AF333906

Singular Integral Equations' Methods
for the Analysis of Microwave Structures

Singular Integral Equations' Methods for the Analysis of Microwave Structures

Liudmila Nickelson and Victor Shugurov

Edited by L.V. Nickelson and T.E. Nickelson

Leiden • Boston, 2005

VSP
an imprint of Brill Academic Publishers
P.O. Box 9000
2300 PA Leiden
The Netherlands

Tel: +31 71 535 3500
Fax: +31 71 531 7532
vsppub@brill.nl
www.brill.nl
www.vsppub.com

First published in 2005

ISBN 90-6764-410-2

A C.I.P. record for this book is available from the Library of Congress

Printed in The Netherlands by Ridderprint bv, Ridderkerk.

Contents

INTRODUCTION ix

PREFACE xi

1. CABLE WAVES 1
    1.1. Cylindrical waves 1
    1.2. The cable wave of coaxial waveguides 4

2. THE SIMPLEST STRIPLINES 10
    2.1. Presentation of an electric field 10
    2.2. The application of conformal mapping 16
    2.3. Edge effects of symmetrical striplines 18
        2.3.1. Short–circuiting metal plane 18
        2.3.2. The open end 20
        2.3.3. The flange end 22
    2.4. Screened lines 24
        2.4.1. Capacitance of a stripline in a circular cylinder 24
        2.4.2. Capacitance of a stripline in a rectangular cylinder 26

3. CAUCHY TYPE INTEGRAL AND SOLUTION OF
SINGULAR INTEGRAL EQUATIONS 29
    3.1. Sokhotsky–Plemelj formulae 29
    3.2. Behavior of the Cauchy type integral at the integration contour ends ... 32
    3.3. Solution of singular integral equations 39

4. ANISOTROPIC MEDIA 41
    4.1. The main parameters of ferrite media 42
    4.2. Magnetic susceptibility tensor taking into account crystallographic
        anisotropy 47
    4.3. The permittivity tensor of semiconductor media 54

5. COMMON DEPENDENCIES FOR TRANSMITION LINES IN TEM - APPROXIMATION 58
5.1. Integral representation 58
5.2. Derivation of singular integral equations 61
5.3. Reciprocity theorem 65
5.4. Connection between matrices of capacitances and inductances 69
5.5. Eigenwaves of a lossless multiconductor transmission line 71
5.6. Eigenwaves of a lossy transmission line 74
5.7. Long line circuits 78
5.8. Realization of the Bogoliubov–Krylov method 81
5.9. Consideration of geometric symmetry 83

6. ANALYSIS OF MICROSTRIP LINES WITH ISOTROPIC AND ANISOTROPIC SUBSTRATES IN TEM–APPROXIMATION BY THE SIE METHOD 86
6.1. Microstrip lines with isotropic substrates 86
6.2. Microstrip lines with n–Si substrates and a comparison of our calculated results with the experimental data 92
    6.2.1. Calculated results for MSLs with two–layer n–Si–dielectric substrates 97
6.3. Microstrip lines with GaAs and GaAs–SiO$_2$ substrates 100
    6.3.1. The calculated results of MSLs with GaAs substrates 101
    6.3.2. The calculated results of MSLs with GaAs–SiO$_2$ substrates 105
6.4. Microstrip lines with Al$_2$O$_3$–SiO$_2$ substrates 111
6.5. Microstrip lines with LiNbO$_3$ substrates 117

7. ANALYSIS OF MICROSTRIP LINES WITH GYROTROPIC SUBSTRATES IN TEM–APPROXIMATION BY THE SIE METHOD 133
7.1 Microstrip lines with longitudinally magnetized semiconductor substrates 133
    7.1.1. One metal strip MSL 135
    7.1.2. Comparison of our calculated results with experimental data 137
    7.1.3. Coupled MSLs 138
    7.1.4. Losses in geometrically asymmetric MSLs 140
7.2. Meander line on a ferrite substrate 141
7.3. Microstrip lines with transversally magnetized ferrite substrates in the approximation of harmonic functions 150
    7.3.1. The differential and integral equations. 151
    7.3.2. The numerical solution of the integral equations 154
7.4. Microstrip lines with the longitudinally magnetized layer $\ddot{\varepsilon}$ – gyrotropic substrates 158
    7.4.1. A one metal strip MSL with a layer semiconductor–dielectric substrate 158
    7.4.2. Comparison of our calculated results with the experimental data 161
    7.4.3. The resonance wave attenuation in MSLs on the metal strip thickness and comparison of our calculated results with experimental data 162

7.4.4. Investigation of coupled MSLs with semiconductor–
    magnetodielectric substrates 166
7.4.5. Coupled MSLs with layer semiconductor substrates. 169
7.5. Microstrip lines with longitudinally magnetized substrates
having double $(\ddot{\varepsilon}, \ddot{\mu})$ – gyrotropy 172
    7.5.1. MSLs with layer semiconductor–ferrite substrates 173
    7.5.2. Coupled MSLs with the substrate having double
    $(\ddot{\varepsilon}, \ddot{\mu})$ –gyrotropy 178

8. SOLUTION OF MAXWELL'S EQUATIONS BY THE SIE METHOD FOR ISOTROPIC WAVEGUIDES

8. SOLUTION OF MAXWELL'S EQUATIONS BY THE SIE
    METHOD FOR ISOTROPIC WAVEGUIDES 185
8.1. The integral representation for the solution of Maxwell's equations 185
8.2. Examples of testing the algorithm 191
8.3. Microstrip line dispersion dependences and comparison of our
calculated results with data from references 193
8.4. The dispersion dependences of planar slot lines and the comparison
of our calculated results with experimental data 196
8.5. The numerical investigation of planar slot lines with asymmetrically
placed metal strips and with micron slots between them 200
8.6. The dispersion dependences of the nonplanar slot lines and a
comparison of our calculated results with the experimental data 206
8.7. The numerical investigation of nonplanar slot lines with
asymmetrically placed strips and micron thicknesses of substrates 209
8.8. The dispersion dependences of dielectric waveguides and the
comparison of our calculated results with reference data 212

9. SOLUTION OF MAXWELL'S EQUATIONS BY THE SIE METHOD FOR LONGITUDINALLY MAGNETUZED $\ddot{\varepsilon}$ –, $\ddot{\mu}$ – AND $\ddot{\varepsilon}$ & $\ddot{\mu}$ – GYROTROPIC WAVEGUIDES

9. SOLUTION OF MAXWELL'S EQUATIONS BY THE SIE
METHOD FOR LONGITUDINALLY MAGNETUZED
$\ddot{\varepsilon}$ –, $\ddot{\mu}$ – AND $\ddot{\varepsilon}$ & $\ddot{\mu}$ – GYROTROPIC WAVEGUIDES 216
9.1. The solution for an open circular longitudinally magnetized waveguide 216
    9.1.1. Comparison of our calculated results with experimental data 219
9.2. Analysis of longitudinally magnetized $\ddot{\varepsilon}$ – & $\ddot{\mu}$ – gyrotropic waveguides 224
9.3. Some examples of testing our algorithm 236
9.4. Electrodynamical characteristics for $\ddot{\mu}$ –gyrotropic slot lines 239
9.5. Electrodynamical characteristics for square–shaped
$\ddot{\varepsilon}$ – gyrotropic waveguides 241
9.6. Electrodynamical characteristics for the $\ddot{\varepsilon}$ – & $\ddot{\mu}$ – gyrotropic
one–comb and two–comb waveguides 243
9.7. Electrodynamical characteristics of mirror magnetized ferrite waveguides
and comparison of calculated results with experimental data 248

10. SOLUTION OF MAXWELL'S EQUATIONS BY THE SIE METHOD FOR OPEN TRANSVERSALLY MAGNETIZED GYROTROPIC WAVEGUIDES 253

10.1. Introduction.........253

10.2. The integral representation of Maxwell's equations solution 254

10.3. The choice of a one-valued branch for the function $\eta(\chi_x)$ 257

10.4. Integration in the complex plane of the variable χ_x and examples 261

10.5. Singling out of the singularity.........266

11. SOLUTION OF MAXWELL'S EQUATIONS BY THE SIE METHOD FOR THREE-DIMENSIONAL SCATTERING PROBLEMS 268

11.1. The fundamental solution of Maxwell's equations for constructing the SIE method 268

11.2. A comparison of calculated results with experimental data of the back scattering cross-section dependence for a circular metal disc 273

11.3. Diffraction parameters of a microstrip reflecting structure.........274

11.4. Numerical investigations of three-dimensional semiconductor structures such as a microwave sensor.........277

11.5. Electrodynamical characteristics of a three-dimensional human heart model 279

11.5.1. The surface integrals over triangular areas 279

11.5.2. Electromagnetic wave scattering by a heart model.........280

11.5.3 The computation of electric fields on a heart model with a microwave catheter placed inside 283

REFERENCES 288

APPENDIXES.........298

A–1. The electromagnetic spectrum 298

A–2. Some basic equations for electrodynamics 299

A–3. Vector differential operations 300

A–4. Relative permittivities.........302

A–5. Fundamental constants.........303

A–6. Index of tables.........304

A–7. Index of figures 305

A–8. Index.........324

A–9. Acronims.........330

INTRODUCTION

This book, the result of more than 30 years of research, presents the accumulation of our methods in solving electrodynamical problems. In solving these problems we have developed unique methods based on the theory of Singular Integral Equations (SIE) as well as related computer programs, which we used for numerical analysis.

The subject matter of this book describes the chronological sequence of solving boundary problems and the numerical investigation of proposed microwave structures. Particularly concerned here are those developments in passive devices, such as those controllable through magnetic or electric fields as well as electromagnetic energy transmission lines.

In the early 1970's theoreticians concentrated their efforts on calculations of microstrip lines (MSLs) on infinite dielectric lossless substrates with infinitely thin metal strip conductors of the TEM–approximation (Transverse Electromagnetic). In the past, solutions of these MSL calculations were limited to one-dimensional (1D) or two–dimensional (2D) problems, meaning that most problems could only be solved analytically using conformal mapping. Currently, however, new computer resources designed specifically for complex three-dimensional (3D) electrodynamical problems are able to analyze complex problems such as the electromagnetic field distribution in the human body.

This book will provide many solutions to solving electrodynamical problems. Beginning with calculating simple striplines by a conformal mapping method in chapter two and ending with our numerically investigating electrodynamical characteristics of a heart model that was under microwave radiation. Our SIE methods created would allow anyone to determine the electrodynamical characteristics of certain 2D or 3D–structures.

To see how one might solve an electrodynamical problem using the SIE methods we would start with differential equations having a certain "point source". Then the fundamental solution to the differential equations is used in the integral presentation of an electromagnetic field for each electrodynamical problem that must be solved. This integral presentation would automatically satisfy the differential equations and the density functions would be determined from the boundary conditions.

The solutions of differential equations, obtained by the SIE method were electrodynamically rigorous, as they satisfied the differential equations and the proper boundary conditions. The edge conditions were satisfied due to the integral presentation of the solution. The practical applications of the SIE methods lead to numerical solutions of the well–determined system of the linear equations that simplified and accelerated the computations. The false roots did not occur applying our SIE methods. The boundary conditions had to be satisfied only on the existing surfaces dividing different materials.

Our SIE methods are universal in that: the cross–section shape can be of any form, the waveguide can be screened or open, the partially homogeneous regions can be dielectric, anisotropic or gyrotropic. These SIE methods are especially effective when the 2D– or 3D–structures have an intricate form and these structures may even include high lossy materials. What we mean by effective is this, we are able to decide very complicated problems with the SIE methods that maybe impossible to solve by other methods.

The SIE methods used to calculate electrodynamical problems enabled us to develop and optimize a number of the microwave devices. Our SIE methods were substantiated through the creation of five devises (patented in the former USSR).

The author's belief is that the SIE methods described in this book will find the widest practical application with experts in the field of microwave techniques in medical, industrial and domestic facilities.

PREFACE

Throughout the book we present examples of using SIE methods in calculating microwave structures. Briefly, in the preface we will present different computation algorithms that were created by our SIE methods. We demonstrate several main formulae for electric and magnetic fields that were used in this book. Expressions of these electric and magnetic fields will be substituted into boundary conditions of certain electrodynamical problems and in this way we will determine the proper solutions. In the following sections we present several SIE methods.

The problems of electrostatics and magnetostatics (chapters 4–7) and the solution of Laplace's equation. In these chapters problems were decided in TEM–approximation. Here the electrostatic field has the integral representation:

$$\dot{E}(z) = -\int_L \frac{\dot{\mu}(t)ds}{t-z} - \sum_{k=1}^{m} \frac{\dot{A}_k}{z_k - z}, \tag{1}$$

where $z = x_0 + jy_0$ is the coordinate of the point when the boundary conditions are written for an electric field component or when the value of the field component is calculated. Here $t = x + jy$ is the coordinate at the point of contour when the value of the unknown function $\dot{\mu}(t)$ is found. Here $ds = |dt|$ is the contour *arc* element. The unknown values $\dot{\mu}(t)$ and $\dot{A}_k$ are found after applying the boundary conditions. The value $z_k = x_k + jy_k$ is the coordinate of the point inside of the k –*th* metal conductor and j is the imaginary unit for the coordinates.

The contours L_ε and L_m separate different media in the cross–section of the structure (Fig.1), which are arbitrarily divided into segments. Along these segments the integration is carried out from the lower limit (x_{a_j}, y_{a_j}) to the upper limit (x_{b_j}, y_{b_j}) with an index "j" running from 1 to $n_\varepsilon + n_m$. Here n_m is the number of segments of the contours, which is limited to metal. Then n_ε is the number of segments of other contours, which is limited to dielectric, semiconductor, ferrite and other materials.

All the contours separating different media are arbitrarily divided into segments and they can be equal or unequal in length. The coordinates of the point "i" (when we formulate the boundary conditions) could be chosen at any place on the segment but the center of the segment is usually the best choice.

Let the structure have in the cross–section several regions bounded by contours L_ε and L_m. Then the boundary conditions have this form:

$$\ddot{\varepsilon}^{+}E_n^{+}(z)\Big|_{L_\varepsilon} = \ddot{\varepsilon}^{-}E_n^{-}(z)\Big|_{L_\varepsilon}, \tag{2}$$

where the signs "+" and "-" single out the regions on the left and right side of the contours.

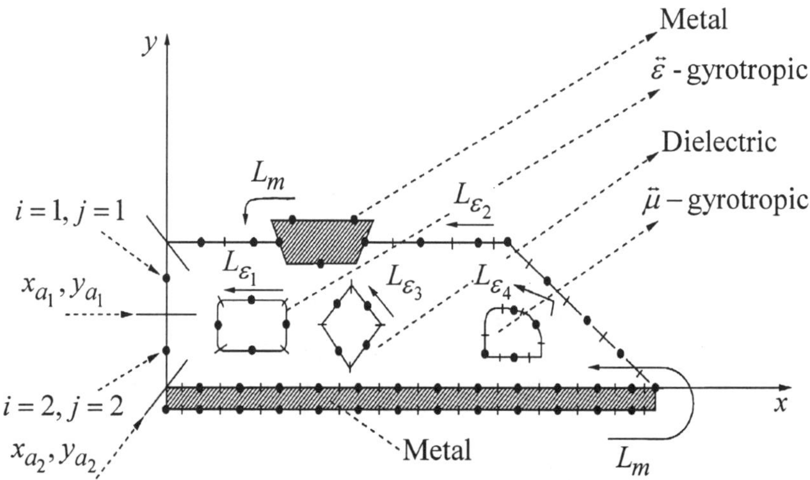

Fig. 1. *Cross–section of a strip waveguide line, where L_ε are contours bordering isotropic, anisotropic and gyrotropic media. L_m are contours bordering metal conductors and other designations.*

For the contours L_m bordering the perfect metal, the boundary conditions are simpler:

$$E_\tau^{-}(z)\Big|_{L_m} = 0. \tag{3}$$

The boundary conditions are formulated for every point "i" when "i" is running from 1 to $n_\varepsilon + n_m$. Thus the boundary conditions (2) and (3) form the system of linear equations of the order $n_\varepsilon + n_m$. The system is not full because the electric field expression (1) has the unknowns $\dot{A}_k$ and we must add an expression, which gives the values of the potential of each metal conductor (chapter 5). Having the solutions $\dot{\mu}(t_i)$ and $\dot{A}_k$ it is easy to find the electrodynamical characteristics of the MSL (chapters 6–7).

The electrodynamical rigorous solution of Maxwell's equations to determine the dispersion dependence of the main and higher modes for the regular waveguides, with a cross–section of arbitrary shapes having partial homogeneous materials. We solved Helmholtz's equation in chapter 8. The longitudinal components of the electric and magnetic fields are presented in these integral forms:

$$E_z(\vec{r}) = \int_{L_\varepsilon + L_m} \mu_e(\vec{r}_s) H_0^{(2)}(k_\perp r') ds , \tag{4}$$

$$H_z(\vec{r}) = \int_{L_\varepsilon + L_m} \mu_h(\vec{r}_s) H_0^{(2)}(k_\perp r') ds , \tag{5}$$

where $E_z(\vec{r})$, $H_z(\vec{r})$ are the longitudinal components of the magnetic and electric fields. Here $\vec{r}$ is the radius vector of the point in the xy plane. Here $\vec{r}_s$ is the radius vector of the contour point (Fig.2). The unknown functions $\mu_h(\vec{r}_s)$ and $\mu_e(\vec{r}_s)$ are determined by the boundary conditions. The $H_0^{(2)}(k_\perp, r')$ is the second kind of Hankel's function of the zeroth order, where $r' = |\vec{r} - \vec{r}_s|$ and $k_\perp = \sqrt{k^2 \varepsilon_r \mu_r - h^2}$ is the transversal wave number.

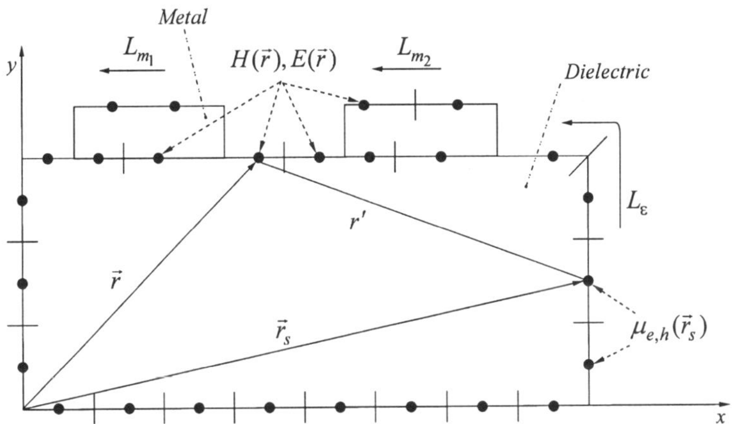

Fig. 2. Cross–section of a slot line, where designations are shown.

From Maxwell's equations, the transversal components H_x, H_y, E_x, E_y of the electromagnetic field can be expressed through two longitudinal components. Substituting the formulae (4) and (5) into the expressions for the field transversal components on the contours bordering isotropic, anisotropic and gyrotropic media:

$$E_\tau^+\big|_L = E_\tau^-\big|_L , \quad H_\tau^+\big|_L = H_\tau^-\big|_L , \tag{6}$$

on contours bordering metal conductors:

$$E_\tau^+\big|_L = 0 . \tag{7}$$

These terms will contain the Cauchy's type integrals. These integrals become singular at $r' \to 0$. We applied Sokhotsky–Plemelj formulae to write singular integrals on the contours. The boundary conditions lead to a system of SIE equations. After applying Bogoliubov–Krylov method we changed the integral equations into the proper system of linear algebraic equations. Setting the system determinant to zero we get certain dispersion equation that permits us to define propagation constants h.

The electrodynamical rigorous solution of Maxwell's equations to determine the dispersion dependence of the main and higher modes for the regular waveguide with an arbitrary shape of the cross–section having partial homogeneous isotropic, anisotropic and gyrotropic materials. The longitudinal components are presented in these forms (chapter 9)

$$E_z(\vec{r}) = \int_L \mu_e(\vec{r}_s)H_0^{(2)}(k_{\perp 2}r')ds + a\int_L \mu_h(\vec{r}_s)H_0^{(2)}(k_{\perp 1}r')ds ,$$

$$H_z(\vec{r}) = \int_L \mu_h(\vec{r}_s)H_0^{(2)}(k_{\perp 1}r')ds + b\int_L \mu_e(\vec{r}_s)H_0^{(2)}(k_{\perp 2}r')ds . \tag{8}$$

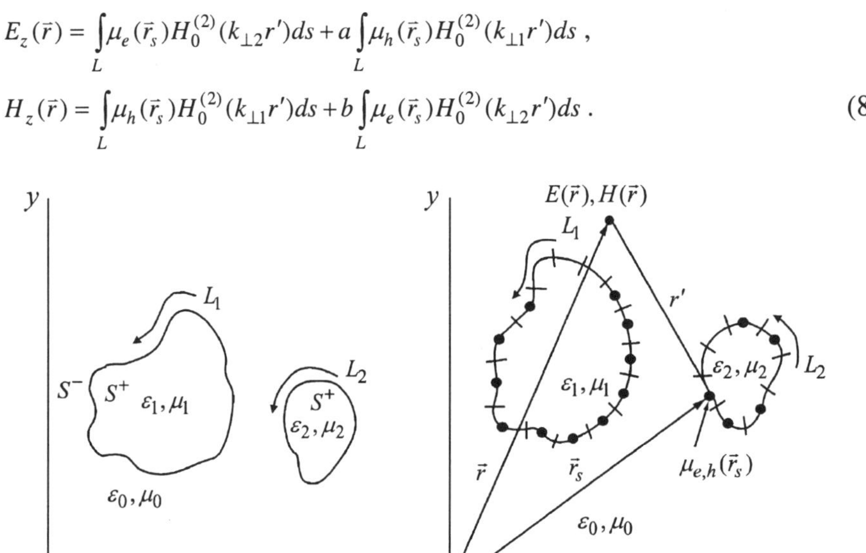

Fig. 3. *(a) Cross–section of two coupled $\ddot{\varepsilon}$ – gyrotropic rods;*
(b) designations.

The approach to solving electrodynamical problems for regular waveguides with arbitrary shapes (Fig.3) having partly homogeneous isotropic, anisotropic, gyrotropic is the same as was described earlier in the preface. These calculations enable one to develop and optimize a number of the microwave devices. Also to propose new constructions of waveguide transmission lines and different devices on their base.

Solution of Maxwell's equations by the SIE method for open transversally magnetized gyrotropic waveguides (chapter10). In this chapter, we describe a new method based on the SIE theory. By this, one is enabled to theoretically investigate open waveguides of arbitrary (complex) cross–section geometry having transversally magnetized ferrite or semiconductor material. In this chapter the SIE method yields the solution to the problem in the rigorous electrodynamical formulation. This method enables one to analyze the main and higher modes propagating in the investigated waveguide.

The electrodynamical rigorous solution of Maxwell's equations for a 3D isotropic structure of arbitrary form placed in free space (chapter 11). The SIE method was verified when we compared our calculations for 3D isotropic structures with experimental data and with results from science publications. Also in this chapter we solved certain diffraction problems, which enabled us to calculate the back scattering cross–section and other electrodynamical characteristics of different 3D structures (for example Figs.4-6). At the end of this chapter we determined the electrodynamical characteristics of a biological heart model that was under the influence of microwaves. The heart model was illuminated by electromagnetic waves from inside and outside sources.

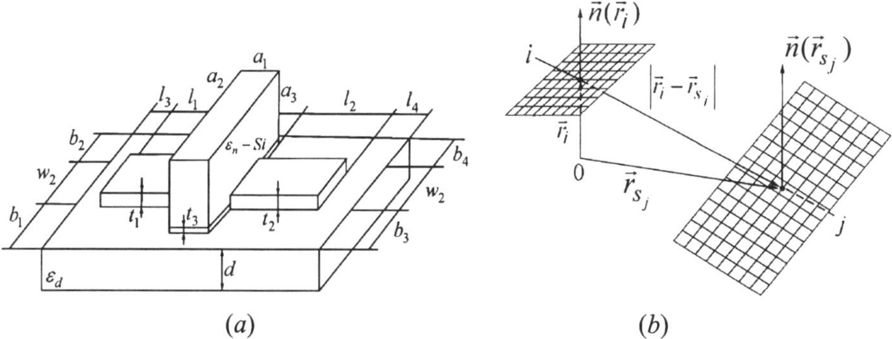

(a) (b)

Fig. 4. *(a) The configuration of a 3D structure, which is illuminated by outside microwaves; (b) designations.*

This 3D structure Fig.4a is a semiconductor sensor and is one kind of dipole probe. It has a sensitive element for detecting a microwave electric field. The sensor is constructed as a symmetric vibrator made of two microstrip conductors with a semiconductor sample of a cuboid form placed between them.

The 3D structure in Fig.5 is a reflector of microwaves containing two–microstrip metallic frames. The ratio of the back reflected signal (toward the microwave source) and the incident microwave field was computed as a function of the frequency.

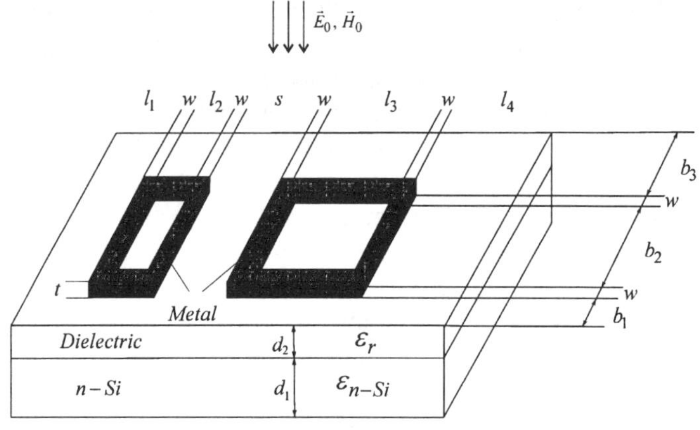

Fig. 5. *A structure having a semiconductor–dielectric substrate with two metallic frames placed on it.*

A human heart may be under the influence of microwave radiation in the medical examination and therapeutic treatment of patients. When this accrues we can theoretically investigate different electrodynamical characteristics of the heart.

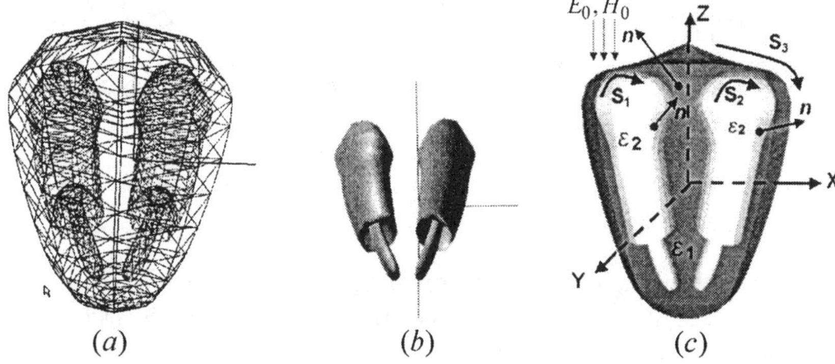

(*a*)   (*b*)   (*c*)

Fig. 6. *The model of the heart used in calculations: (a) the surfaces are approximated by triangles, (b) a view of ventricles and atria, (c) certain designations.*

An electrodynamical accurate solution of Maxwell's equations based on the SIE method was used to calculate 3D structures. We investigated 3D heart models when the source of the electromagnetic fields was inside or outside of the heart.

One can determine the components of the electromagnetic field inside and outside of a 3D structure depending on the polarization, wave incidence angle and the electrophysical parameters of the material. In order to calculate the 3D structure and electromagnetic waves interaction it is necessary to solve a diffraction problem. The known incident electromagnetic wave has components $E_0(\vec{r})$ and $H_0(\vec{r})$, where $\vec{r}$ is the radius vector of a point in which we calculate an electromagnetic field. When solving electrodynamical problems for these 3D structures in Figs.4–6 we wanted to find the scattering or transmitted field. And to present the electromagnetic fields in integral form we used the solutions of Maxwell's equations with electric $\vec{j}_e$ and magnetic $\vec{j}_m$ point sources:

$$rot\vec{H} = i\omega\varepsilon_0\varepsilon E + \vec{j}_e,$$

$$rot\vec{E} = -i\omega\varepsilon_0\varepsilon\vec{H} - \vec{j}_m.$$

These equations are linear when the general solution is taken as a sum of solutions for an electric wave (when $\vec{j}_e \neq 0$ and $\vec{j}_m = 0$) and as a sum of solutions for a magnetic wave (when $\vec{j}_e = 0$ and $\vec{j}_m \neq 0$).

For the electric $\vec{E}(\vec{r})$ and magnetic $\vec{H}(\vec{r})$ fields the following expressions were derived:

$$\vec{E}(\vec{r}) = \int_s \mu_e(\vec{r}_s)\left(\frac{1}{k^2\varepsilon\mu}\nabla\left(\vec{n}(\vec{r}_s),\nabla\right)+\vec{n}(\vec{r}_s)\right)h_0(z)ds -$$

$$-iZ_0\sqrt{\frac{\mu}{\varepsilon}}\int_s \mu_h(\vec{r}_s)\left[\vec{n}(\vec{r}_s),\hat{r}\right]h_1(z)ds,$$

$$\vec{H}(\vec{r}) = \int_s \mu_h(\vec{r}_s)\left(\frac{1}{k^2\varepsilon\mu}\nabla\left(\vec{n}(\vec{r}_s),\nabla\right)+\vec{n}(\vec{r}_s)\right)h_0(z)ds +$$

$$+\frac{i}{Z_0}\sqrt{\frac{\varepsilon}{\mu}}\int_s \mu_e(\vec{r}_s)\left[\vec{n}(\vec{r}_s),\hat{r}\right]h_1(z)ds,$$

where $h_0(z)$ is a spherical Hankel function of the zeroth order, and the second kind $h_1(z) = -\dfrac{d}{dz}h_0(z)$ and $z = k\sqrt{\varepsilon\mu}|\vec{r} - \vec{r}_s|$. The wave number $k = \omega/c$, where $\omega = 2\pi f$ and $c = 1/\sqrt{\mu_0\varepsilon_0}$ is the velocity of light in a vacuum. The value $Z_0 = \sqrt{\mu_0/\varepsilon_0} \sim 120\pi$. Values $\mu_e(\vec{r}_s)$, $\mu_h(\vec{r}_s)$ are the electric and magnetic source densities in the point $\vec{r}_s$ of the surface. Here $\vec{n}(\vec{r}_s)$ is the unit normal vector to the surface in the same point $\vec{r}_s$ of the surface. Here ds is an infinitesimal patch of aria with a direction that is perpendicular to the surface. The expression $\hat{r} = (\vec{r} - \vec{r}_s)/|\vec{r} - \vec{r}_s|$ is a unit vector in the direction from $\vec{r}_s$ to $\vec{r}$. This symbol ∇ is a vector operator *del*. The $(\vec{n}(\vec{r}_s),\nabla)$ designation is a dot product and $\left[\vec{n}(\vec{r}_s),\hat{r}\right]$ designation is a cross product of two vectors. The densities $\mu_e(\vec{r}_s)$, $\mu_h(\vec{r}_s)$ will be found from the boundary conditions for the electric $\left[\vec{n}(\vec{r}_1),\vec{E}(\vec{r}_1)^+\right]=\left[\vec{n}(\vec{r}_1),\vec{E}^i + \vec{E}(\vec{r}_1)^-\right]$ and magnetic

$\left[\vec{n}(\vec{r}_1),\vec{H}(\vec{r}_1)^+\right]=\left[\vec{n}(\vec{r}_1),\vec{H}^i+\vec{H}(\vec{r}_1)^-\right]$ fields on the boundary surfaces, where $\vec{r}\to\vec{r}_1$ ($\vec{r}_1$ is a radius vector of a point on the boundary surfaces). These SIE methods can be used to analyse the electrodynamical characteristics of different biological organ models and can be applied to 3D object of any shape or size.

We would like to draw the reader's attention to the fact that value $\mu(t_j)$ takes two values for every point "j". The value $\mu(t_j)$ depends on whether the point is nearing the contour from its left or right side (chapters 8-11). For example (Fig.7):

$$E_z^-(\vec{r}_i)=\sum_j \mu^-(r_{s_j})\int_{a_j}^{b_j}H_0^{(2)}\left(k_{\perp 2}\left|\vec{r}_i-\vec{r}_{s_j}\right|\right)\,ds\ ,$$

$$E_z^+(\vec{r}_i)=\sum_j \mu^+(r_{s_j})\int_{a_j}^{b_j}H_0^{(2)}\left(k_{\perp 2}\left|\vec{r}_i-\vec{r}_{s_j}\right|\right)\,ds$$

on the contour which separate media.

But in electrostatic and magnetostatic problems the values of $\mu(t)$ on the right and left contour sides are equal $\mu^+(t)=\mu^-(t)$.

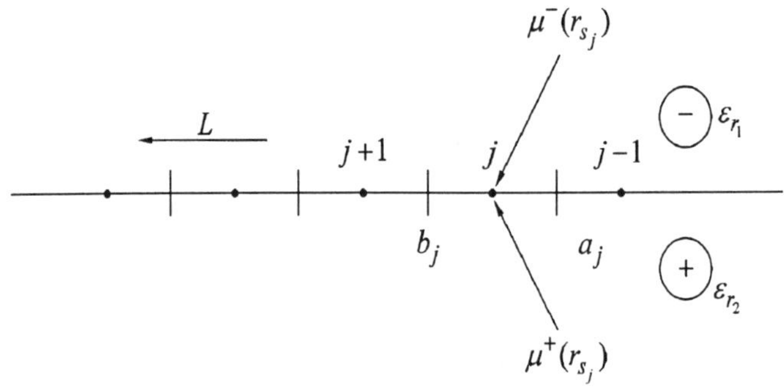

Fig. 7. Designations used in the expressions for the field components.

We have attempted to show some essential points in this preface, which would enable the reader to comprehend our SIE methods and the great diversity in which they can be used. We know that our SIE methods of calculating problems will be extremely helpful in investigating and designing new and unique microwave devices for the future.

1. CABLE WAVES

Here we will study waveguide systems surrounded by cylindrical metal surfaces, which are filled with isotropic or anisotropic media (Fig.1.1). We used a cylindrical coordinate system with a z–axis directly parallel to the straight line, which forms the cylindrical surfaces.

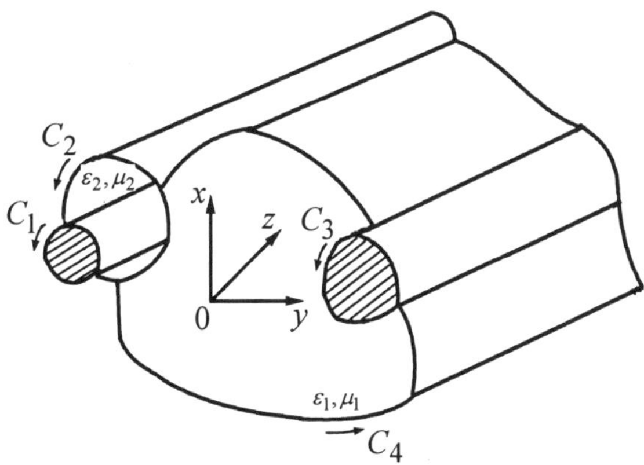

Fig. 1.1. Waveguide transmission line with inhomogeneous fillings.

The cross–section of the waveguide transmission line (when the plane $z =$ const) consist of the contours C_1, C_2, C_3 .and. C_4. Electromagnetic waves propagating along such a transmission line are called cylindrical waves. Because of the uniformity of the waveguide along axis, oz (regular waveguide) dependence on coordinate z is described by an exponent e^{-ihz}, where h is a propagation constant. Dependence on time is taken as an exponent $e^{i\omega t}$.

1.1. Cylindrical Waves

Here Maxwell's equations are:

$$rot\ \vec{H} = \frac{\partial \vec{D}}{\partial t},$$

$$rot\ \vec{E} = -\frac{\partial \vec{B}}{\partial t}, \qquad (1.1)$$

where constitutive relations:

$$\vec{D} = \varepsilon\varepsilon_0\vec{E},$$

$$\vec{B} = \mu\mu_0\vec{H}.$$

When we will replace the derivatives $\partial/\partial z \to -ih$ and $\partial/\partial t \to i\omega$ then the equations (1.1) for the complex amplitudes (depending on x, y) take the form in Cartesian coordinates:

$$\frac{\partial H_z}{dy} + ihH_y = i\omega D_x,$$

$$-ihH_x - \frac{\partial H_z}{dx} = i\omega D_y,$$

$$\frac{\partial H_y}{\partial x} - \frac{\partial H_x}{\partial y} = i\omega D_z,$$

$$\frac{\partial E_z}{dy} + ihE_y = -i\omega B_x,$$

$$-ihE_z - \frac{\partial E_z}{dx} = -i\omega B_y,$$

$$\frac{\partial E_y}{\partial x} - \frac{\partial E_x}{\partial y} = -i\omega B_z. \tag{1.2}$$

From this system of equations (1.2) (for the isotropic medium the quantities ε and μ are scalars) transverse field components are expressed:

$$E_x = \frac{1}{k^2\varepsilon\mu - h^2}\left(-ih\frac{\partial E_z}{\partial x} - i\omega\mu\mu_0\frac{\partial H_z}{\partial y}\right),$$

$$E_y = \frac{1}{k^2\varepsilon\mu - h^2}\left(-ih\frac{\partial E_z}{\partial y} + i\omega\mu\mu_0\frac{\partial H_z}{\partial x}\right),$$

$$H_x = \frac{1}{k^2\varepsilon\mu - h^2}\left(i\omega\varepsilon\varepsilon_0\frac{\partial E_z}{\partial y} - ih\frac{\partial H_z}{\partial x}\right),$$

$$H_y = \frac{1}{k^2\varepsilon\mu - h^2}\left(-i\omega\varepsilon\varepsilon_0\frac{\partial E_z}{\partial x} - ih\frac{\partial H_z}{\partial y}\right), \quad k = \omega\sqrt{\varepsilon_0\mu_0} = \omega/c, \tag{1.3}$$

through the derivatives of the longitudinal components, which satisfy the differential equations:

$$\left(\frac{\partial^2}{\partial x^2} + \frac{\partial^2}{\partial y^2}\right)E_z + \left(k^2\varepsilon\mu - h^2\right)E_z = 0,$$

$$\left(\frac{\partial^2}{\partial x^2}+\frac{\partial^2}{\partial y^2}\right)H_z+\left(k^2\varepsilon\mu-h^2\right)H_z=0. \tag{1.4}$$

Since the equations (1.4) are independent, the solutions can be divided into two groups: 1) $E_z\neq 0$, $H_z=0$ and 2) $E_z=0$, $H_z\neq 0$. The first waves are called E–waves and the second ones are called H–waves. Solutions to the equations (1.4) and transversal components (1.3) were chosen in such a way that certain boundary conditions were satisfied on the surface of the divided media.

On the surface of ideal metal the tangential components of an electric field are equal to zero. Here in this case when we have idea metal:

$$E_z\big|_C=0,$$
$$E_s\big|_C=0, \tag{1.5}$$

where index $\bar{s}$ is a vector tangential to the contour C (Fig.1.2).

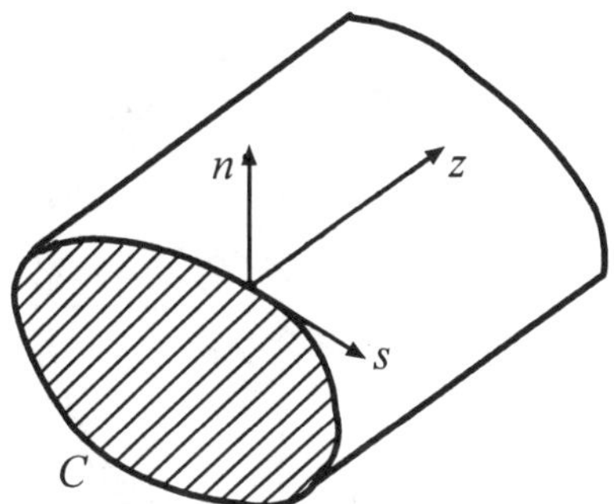

Fig. 1.2. *Boundary conditions on the surface of ideal metal.*

For other media (not metal) the tangential components of electric and magnetic fields from different sides of the surfaces will be equal:

$$E_{z_1}=E_{z_2},$$
$$H_{z_1}=H_{z_2},$$
$$E_{s_1}=E_{s_2},$$
$$H_{s_1}=H_{s_2}, \tag{1.6}$$

when there are no currents on the surfaces.

Here boundary conditions can "mix up" E–wave and H–waves. The boundary conditions will be satisfied only by the combination of both types of waves.

In the case of two or more metal conductors in a waveguide there exists a wave having zero cutoff frequency. Such a wave is called a cable wave. At zero frequency the electric field of the wave has the configuration of a static field. This field was obtained after applying different constant potentials to metal conductors. The field has only transverse components and the wave is called transverse electromagnetic (TEM). When filling is inhomogeneous and frequency is not zero some longitudinal components occur. It is usually assumed that the longitudinal components are much smaller than transverse ones (TEM–approximation) and so the simpler static problem is solved.

1.2. The Cable Wave of Coaxial Waveguides

Here we describe the cable wave that is propagating in a coaxial waveguide in greater detail. It is known that a hollow waveguide cannot support TEM–waves. But a coaxial waveguide (Fig.1.3), consisting of a long straight strip of r_1 and surrounded by a cylindrical–conducting sheath of r can.

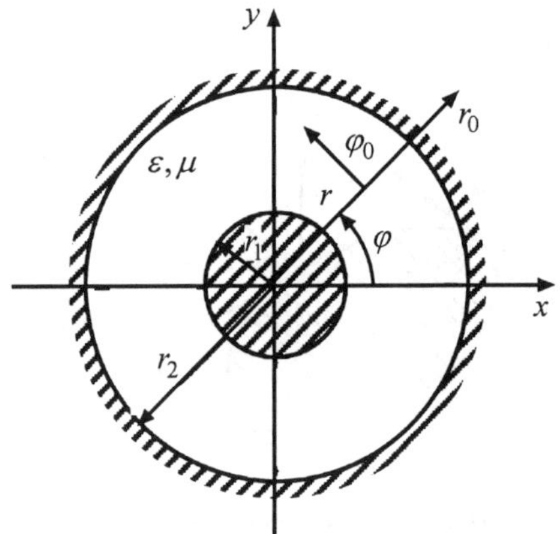

Fig. 1.3. *Cross–section of a coaxial waveguide.*

The geometry of this coaxial waveguide dictates that we use the cylindrical coordinate system. The transverse components are vectors projections on the directions of unit basic vectors r_0 and φ_0. Similarly (1.3), the transversal components:

$$E_r = \frac{1}{k^2\varepsilon\mu - h^2}\left(-ih\frac{\partial E_z}{\partial r} - i\omega\mu\mu_0\frac{1}{r}\frac{\partial H_z}{\partial \varphi}\right),$$

$$E_\varphi = \frac{1}{k^2\varepsilon\mu - h^2}\left(-ih\frac{1}{r}\frac{\partial E_z}{\partial \varphi} + i\omega\mu\mu_0\frac{\partial H_z}{\partial r}\right),$$

$$H_r = \frac{1}{k^2\varepsilon\mu - h^2}\left(i\omega\varepsilon\varepsilon_0\frac{1}{r}\frac{\partial E_z}{\partial \varphi} - ih\frac{\partial H_z}{\partial r}\right),$$

$$H_\varphi = \frac{1}{k^2\varepsilon\mu - h^2}\left(-i\omega\varepsilon\varepsilon_0\frac{\partial E_z}{\partial r} - ih\frac{1}{r}\frac{\partial H_z}{\partial \varphi}\right) \tag{1.7}$$

are expressed through the longitudinal components. The solutions of the differential equations (1.4) are cylindrical functions of the order m multiplied by $\sin m\varphi$ or $\cos m\varphi$. In the case where the waveguide has a uniform filling, the fields of the main wave ($m = 0$) are important. These waves are independent of coordinate angle φ and the solutions are:

$$E_z = AJ_0(gr) + BN_0(gr),$$

$$H_z = CJ_0(gr) + DN_0(gr),$$

$$g^2 = k^2 \varepsilon\mu - h^2, \tag{1.8}$$

where $J_0(z)$ is the Bessel cylindrical function of the zeroth order and $N_0(z)$ is the Neumann cylindrical function of the zeroth order. Then $C = D = 0$, $A \neq 0$ and $B \neq 0$ are the amplitudes for the E-wave. Component $E_\varphi = 0$ is for any r and $r_1 \leq r \leq r_2$, therefore the second boundary condition (1.5) is satisfied. The first leads to the system:

$$AJ_0(gr_1) + BN_0(gr_1) = 0,$$

$$AJ_0(gr_2) + BN_0(gr_2) = 0. \tag{1.9}$$

From this system one gets eigenvalues $g \neq 0$ as roots of the transcendental equation:

$$\begin{vmatrix} J_0(gr_1) & N_0(gr_1) \\ J_0(gr_2) & N_0(gr_2) \end{vmatrix} = 0. \tag{1.10}$$

The ratio of the longitudinal and transversal components of the electric fields:

$$\frac{E_z}{E_r} = \frac{g}{ih} \frac{J_0(gr_1)N_0(gr) - J_0(gr)N_0(gr_1)}{J_0'(gr)N_0(gr_1) - J_0(gr_1)N_0'(gr)} \tag{1.11}$$

equals zero at $r = r_1$, because the numerator of formula (1.11) becomes zero. Also the ratio of the longitudinal and transversal components of the electric fields equals zero at $r = r_2$ because the numerator of formula (1.11) equals the equation (1.10), which is equal to zero. When a solution of the transcendental equation (1.10) exists at the eigenvalues $g = 0$ then the ratio (1.11) also becomes equals zero for any value of r, when $r_1 \leq r \leq r_2$. Everyone will get this solution separating the factor g^2. Then we can write:

$$A = g^2 A',$$

$$B = g^2 B'.$$

Then the conditions (1.9) are satisfied when $g \to 0$. The longitudinal component is $E_z = 0$ and the amplitude of the transversal component:

$$E_r = -ih \lim_{g \to 0} \left(\frac{\partial \left(A' J_0 \left(gr \right) + B' N_0 \left(gr \right) \right)}{\partial r} \right) \approx -ihB' \frac{2}{\pi r} \tag{1.12}$$

is proportional to the static field in a coaxial capacitor described by the potential $\varphi \sim \ln r$. This potential satisfies Laplace's equation in the range $r_1 \leq r \leq r_2$. The only nonvanishing component of the magnetic field is:

$$H_\varphi = \frac{\omega \varepsilon \varepsilon_0}{h} E_r = \sqrt{\frac{\varepsilon \varepsilon_0}{\mu \mu_0}} E_r . \tag{1.13}$$

Thus, the wave impedance is defined by the ratio:

$$Z = \frac{E_r}{H_\varphi} \qquad \text{and}$$

$$Z = \sqrt{\frac{\mu \mu_0}{\varepsilon \varepsilon_0}} .$$

This wave impedance is constant.

The relationship to the static field allows us in a natural way to introduce the capacitance C and inductance L per unit length. From (1.12), the density of the surface charge is $\sigma = -i \varepsilon \varepsilon_0 k \sqrt{\varepsilon \mu} \ B' \frac{2}{\pi} \frac{1}{r_2}$.

The charge per unit length is $Q = -4 i \varepsilon \varepsilon_0 k \sqrt{\varepsilon \mu} \ B'$.

The difference of the potentials of the metal surfaces is:

$$V_{12} = -ik \sqrt{\varepsilon \mu} \ E' \frac{2}{\pi} \ln \left(\frac{r_2}{r_1} \right) .$$

So $C = \dfrac{2\pi \varepsilon \varepsilon_0}{\ln \left(r_2 / r_1 \right)}$ and similarly $L = \dfrac{\mu \mu_0 \ln \left(r_2 / r_1 \right)}{2\pi}$.

We now consider the coaxial waveguide (Fig.1.4) with a nonuniform filling along the coordinate $r = const$. The solutions of the differential equations (1.4) are:

$$E_{z_1} = A_1 J_0 \left(g_1 r \right) + B_1 N_0 \left(g_1 r \right),$$

$$H_{z_1} = C_1 J_0 \left(g_1 r \right) + D_1 N_0 \left(g_1 r \right), \qquad r_1 \leq r \leq r_2 ,$$

$$E_{z_2} = A_2 J_0 \left(g_2 r \right) + B_2 N_0 \left(g_2 r \right),$$

$$H_{z_2} = C_2 J_0 \left(g_2 r \right) + D_2 N_0 \left(g_2 r \right), \qquad r_2 \leq r \leq r_3 ,$$

$$g_i^2 = k^2 \varepsilon_i \mu_i - h^2 , \ i = 1, \ 2... \tag{1.14}$$

for the waves that are independent of the angle φ.

The solutions (1.14) are the linear combinations of Bessel and Neumann functions of the zeroth order like those in the previous example.

The electric field component E_z is equal to zero at $r = r_1$:

$$A_1 J_0(g_1 r_1) + B_1 N_0(g_1 r_1) = 0,\tag{1.15}$$

the electric field component E_z is equaled to zero at $r = r_3$:

$$A_2 J_0(g_2 r_3) + B_2 N_0(g_2 r_3) = 0.\tag{1.16}$$

The continuity condition for the component E_z at $r = r_2$ takes the form:

$$A_1 J_0(g_1 r_2) + B_1 N_0(g_1 r_2) = A_2 J_0(g_2 r_2) + B_2 N_0(g_2 r_2).\tag{1.17}$$

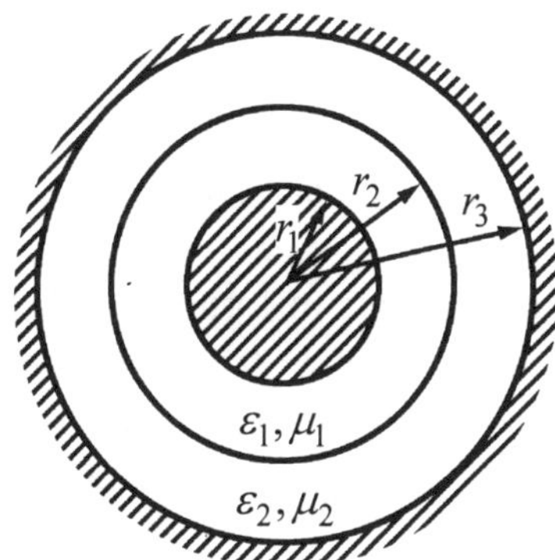

Fig. 1.4. *Cross–section of a coaxial waveguide with layer fillings.*

Similarly, for the magnetic field component H_φ at $r = r_2$ we have:

$$\frac{\varepsilon_1}{g_1} A_1 J_0'(g_1 r_2) + B_1 N_0'(g_1 r_2) = \frac{\varepsilon_2}{g_2} A_2 J_0'(g_2 r_2) + B_2 N_0'(g_2 r_2).\tag{1.18}$$

The components E_φ at $r = r_1$ and $r = r_3$ must vanish:

$$\frac{1}{g_1}\left(C_1 J_0'(g_1 r_1) + D_1 N_0'(g_1 r_1)\right) = 0,\tag{1.19}$$

$$\frac{1}{g_2}\left(C_2 J_0'(g_2 r_3) + D_2 N_0'(g_2 r_3)\right) = 0.\tag{1.20}$$

The continuity of components H_z and E_φ at $r = r_2$ allows us to arrive at the last two equations:

$$C_1 J_0(g_1 r_2) + D_1 N_0(g_1 r_2) = C_2 J_0(g_2 r_2) + D_2 N_0(g_2 r_2),\tag{1.21}$$

$$\frac{\mu_1}{g_1}\left(C_1 J_0'(g_1 r_2) + D_1 N_0'(g_1 r_2)\right) = \frac{\mu_2}{g_2}\left(C_2 J_0'(g_2 r_2) + D_2 N_0'(g_2 r_2)\right).\tag{1.22}$$

These are the unknowns magnitudes A_1, B_1, A_2 and B_2 in the equations (1.15)–(1.18) and the unknowns magnitudes C_1, D_1, C_2 and D_2 in the equations (1.19)–(1.22). Therefore the boundary conditions (1.15)–(1.22) split into two uncoupled systems. The coaxial waveguide can propagate $E-$ waves and $H-$ waves independent of each other (Fig.1.4). Here there is an $E-$ wave in the waveguide and the amplitudes are $C_i = D_i = 0$, where the index is $i = 1$ or 2. So it follows that from formulae (1.7), (1.14)–(1.16) the ratio of the longitudinal and transversal electric field components:

$$\frac{E_z}{E_r} = -\frac{g_1}{ih}\frac{J_0(g_1 r)N_0(g_1 r_1) - J_0(g_1 r_1)N_0(g_1 r)}{J_0'(g_1 r)N_0(g_1 r_1) - J_0(g_1 r_1)N_0'(g_1 r)}, \qquad r_1 \leq r \leq r_2,$$

$$\frac{E_z}{E_r} = -\frac{g_2}{ih}\frac{J_0(g_2 r)N_0(g_2 r_3) - J_0(g_2 r_3)N_0(g_2 r)}{J_0'(g_2 r)N_0(g_2 r_3) - J_0(g_2 r_3)N_0'(g_2 r)}, \qquad r_2 \leq r \leq r_3. \qquad (1.23)$$

Obviously values g_1 and g_2 can not vanish at the same time. Therefore, the component E_z can be small in comparison to E_r because it does not turn into zero.

The arguments for the functions in formula (1.23) are so small that one can confine himself to the first member of their expansions. It is justified for $(g_i r/2)^2 \prec\prec 1$ when $r_1 \leq r \leq r_3$. If an error is about 10% the magnitude is approximately $gr \sim 0.6$. When the arguments for the functions are small formula (1.23) changes to:

$$\frac{E_z}{E_r} \cong \frac{g_1^2}{ih} r \ln\left(\frac{r_1}{r}\right), \qquad r_1 \leq r \leq r_2,$$

$$\frac{E_z}{E_r} \cong \frac{g_2^2}{ih} r \ln\left(\frac{r_3}{r}\right), \qquad r_2 \leq r \leq r_3. \qquad (1.24)$$

So if the thickness of every dielectric layer does not exceed quantity Δ then:

$$\frac{E_z}{E_r} \prec \frac{\left|g_{1,2}^2\right| \cdot \Delta}{h}, \qquad (1.25)$$

therefore by diminishing Δ one always gets a small ratio of the longitudinal and transversal components.

The simple configuration of the waveguide (Fig.1.4) allows us to arrive at the exact formulae (1.23) and to calculate the ratio of the longitudinal and transversal electric field components. In the case of intricate waveguide cross–sections as MSLs it is difficult to receive simple and exact formulae like (1.23). However, the qualitative picture of electric field distributions remains the same in both cases: when the cross–sizes of a waveguide are small in comparison to the wavelength of a wave in this waveguide and when the ratio of longitudinal and transversal components is small for the low frequencies. This means that the main wave fields can be looked at with accuracy in the static approximation, which is often called the TEM–approximation.

Here the application of TEM–approximation means deleting the last few members in the differential equations (1.4). In this case instead of the solution to the Helmholtz's equation, it is enough to solve a simpler Laplace's equation for the problem when the function sought depends on two coordinates. The last circumstance allows us to formulate criterion of the TEM–approximation when the corresponding solution of the boundary problem for Laplace's equation is found. The criterion of the TEM–approximation is:

$$\left|\left(k^2\varepsilon\mu-h^2\right)E_z\right| \prec\prec \left|\frac{\partial^2 E_z}{\partial x^2}\right|. \tag{1.26}$$

Finally because the cross–section shape of a coaxial waveguide is simple it allows us to receive an electrodynamical rigorously solution to a waveguide problem in an analytical form. Analysis of this electrodynamical rigorous solution allows us to formulate a criterion for the use of TEM–approximation in solving certain waveguide problems, an example would be for simple striplines. As we will see in the next chapter.

2. THE SIMPLEST STRIPLINES

In the previous chapter it was shown that in case of the uniform filling of a waveguide with an isotropic medium, the main wave is characterized by the propagation constant:

$$h = k\sqrt{\varepsilon\mu}$$

and the electromagnetic field is transversal. The electromagnetic field can be described by the potential $\varphi(x, y)$ so that:

$$E_x = -\frac{\partial\varphi}{\partial x}, \quad E_y = -\frac{\partial\varphi}{\partial y}, \qquad E_z = 0,$$

$$H_x = -\frac{E_y}{Z}, \quad H_y = \frac{E_x}{Z}, \qquad H_z = 0. \tag{2.1}$$

Here $Z = \sqrt{\mu\mu_0/\varepsilon\varepsilon_0}$ is a wave impedance of the medium, which fills the waveguide. The potential φ satisfies Laplace's equation:

$$\frac{\partial^2\varphi}{\partial x^2} + \frac{\partial^2\varphi}{\partial y^2} = 0 \tag{2.2}$$

and certain boundary conditions. The equation (2.2) shows that the potential φ is a harmonic function. One can represent this potential φ as a real or an imaginary part of the corresponding analytic function.

In this chapter, we will present our calculation of capacitances for several different striplines that can be expressed through simple formulae.

2.1. Presentation of an Electric Field

Here we see the complex function:

$$\Phi(z) = \varphi(x, y) + j\psi(x, y)$$

of a complex coordinate:

$$z = x + jy$$

(at the point with coordinates x, y) has to be an analytic function in a certain region. In the analyticity region, the derivative $\Phi'(z)$ does not depend on the direction of approach to point z. This is expressed by Cauchy–Riemann conditions:

$$\frac{\partial \varphi}{\partial x} = \frac{\partial \psi}{\partial y}, \qquad \frac{\partial \varphi}{\partial y} = -\frac{\partial \psi}{\partial x}, \tag{2.3}$$

for partial derivatives of the real and imaginary parts. We write here the derivative $\Phi'(z)$ while approaching the point z parallel to axis x then:

$$\Phi'(z) = \frac{\partial \varphi}{\partial x} + j\frac{\partial \psi}{\partial x}.$$

We notice that another complex unit j is introduced here. Unit j is independent of the complex unit i (introduced earlier in chapter one).

The expression for $\Phi'(z)$, after using the second formula of the Cauchy–Riemann conditions (2.3), will become:

$$\Phi'(z) = \frac{\partial \varphi}{\partial x} - j\frac{\partial \varphi}{\partial y} = -E_x + jE_y \cong \hat{E}, \tag{2.4}$$

this means that the derivative of the complex potential $\Phi'(z)$ describes the plane electric field:

$$\mathrm{Re}\,\Phi'(z) = -E_x,$$

$$\mathrm{Im}\,\Phi'(z) = E_y.$$

Now we write the product:

$$\Phi'(z)dz = \left(-E_x + jE_y\right)\left(dx + jdy\right) = -\left(E_x dx + E_y dy\right) - j\left(E_x dy - E_y dx\right). \tag{2.5}$$

Bearing this in mind the surface vector element has projections as is shown in Fig.2.1:

$$dS_x = \ell dy, \qquad dS_y = -\ell dx \tag{2.5a}$$

and we represent the formula (2.5) in this form:

$$\Phi'(z)dz = -(Edr) - \frac{j}{\ell}(EdS). \tag{2.6}$$

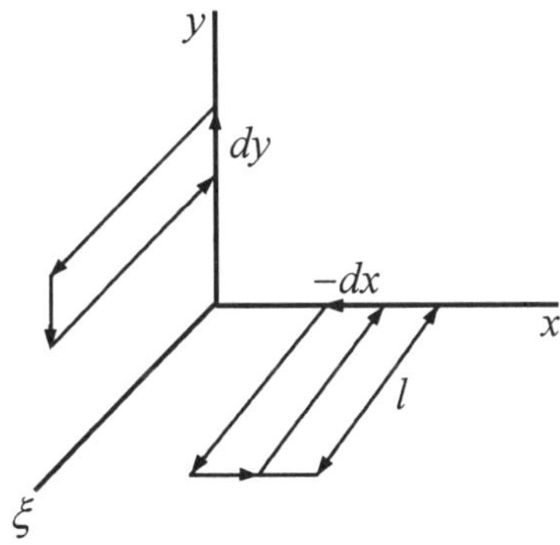

Fig. 2.1. *Projections of a vector surface element.*

Integrating this expression in the plane x and y from the point r_1 to the point r_2 and separating the real part, we find that the real part of the integral:

$$\mathrm{Re} \int_{r_1}^{r_2} \Phi'(z)dz = \varphi(r_2) - \varphi(r_1).$$ (2.7)

The formula (2.7) gives the potential difference at the points r_1 and r_2. Integrating the expression (2.6) over the closed contour C and separating the imaginary part we get the value:

$$\mathrm{Im} \oint_C \Phi'dz = -\frac{1}{i\varepsilon\varepsilon_0 \ell} \oint_S (DdS) = -\frac{Q}{\varepsilon\varepsilon_0 \ell},$$ (2.8)

which is proportionate to the charge per unit length. The closed surface S in the formula (2.8) consists of the side surface of the length ℓ of the cylinder. This cylinder is made by a straight line (normal to the plane xy) that slides along the contour C and has two butt–end lids. The flux through the lids is equal to zero because the electric field is plane.

In this first example, we determine the electric field of an infinitely thin ideal metal strip when the width is $2a$ (Fig.2.2) and has a carrying charge of Q/ℓ per unit length. Here Q is the charge of the metal strip of the length ℓ.

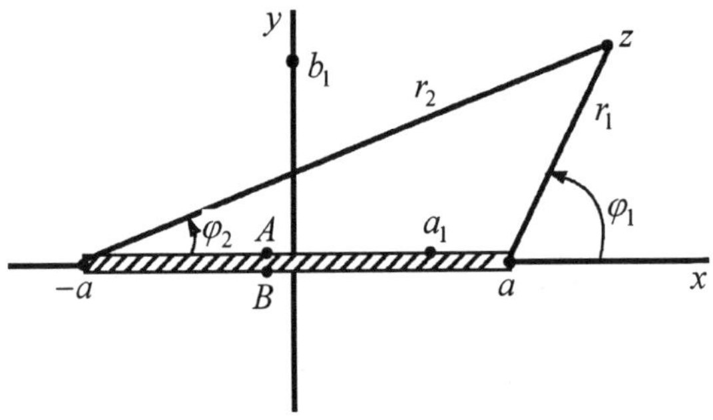

Fig. 2.2. *Designations for determining the electric field of a metal strip.*

The qualitative picture of this electric field is obvious. In the points $y = 0$ of the metal surface the direction of vector $\vec{E}$ is normal to the surface. The field directions on the upper and low strip sides (in the points A and B) are opposite. The potential of the point with these coordinates $(a_1, 0)$ for $|a_1| \leq a$ does not depend on a_1 value. When the distant between the metal strip and the point, which we observe increases then the electric field strength $|\vec{E}|$ diminishes. When the distance increased from this metal strip, the modulus $|\vec{E}|$ of the electric field is inversely proportional to the distance directed along the radius vector. At very great distances from the metal strip the electric field coincides with the electric field of a linear charge having the potential $\varphi \sim \ln r$. So in this case the electric field strength is $\bar{E} \sim \bar{r}/r^2$.

Thus, the function $\Phi'(z)$ describing an electric field according to the formula (2.4) must be two valued at the points $-a \le x \le a$ and $y = 0$. The function values $\Phi'(z_A) = -\Phi'(z_B)$ have opposite signs at the points A, B and $\lim\limits_{z \to \infty} |\Phi'(z)| \sim 1/|z|$.
Two–valuedness of the function occurs taking the square root. Therefore:

$$\Phi'(z) = \frac{A_0}{\sqrt{z^2 - a^2}} \qquad (2.9)$$

and the cut defining the one–valued function branch joins the points $(-a,0)$ and $(a,0)$ along the real axis (in the cut points one can enter the second sheet of Riemann surface).

We now substitute in the expression (2.9) $|x| \prec a$, $y \to +0$ and we determine the field $\hat{E}$ at the point A. We receive the expression:

$$\Phi'(x,+0) = -j\frac{A_0}{\sqrt{a^2 - x^2}}, \qquad (2.10)$$

so here one takes the arithmetic value of the square root (positive or zero). Comparing the expressions (2.10) and (2.4) and bearing in mind that the component $E_x = 0$ in the point A one can see that the constant A_0 must be the real number:

$$E_y(A) = -E_y(B) = -\frac{A_0}{\sqrt{a^2 - x^2}}. \qquad (2.11)$$

Now we find the integral (2.8) over the closed contour around the strip. The integral value does not depend on the contour shape. Choosing the contour as a circle with the center in the origin and with a large enough radius we find that $|z \succ\succ a|$. We found the expression:

$$\oint_C \Phi'(z)dz = A_0 \oint_C \frac{dz}{z} = 2\pi jA_0$$

and the constant A_0 is determined according to the formula (2.8) and A_0 is expressed like this:

$$A_0 = -\frac{Q}{2\pi\varepsilon\varepsilon_0\ell}. \qquad (2.12)$$

The relationship (2.12) between the constant A_0 and the charge per unit length Q/ℓ can be found in another way. We use the fact that the surface charge density on the conductor surface $\sigma = D_n$ coincides with the normal component of electric inductance vector $D_n = \varepsilon\varepsilon_0 E_n$. Using (2.11) we express the charge of the metal strip which has the length ℓ:

$$Q = 2\ell \int\limits_{-a}^{+a} \sigma \, dx = -4 A_0 \varepsilon \varepsilon_0 \ell \int\limits_{0}^{a} \frac{dx}{\sqrt{a^2 - x^2}} = -2\pi \varepsilon \varepsilon_0 \ell A_0 \,.$$

This formula coincides with the value (2.12) because the calculation that was made means integrating the formula (2.8) over the contour closely adjoining the cut (Fig.2.2).

By using this formula (2.7) we can determine the potential difference of φ_{12} between these points $z_1 = a_1$ (when $|a_1| \prec a$) and $z_2 = jb_1$. The potential difference is:

$$\varphi_{12} = A_0 \operatorname{Re} \int\limits_{a_1}^{jb_1} \frac{dz}{\sqrt{z^2 - a^2}} = A_0 \operatorname{Re}\left[\ln\left(z + \sqrt{z^2 - a^2}\right)\Big|_{z = a_1}^{z = jb_1}\right] =$$

$$= A_0 \operatorname{Re}\left[\ln \frac{j\left(b_1 + \sqrt{a^2 + b_1^2}\right)}{a_1 + j\sqrt{a^2 - a_1^2}}\right] = A_0 \ln \frac{b_1 + \sqrt{a^2 + b_1^2}}{a} \,.$$

The result does not depend on the value a_1. This means that the potentials of all the conductor points are equal.

Finally, we will find the dependence of the electric field strength modulus near the strip edge on the distance r_1. According to the expression (2.4) we will place into the formula (2.9) these magnitudes:

$$z - a = r_1 e^{j\varphi_1} \,,$$

$$z + a = r_2 e^{j\varphi_2} \,.$$

After taking into account the magnitudes $r_1 \to 0$ and $r_2 = 2a$ we will obtain:

$$|\vec{E}| = \frac{|A_0|}{\sqrt{2 a r_1}} \sim \frac{1}{\sqrt{r_1}} \,.$$

The result in this particular problem displays the general rule. The electric field behavior near the edge points is determined by the behavior of the electrostatic fields.

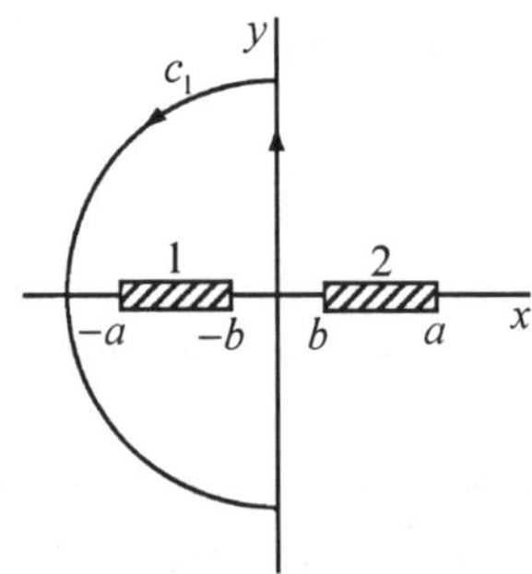

Fig. 2.3. *Designations for determining the electric field of two interacting metal strips.*

In this second example, we consider the electric field of two infinitely thin metal strips lying in the same plane as shown in Fig.2.3. Here the field is described by the function:

$$\hat{E} = \frac{Az + B}{\sqrt{\left(z^2 - a^2\right)\left(z^2 - b^2\right)}},$$ (2.13)

with the real constants A, B, a, b and $b \prec a$. The cuts in the complex plane $z = x + jy$ coincide with strips plane. Integrating over the circle of a large radius is:

$$\oint_C \hat{E}dz = 2\pi jA.$$

By the formula (2.8) we determine the constant:

$$A = -\frac{Q}{2\pi\varepsilon\varepsilon_0\ell},$$ (2.14)

where $Q = Q_1 + Q_2$, Q_1 and Q_2 are the charges of the first and second strips of the length ℓ. We now integrate the expression (2.13) over the contour C_1 (Fig.2.3) that is made of a semicircle with an infinite radius and with a part of axis y. We have:

$$\oint_{C_1} \hat{E}dz = j\pi A + \int_{-\infty}^{\infty} \frac{(jB - Ay)dy}{\sqrt{\left(y^2 + a^2\right)\left(y^2 + b^2\right)}} = j\pi A + 2jB \int_0^{\infty} \frac{dy}{\sqrt{\left(y^2 + a^2\right)\left(y^2 + b^2\right)}}.$$ (2.15)

The last integral according to the formulae (3.152.1) and (8.112.1) of ref. [2.1]:

$$\int_0^{\infty} \frac{dy}{\sqrt{\left(y^2 + a^2\right)\left(y^2 + b^2\right)}} = \frac{1}{a} K\left(\frac{\sqrt{a^2 - b^2}}{a}\right),$$

is proportional to the full elliptic integral K of the first kind. The tables of its values can be found in ref. [2.2]. Therefore, from (2.8), (2.14) and (2.15), we found the constant:

$$B = \frac{Q_1 - Q_2}{4\varepsilon\varepsilon_0\ell} \frac{a}{K(k)}, \qquad k = \frac{\sqrt{\left(a^2 - b^2\right)}}{a}.$$ (2.16)

If the whole charge Q is zero then constants is expressed:

$$A = 0, \qquad B = \frac{Q_1}{2\varepsilon\varepsilon_0\ell} \frac{a}{K(k)}.$$

The potential difference between strips:

$$\varphi_{12} = \int_{-b}^{+b} E_x dx = 2B \int_0^{+b} \frac{dx}{\sqrt{\left(a^2 - x^2\right)\left(b^2 - x^2\right)}} = \frac{2B}{a} K\left(\frac{b}{a}\right),$$ (2.17)

is calculated by the formula (2.7) using the formulae (3.152.7) and (8.112.1) from ref. [2.1]. So the potential difference between strips is expressed:

$$\varphi_{12} = \frac{Q_1}{2\varepsilon\varepsilon_0\ell}\frac{K(k')}{K(k)}, \text{ where } k' = \sqrt{1-k^2}.$$

The capacitance per unit length will be expressed:

$$C \equiv \frac{Q_1}{\ell\varphi_{12}} = \varepsilon\varepsilon_0\frac{K(k)}{K(k')}, \quad \text{where} \quad k = \frac{\sqrt{a^2-b^2}}{a}. \tag{2.18}$$

The capacitance C is the most important integral characteristic of the main wave propagating along the waveguide. In the same way one can calculate characteristics of a stripline having two or more strips of different widths.

2.2. The Application of Conformal Mapping

The electric field strength $\vec{E}$ of the main wave of the multi-conductor line is described by an analytic function $\Phi'(z)$. This allows us to calculate the field, capacitances and other electrodynamical characteristics of a stripline with a cross–section configuration, obtained by conformal mapping of a simple stripline. The capacitances of both lines are the same, because calculating charges and potential differences of the corresponding integrals coincide for both lines. This is verified by changing the integration variable according to the conformal transformation. In this chapter we will give some simple examples of the usage of one region conformal mapping into another.

From the symmetry of Fig.2.3 with respect to the axis y, it is clear that for $Q_1 = -Q_2$ the component $E_y = 0$ on the straight line $x = 0$ is a metal wall along the axis y, as shown in the Fig.2.4. We will get the same results from formula (2.13) if we allow that the constant $A = 0$.

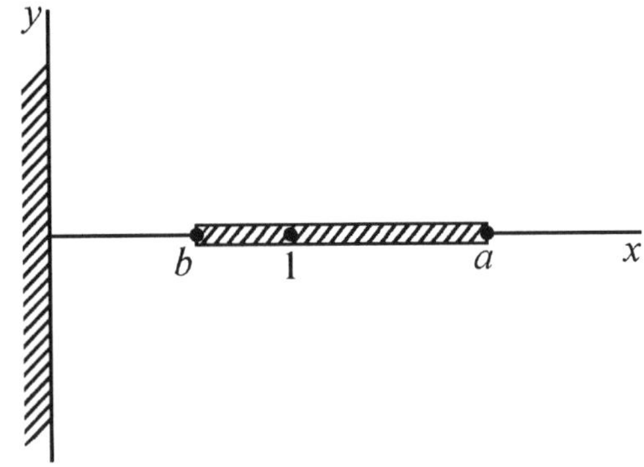

Fig. 2.4. For determination of the capacitance of a metal strip.

The capacitance per unit length of the metal strip shown in Fig.2.4:

$$C = 2\varepsilon\varepsilon_0 \frac{K\left(\sqrt{1 - b^2/a^2}\right)}{K(b/a)}. \tag{2.19}$$

This capacitance per unit length is exactly twice as large as the metal strip shown in Fig.2.3.

The conformal mapping:

$$w = d \ln z/\pi, \quad w = u(x, y) + iv(x, y), \tag{2.20}$$

transfers the axis $y \geq 0$ (Fig.2.4) into the straight line $v = d/2$. The negative semi–axis $y \leq 0$ into $v = -d/2$ and the real axis x is transferred into the real axis u. Because of this the point $z = 1$ has as its image the origin $w = 0$. These metal strip's (Figs.2.4 and 2.5) geometric parameters are interconnected to the formulae:

$$u_1 = d \ln a/\pi, \qquad a \succ 1,$$
$$-u_1 = d \ln b/\pi, \qquad b \prec 1, \quad u_1 \succ 0.$$

Then by the formula (2.19), we will determine the capacitance:

$$C = 2\varepsilon\varepsilon_0 \frac{\left(\sqrt{1 - e^{-4\pi u_1/d}}\right)}{K\left(e^{-2\pi u_1/d}\right)} \tag{2.21}$$

for the line shown in Fig.2.5. The electric field lines deviate more from the straight lines $u = const$ near the strip edges $u = \pm u_1$.

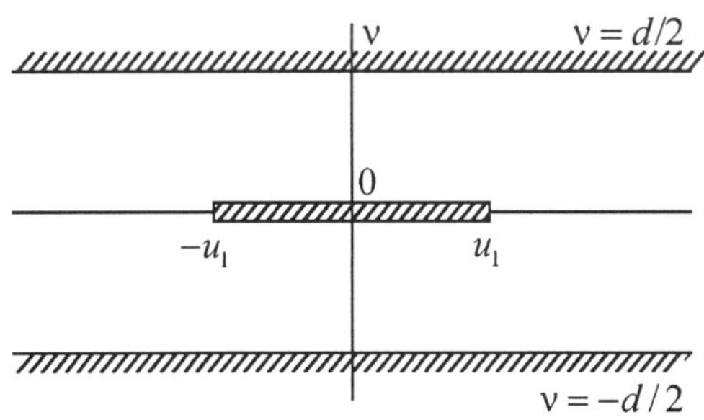

Fig. 2.5. *Cross–section of the symmetric stripline.*

If the strip width is $2u_1 \succ\succ d$ then the influence of width deviations on the capacitance value (per unit length) is small and the capacitance can be calculated by the formula:

$$C = 2 \frac{\varepsilon\varepsilon_0 \cdot 2u_1}{d/2}.$$

The edge effect correction for $u_1/d \ggg 1$ can be determined by the formula (2.21), which uses the expansion (ref. [2.2] pages 109 and 114) of an elliptic integral K :

$$K\left(e^{-2\pi u_1/d}\right) \approx \frac{\pi}{2}\left(1+\frac{1}{4}e^{-4\pi u_1/d}+\ldots\right),$$

$$K\left(\sqrt{1-e^{-4\pi u_1/d}}\right) \approx \ln 4 + \frac{2\pi u_1}{d} + \frac{1}{4}\left(\ln 4 - 1 + \frac{2\pi u_1}{d}\right)\cdot e^{-4\pi u_1/d} + \ldots .$$

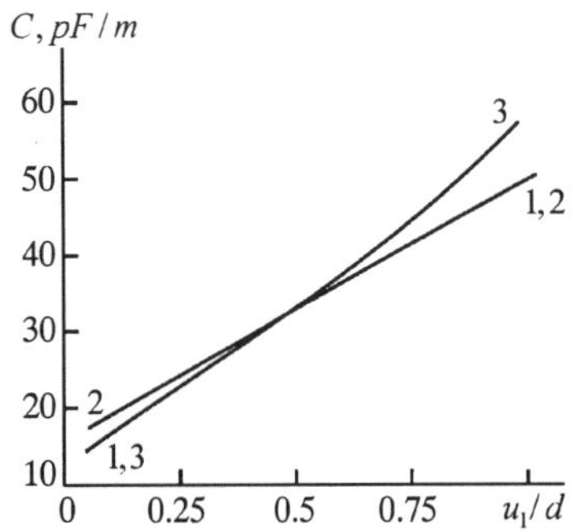

Fig. 2.6. *The symmetric stripline capacitance dependence on the ratio u_1/d : curves 1, 2 and 3 were calculated by the formulae (2.21), (2.22) and (2.23) respectively.*

Here we write the approximate expression of the stripline capacitance when considering the edge effect correction:

$$C \approx \frac{8\varepsilon\varepsilon_0 u_1}{d} + \frac{4\varepsilon\varepsilon_0}{\pi}\left(\ln 4 - \frac{1}{4}e^{-4\pi u_1/d}\right), \qquad u_1/d \ggg 1. \tag{2.22}$$

For $2u_1/d \prec 0.35$ the approximate formula from ref. [2.3]:

$$C = \frac{2\pi\varepsilon\varepsilon_0}{\ln(5.1 u_1/d)} \tag{2.23}$$

makes an error of less than 1%. A comparison of capacitances per unit length calculated by the exact and approximate formulae (Fig.2.6) show that they concur with each other.

2.3. Edge Effects of Symmetrical Striplines

2.3.1. Short–circuiting metal plane. The formula:

$$w = \frac{d}{\pi}\,arcsh\ z\,,$$

transforms the region shown in Fig.2.4 into the region of Fig.2.7. The simplest way to see this is to consider the inverse transformation:

$$z = sh\left(\frac{\pi w}{d}\right).$$

(2.24)

When separating the real and imaginary parts of the expression (2.24) we receive:

$$x = sh\left(\frac{\pi u}{d}\right)\cos\left(\frac{\pi v}{d}\right),$$

$$y = ch\left(\frac{\pi u}{d}\right)\sin\left(\frac{\pi v}{d}\right).$$

(2.25)

The formulae show that the contour *ABCD* in Fig.2.7 is transformed (after applying (2.24)) into the axis y in Fig.2.4 and the real axes are transformed into each other. From the formula (2.24) we find the magnitudes:

$$b = sh\left(\frac{\pi u_1}{d}\right), \qquad a = sh\left(\frac{\pi u_2}{d}\right), \qquad u_2 = u_1 + \ell.$$

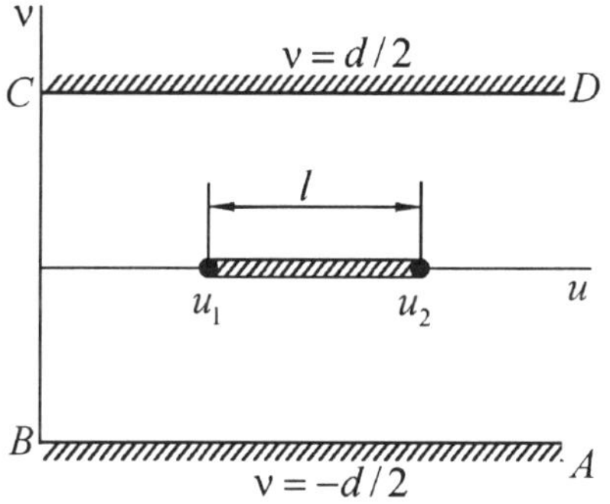

Fig. 2.7. *The designations on conformal mapping of one region into another.*

The mutual magnitudes correspond to the geometrical parameters of striplines in Figs.2.7 and 2.4. Therefore, the capacitance per unit length of the stripline (Fig.2.7) is:

$$C = 2\varepsilon\varepsilon_0 \frac{K\left(\sqrt{1 - \dfrac{sh^2\left(\pi u_1/d\right)}{sh^2\left(\pi u_2/d\right)}}\right)}{K\left(\sqrt{\dfrac{sh\left(\pi u_1/d\right)}{sh\left(\pi u_2/d\right)}}\right)}.$$

(2.26)

Fig.2.8 shows the dependences of the capacitance per unit length on the ratio u_1/d for three different values ℓ/d. The values of ℓ/d are 0.1, 0.5 and 1.0 for the curves 1, 2 and 3 respectively. When $u_1/d \geq 0.7$ the capacitance calculated by the formula (2.26) coincides with the capacitance of the boundless symmetric stripline with an error of less than 1%. We can see in Fig.2.8 that the capacitance C grows rapidly when the values of u_1/d is less than 0.1.

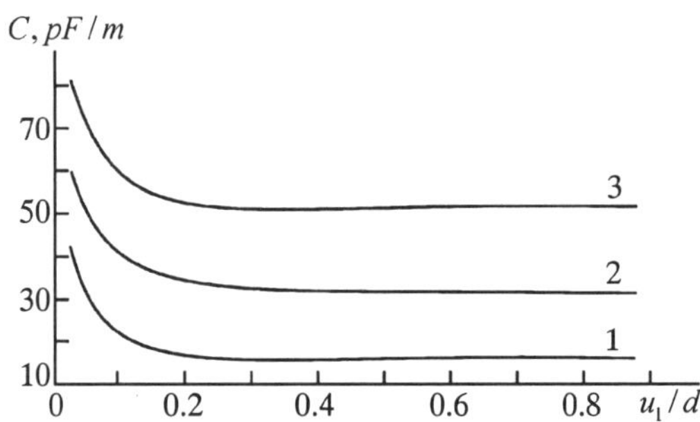

Fig. 2.8. *The dependences of the capacitance per unit length on the ratio u_1/d for a stripline with the shot–circuited end at the values ℓ/d : where curves 1, 2 and 3 are 0.1, 0.5 and 1.0 respectively.*

2.3.2. The open end. We now consider the transformation:

$$w = z + \frac{d}{2\pi} e^{2\pi z/d} , \tag{2.27}$$

this transforms the region of Fig.2.9 into the region Fig.2.10. To verification this, we must separate the real and imaginary parts:

$$u = x + \frac{d}{2\pi} e^{2\pi x/d} \cos\left(2\pi y/d\right) , \tag{2.28}$$

$$v = y + \frac{d}{2\pi} e^{2\pi x/d} \cos\left(2\pi y/d\right) \tag{2.29}$$

in the formula (2.27).

The relationship of (2.29) shows that the straight lines $y = \pm d/2$ in Fig.2.9 correspond to the half–straight lines $v = \pm d/2$, which are running parallel to the axis u from infinity to the points A and B in Fig.2.10. The formula (2.28) at $y = d/2$ takes the form:

$$u = x - \frac{d}{2\pi} e^{2\pi x/d} ,$$

$$u = x - \frac{d}{2\pi} e^{2\pi x/d} ,$$

where one can see that while the coordinate x is growing, that the coordinate u grows first and reaches the value u_A . Then it diminishes going into negative infinity. The maximum is reached at $x = 0$, when the derivative of the function turns into zero. Therefore, the values are

$$u_A = -d/2\pi , \qquad v_A = d/2 .$$

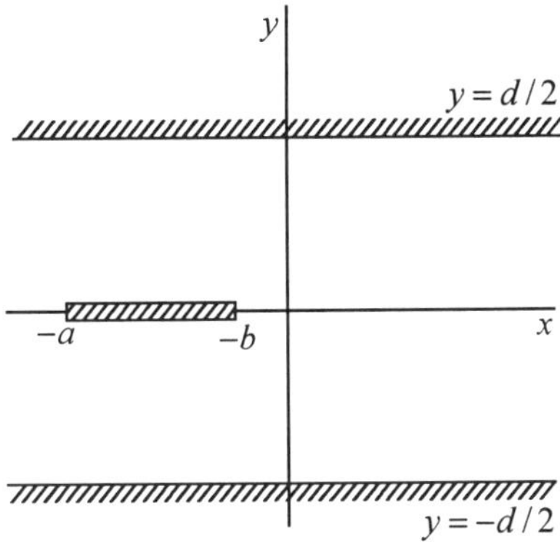

Fig. 2.9. *Topology of the symmetric stripline cross–section.*

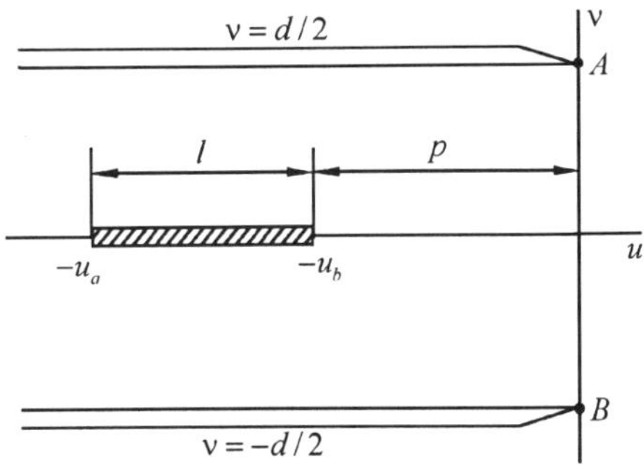

Fig. 2.10. *On conformal mapping of one region into another.*

The capacitance of the striplines shown in Figs.2.09 and 2.10 is expressed by the formula (2.21):

$$C = 2\varepsilon\varepsilon_0\, K(m')/K(m),$$

where the arguments are:

$$m' = \sqrt{1-m^2} \quad \text{and} \quad m = e^{-\pi(a-b)/d}.$$

The equations to find the ratios a/d and b/d are derived from the expression (2.28) at $y = 0$:

$$\frac{u_a}{d} = \frac{a}{d} - \frac{1}{2\pi}e^{-2\pi a/d},$$

$$\frac{u_b}{d} = \frac{b}{d} - \frac{1}{2\pi}e^{-2\pi b/d}.$$

Fig.2.11 shows the dependence of the capacitance per the unit length of the stripline (Fig.2.10) on the ratio p/d when the values of ℓ/d are 0.1 (curve 1), 0.5 (curve 2) and 1.0 (curve 3).

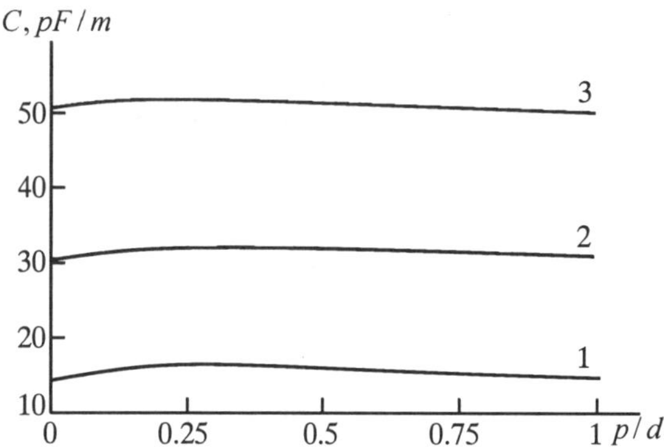

Fig. 2.11. *The dependence of the capacitance per unit length on the ratio p/d for an open end stripline at the following values of ℓ/d : the curves 1, 2 and 3 are 0.1, 0.5 and 1.0 respectively.*

For $p/d \geq 0.6$, the capacitance of this stripline almost coincides with that of the boundless symmetric stripline. The difference is less than one percent.

2.3.3. The flange end. It is possible to receive a symmetric stripline with a flange end (Fig.2.12) from the stripline shown in the Fig.2.4 by use of conformal mapping:

$$w = \frac{d}{\pi}\left(\sqrt{z^2+1} - arcsh(1/z)\right),$$ (2.30)

where the second part of this formula can be written in this form:

$$arcsh(1/z) = \ln\left(\left(1+\sqrt{1+z^2}\right)\Big/z\right).$$ (2.31)

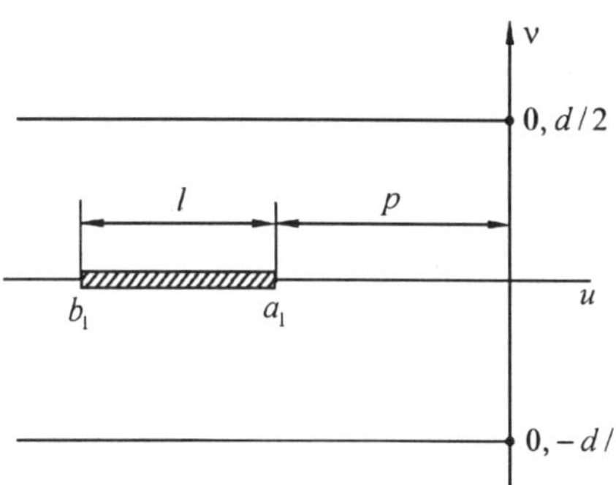

Fig. 2.12. *The designations for determining the capacitance of a symmetric stripline with a flange end.*

Points $(x,0)$ and $x \geq 0$ of a positive real axis transfer into points $(u,0)$ and $-\infty \prec u \prec \infty$ on the whole real axis in the plane of the variable $w = u + iv$. We now consider the images of the points of the imaginary axis $(0,y)$ when $y \geq 0$.

When we substitute the coordinate $z = iy$ in the formulae (2.30) and (2.31) one will obtain this formula:

$$w = \frac{d}{\pi}\left(\sqrt{1-y^2} - \ln\frac{1+\sqrt{1-y^2}}{y} + j\frac{\pi}{2} \right). \tag{2.32}$$

While the quantity y is varying from 0 to 1, the real part u of the expression (2.12) varies from $-\infty$ to 0, and the imaginary part $v = +d/2$ remains fixed. When it is this $y \geq 1$, then in the formula:

$$w = \frac{d}{\pi}\left(j\sqrt{y^2-1} - \ln\frac{1+j\sqrt{y^2-1}}{y} + j\frac{\pi}{2} \right) \tag{2.33}$$

the logarithm's argument has the modulus equal to a unit and it can be presented as $e^{j\varphi}$ and $tg\,\varphi = \sqrt{y^2-1}$, so that the angle φ varies within the first quarter. Here the expression (2.33) is purely imaginary and the quantity $v \geq d/2$ for $y \geq 1$. Therefore, we obtain the image of these points (Fig.2.4) of the imaginary axis $y \leq 1$ in Fig.2.12. The correspondence of the regions which are shown in Fig.2.4 and Fig.2.12 means that the capacitance of the stripline (Fig.2.12) is expressed by the formula:

$$C = 2\varepsilon\varepsilon_0 \frac{K(m')}{K(m)}, \quad m' = \sqrt{1-m^2}, \quad m = b/a, \tag{2.34}$$

where a and b are the roots of the equations:

$$\frac{a_1}{d} = \frac{1}{\pi}\left(\sqrt{1+a^2} - arcsh\frac{1}{a} \right),$$

$$\frac{b_1}{d} = \frac{1}{\pi}\left(\sqrt{1+b^2} - arcsh\frac{1}{b} \right), \qquad a \succ b, \quad a_1 \succ b_1.$$

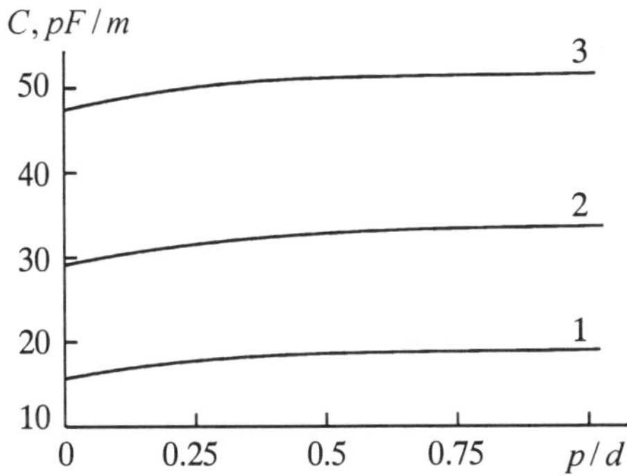

Fig. 2.13. The dependence of the capacitance per the unit length on the ratio p/d for the stripline with a flange end for the following values of ℓ/d : the curves 1, 2 and 3 are 0.1, 0.5 and 1.0 respectively.

The calculated results of the capacitance for three values of ℓ/d : 0.1 (curve 1), 0.5 (curve 2) and 1.0 (curve 3) are presented in Fig.2.13.

One can see that for $p/d = 0.3$ the capacitance of the stripline (Fig.2.12) differs from the capacitance of the boundless symmetric line by less than 1%.

2.4. Screened Lines

2.4.1. Capacitance of a stripline in a circular cylinder. The conformal mapping:

$$z = R \; th\frac{w\pi}{2d} , \tag{2.35}$$

transforms the stripline $|v| \le d/2$ of Fig.2.5 into the circle of the radius R (Fig.2.14). Here the real and imaginary axes (Fig.2.5) are transformed into real and imaginary axes (Fig.2.14) respectively.

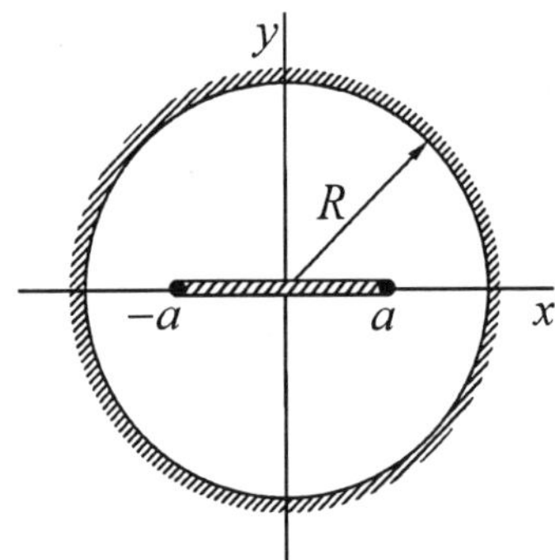

Fig. 2.14. Cross–section of a stripline in a circular screen.

By placing in the formula (2.35), $w = \pm j\,d/2 + u$, we find that the boundaries of the region in Fig.2.5 are transformed into the circumference $|z| = R$.

Using this formula:

$$arcth \; w = \frac{1}{2}\ln\frac{1+w}{1-w} ,$$

we can recalculate the arguments of the functions in the formula (2.21) and determine the capacitance:

$$C = 2\varepsilon\varepsilon_0 \; K\!\left(\sqrt{8a\left(1+a^2\right)}\big/\left(1+a\right)^2 \right)\!\Big/ K\!\left(\left(1-a\right)^2\big/\left(1+a\right)^2\right) \tag{2.36}$$

of the stripline (Fig.2.14) with the radius $R = 1$. Here, the approximate formula [2.3] is:

$$C = 2\pi\varepsilon\varepsilon_0 / \ln(2R/a). \tag{2.37}$$

The calculated results according to the exact and approximate formulae are shown in Fig.2.15 by curves 1 and 2 respectively.

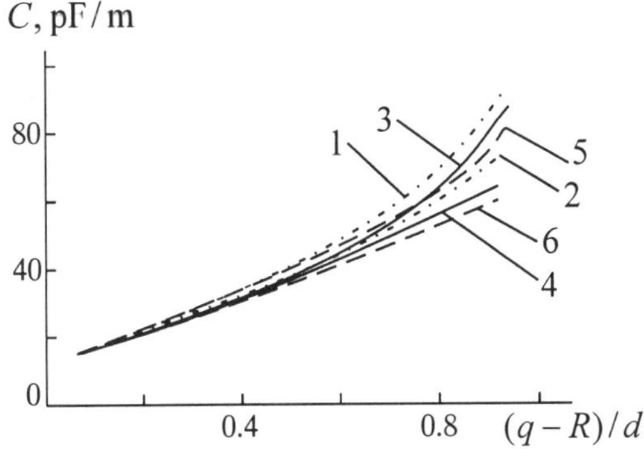

Fig. 2.15. *The dependence of capacitance per unit length on geometric parameters of striplines shown in Fig.2.14 (curves 1, 2) and Fig.2.16 (curves 3, 4).*

The capacitance of the stripline (Fig.2.14) does not change if the stripline is turned by 90°. However, the conformal transformation inversed to (2.35), gives the stripline with a strip oriented normally to metal planes (Fig.2.16). The capacitance of this stripline is expressed by the formula:

$$C = 2\varepsilon\varepsilon_0 \, K\!\left(2\sqrt{\sin(\pi a/d)}\big/(1+\sin(\pi a/d))\right)\!\big/ K\!\left((1-\sin(\pi a/d))/(1+\sin(\pi a/d))\right), \tag{2.38}$$

which can be presented approximately [2.3] in this form:

$$C = 2\pi\varepsilon\varepsilon_0 / \ln(2.25d/2a). \tag{2.39}$$

The calculated results according to the exact and approximate formulae are presented in Fig.2.15 by the curves 5 and 6 respectively.

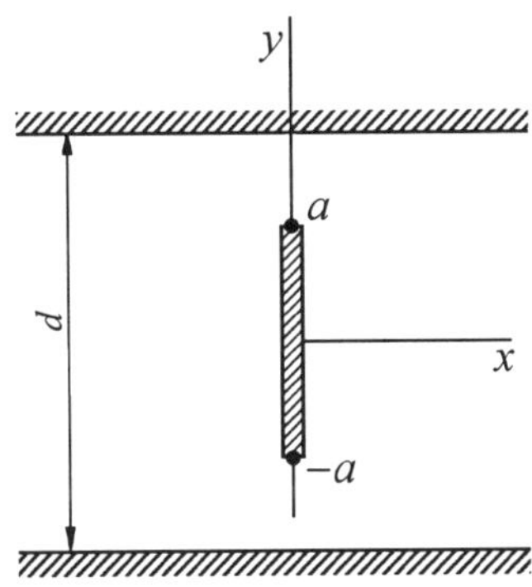

Fig. 2.16. *The capacitance of a symmetric stripline with a vertical strip.*

2.4.2. Capacitance of a stripline in a rectangular cylinder. The conformal mapping of the upper half–plane Im $z \geq 0$ (Fig.2.17) into the polygon of Fig.2.18 in the plane of the complex variable $w = u + iv$ is realized by the Christoffel–Schwarz formula [2.4]:

$$w(z) = w_0 + A \int_0^z (z - b_1)^{-\varphi_1/\pi} (z - b_2)^{-\varphi_2/\pi} \ldots dz \, , \qquad (2.40)$$

when the point b_1 (Fig.2.17) goes over into the vertex with coordinate a_1 (Fig.2.18). We choose the points a_i, b_i and $i = 1,\ldots, 4$, in accordance with Figs.2.19 and 2.20, $n \prec 1$.

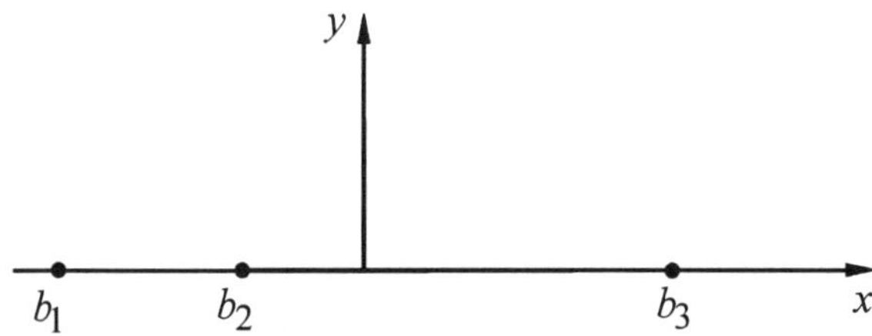

Fig. 2. 17. *The plane of the complex variable* $z = x + jy$.

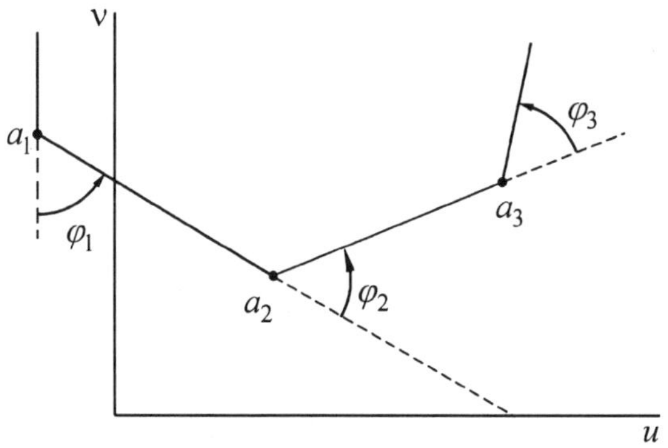

Fig. 2.18. *The polygon in the plane of the complex variable* $w = u + iv$.

The corresponding formula:

$$w(z) = w_1 + A \int_0^z \frac{dz}{\sqrt{\left(1 - z^2\right)\left(1 - n^2 z^2\right)}} \qquad (2.41)$$

or $\quad w(z) = w_1 + AF(z, n)$, $\qquad (2.42)$

where $F(z, n)$ is the elliptic function of the first kind. The formulae (2.41) or (2.42) are determined from the general formula (2.40).

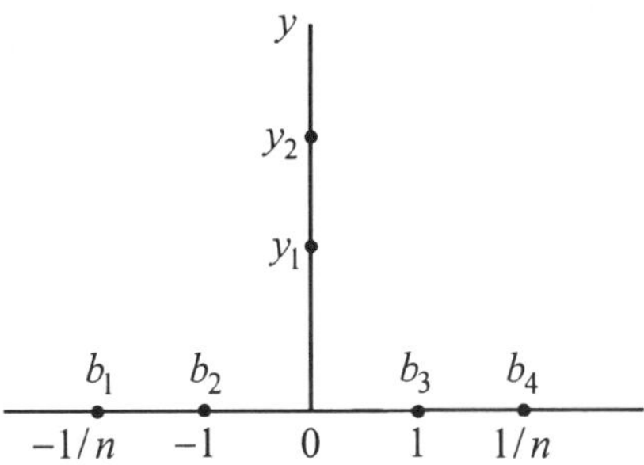

Fig. 2.19. *The designation for determining the capacitance of a stripline in a rectangular waveguide.*

From the definitions (2.41) and (2.42) one can see that the function:

$$F(z,n) = -F(-z,n) \tag{2.43}$$

is uneven. Bearing this in mind the designation

$$F(1,n) = K(n) \equiv K \tag{2.44}$$

of the elliptic integral [2.1] we can write that:

$$F\left(\pm\frac{1}{n}, \ n\right) = \pm K + jK',$$

$$K' = K(n'), \qquad n' = \sqrt{1-n^2} \ . \tag{2.45}$$

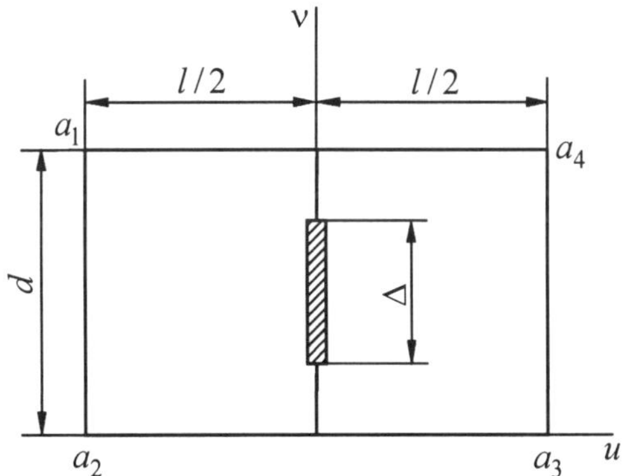

Fig. 2.20. *Screened stripline with a vertical metal conductor.*

We arrive at the expression:

$$\int_0^{\pm\frac{1}{n}} \frac{dz}{\sqrt{(1-z^2)(1-n^2z^2)}} = \int_0^{\pm 1} \frac{dz}{\sqrt{(1-z^2)(1-n^2z^2)}} + \int_{\pm 1}^{\mp\frac{1}{n}} \frac{dz}{\sqrt{(1-z^2)(1-n^2z^2)}} \ .$$

The first integral is $\pm K$ according to the definition (2.44) and the second integral is equal to iK'. This is easy to verify changing the variable

$z = \left(1 - n'^2 t^2\right)^{-1/2}$. Constants w_1, A and n in the formula (2.42) are determined by the correspondence of the points b_1, b_2, b_3, b_4 (Fig.2.19) to the points a_1, a_2, a_3, a_4 (Fig.2.20):

$$\ell/2 = w_1 + A \cdot K, \qquad \ell/2 + jd = w_1 + A \cdot \left(K + jK'\right)$$

and

$$-\ell/2 + jd = w_1 + A \cdot \left(-K + jK'\right).$$

From that:

$$w_1 = 0, \qquad\qquad A = \ell/(2K) = d/K'. \tag{2.46}$$

Therefore, the modulus n is determined by the equation:

$$\ell/(2d) = K/K'. \tag{2.47}$$

We now write the formulae:

$$F(iy_1, n) = j\left(\frac{d - \Delta}{2d}\right) \cdot K',$$

$$F(iy_2, n) = j\left(\frac{d + \Delta}{2d}\right) \cdot K'. \tag{2.48}$$

The function inverted to $F(z, n)$ is usually denoted as $sn\ F$ and is called the elliptic sine. Then:

$$m \equiv \frac{y_1}{y_2} = \frac{sn\left(jK(n') \cdot (d - \Delta)/2d\right)}{sn\left(jK(n') \cdot (d + \Delta)/2d\right)}. \tag{2.49}$$

The value m is the real magnitude, because $F(jy, n)$ is imaginary number for the real number y. Now the stripline capacitance (Fig.2.20) is calculated by the formula (2.19) in which the ratio $b/a = m$ is found from the expression (2.49). The approximate formula for $\ell = d$ has the form [2.3] like this:

$$C = \frac{2\pi\varepsilon\varepsilon_0}{2\ln\left(2.16d/2\Delta\right)}.$$

The calculated results from the exact and approximate formulae are presented in Fig.2.15, curves 3 and 4. In a similar way we found the capacitance of the striplines (Figs.2.14, 2.16 and 2.20) having several strips [2.5].

Our calculations for all striplines considered in this chapter show that at the ratio $p/d \geq 2$ these striplines did not "feel" their edges when the calculations were fulfilled with the relative accuracy of the order 10^{-4}.

As you can see here we calculated the simplest striplines by conformal mapping. In order to calculate striplines or other type of waveguides with complicated shapes of cross-sections we are going to use the theory of Singular Integral Equations (SIE). We now follow up with singular integrals and their behaviors.

3. CAUCHY TYPE INTEGRAL AND SOLUTION OF SINGULAR INTEGRAL EQUATIONS

To determine the characteristics of simple symmetric stripline (in the case of uniform fillings) we must use special functions for their calculation. For striplines having several strips (in case of nonuniform fillings with isotropic or anisotropic medium), one can hardly expect to arrive at the solutions by simple formulae. And so the application of numerical methods becomes essential.

One of these numerical methods we will be using to investigate striplines structures is the SIE methods. The solution of the SIE is closely connected with the theory of Cauchy Type Integrals and the problem of conjugation (determining of an analytical function from the ratio of its values on opposite sides of a cut). In diffraction theory, the conjugation problem was skillfully used, for example, ref. [3.1]. The theory of singular equations was successfully applied in ref. [3.2], where many electrodynamical problems, mainly of rectangular waveguides were solved by this method.

We make no claim to mathematical rigorousness, but for the sake of exposition, we will be describing the main properties of the Cauchy type integral in this chapter. Those wishing to learn this rigorous way of problem solving should turn to the books [3.3] and [3.4].

3.1. Sokhotsky–Plemelj Formulae

The integral:

$$\Phi(z) = \frac{1}{2\pi j} \int_L \frac{\mu(t)dt}{t-z}, \qquad (3.1)$$

along the line L in the plane of the complex variable t is called the Cauchy type integral and the function $\mu(t)$ is the density.

The factor $1/(t-z)$ for the fixed value t with respect to the variable z, is an analytical function everywhere except the point $z=t$, in which there is a pole. Therefore, the function $\Phi(z)$ is analytical everywhere except, the points of the contour L, where it is not single–valued $\Phi(\infty)=0$. We designate $\Phi^+(t_0)$ having the limit value of the integral (3.1) when the point z, is on the left side of the contour L, approaching the point t_0 of the contour. We get the limited value taking the integral (3.1) along the line, which at the finite distances from the point t_0 coincides with the contour L. At a small distance from the point t_0, we take the integral (3.1) along the line of a part of the circumference with a small radius r, when the limit $r \to 0$. The integral along the circumference part is:

$$\int_{ABC} \frac{dt}{t-t_0} = j\left(2\pi - \alpha\right).$$

Here α is the angle between the tangents of contour L in the point t_0. If the contour L is smooth then $\alpha = \pi$. In case of a corner point (Fig.3.1a) the angle α can be any $0 \le \alpha \le 2\pi$. According to the definition, the integral that remains is the principal value integral because $\left|At_0\right| = \left|t_0C\right|$.

Therefore:

$$\Phi^+\left(t_0\right) = \left(1 - \frac{\alpha}{2\pi}\right)\mu\left(t_0\right) + \frac{1}{2\pi j}\int_L \frac{\mu(t)\,dt}{t-t_0}. \tag{3.2}$$

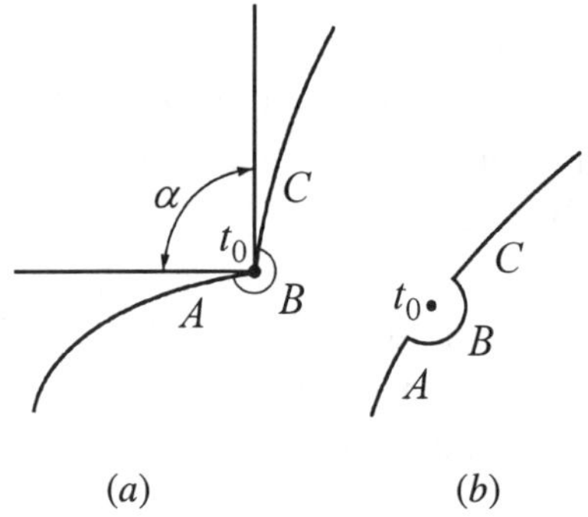

Fig. 3.1. (a) The piecewise arc L with a corner point t_0 and (b) the smooth arc L.

When point z is approaching point t_0 from the right side of the contour L, the corresponding part of the circumference Fig.3.2, is on the left side of the contour L. So that the contour L on the *arc* moves around clockwise with respect of its center.

Therefore, the integral:

$$\int_{ABC} \frac{dt}{t-t_0} = -j\alpha,$$

where for the smooth point $\alpha = \pi$.

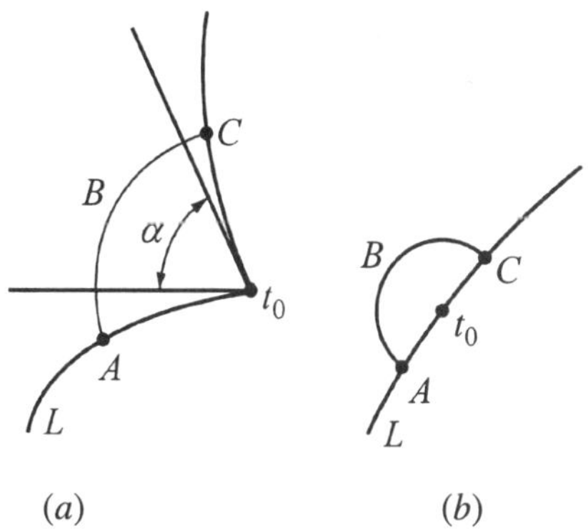

(a)　　　　　　　　　　　*(b)*

Fig. 3.2. *(a) The piecewise smooth arc* L*, (b) the smooth arc* L*.*

We can write:

$$\Phi^-(t_0) = -\frac{\alpha}{2\pi}\mu(t_0) + \frac{1}{2\pi j}\int_L \frac{\mu(t)\,dt}{t-t_0}, \tag{3.3}$$

where the second term of expression is understood, in the sense of the principal value integral. These formulae (3.2) and (3.3) are called the Sokhotsky–Plemelj formulae. We rewrite these formulae for the smooth contour L (Fig.3.1b):

$$\Phi^{\pm}(t_0) = \pm\frac{1}{2}\mu(t_0) + \frac{1}{2\pi j}\int_L \frac{\mu(t)\,dt}{t-t_0}. \tag{3.4}$$

From the formula (3.4), we arrive at the expressions:

$$\Phi^+(t_0) - \Phi^-(t_0) = \mu(t_0), \tag{3.5}$$

$$\Phi^+(t_0) + \Phi^-(t_0) = \frac{1}{\pi j}\int_L \frac{\mu(t)\,dt}{t-t_0} \equiv 2\Phi(t_0), \tag{3.6}$$

showing that the difference of the limited values is equal to the density. The integral value in the point $z = t_0$ of the contour L coincides with the half–sum of the limited values.

The analytic function $\psi(z)$ is given in the plane of the variable z with the cut along the contour L and has the value leap (3.5) on the cut sides. Then the function $\psi(z)$ is different from $\Phi(z)$, given the formula (3.1), by the polynomial, which is an analytical function in the whole plane.
So here:

$$\omega(z) = \Phi(z) - \psi(z),$$

we have:

$$\omega^+(t) = \omega^-(t), \qquad t \in L,$$

the function $\omega(z)$ is analytical in the whole plane and therefore may be presented as a power series.

Let the density $\mu(t)$ be a real function. Then it follows from the formula (3.4) that the function $\operatorname{Im}\Phi(z)$ has no leap on the contour L and the leap of the function $\operatorname{Re}\Phi(z)$, while crossing the contour L according to the formula (3.5) coincides with the $\mu(t)$.

3.2. Behaviour of the Cauchy Type Integral at the Integration Contour Ends

Let the contour L be open with ends in the points a and b (Fig.3.3) and the density satisfies a Hoelder condition:

$$\left|\mu(t_1)-\mu(t_2)\right| \le A\left|t_1-t_2\right|^{\alpha}, \quad 0 < \alpha \le 1 \tag{3.7}$$

and has the finite values $\mu(a)$ and $\mu(b)$.

Now we consider the behavior of a Cauchy type integral:

$$\Phi(z) = \frac{1}{2\pi j} \int_a^b \frac{\mu(t)\,dt}{t-z} \tag{3.8}$$

near the point $z = a$.

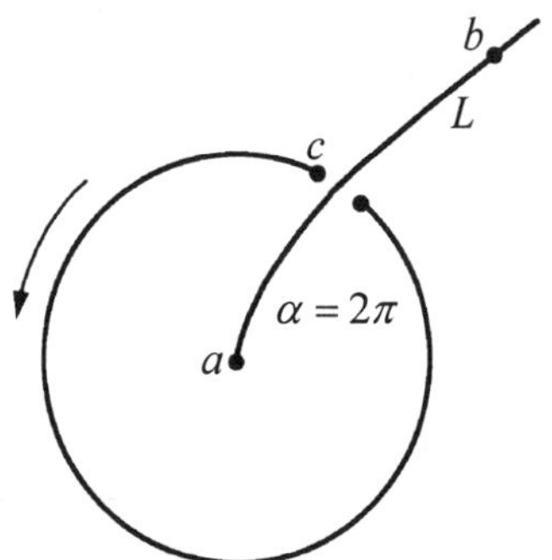

***Fig. 3.3.** The plane cut running from the point a into ∞.*

We can write the function $\Phi(z)$ in the form:

$$\Phi(z) = \frac{1}{2\pi j} \int_a^b \frac{\mu(t)-\mu(a)}{t-z}\,dt + \frac{\mu(a)}{2\pi j} \int_a^b \frac{dt}{t-z}.$$

The first integral near the point $z \sim a$ due to the Hoelder condition (3.7) is a finite quantity and the second integral:

$$\int_a^b \frac{dt}{t-z} = \ln\frac{b-z}{a-z} = \ln\frac{z-b}{z-a}$$

has a growing modulus, while the point z is approaching the low integration limit.

Thus:

$$\Phi(z) \approx -\frac{\mu(a)}{2\pi j}\ln(z-a) \qquad \text{when} \qquad z \sim a. \tag{3.9}$$

And similarly:

$$\Phi(z) \approx +\frac{\mu(b)}{2\pi j}\ln(z-b) \qquad \text{when} \qquad z \sim b. \tag{3.10}$$

In this same way, we consider the case when the density in the point c has the finite leap and $\mu(c-0) \neq \mu(c+0)$. Here $\mu(c\pm0)$ means the limited values of the function $\mu(t)$, while $t \to c$ from one side or the other. Representing the integral (3.8) as a sum of two integrals, we write:

$$\Phi(z) = \frac{1}{2\pi j}\int_a^c \frac{\mu(t)-\mu(a)}{t-z}dt + \frac{\mu(a)}{2\pi j}\int_c^b \frac{dt}{t-z}.$$

According to the formulae (3.9) and (3.10) we have:

$$\Phi(z) \approx \frac{\mu(c-0)-\mu(c+0)}{2\pi j}\ln(z-c) \qquad \text{when} \qquad z \sim c. \tag{3.11}$$

We now consider the case when the density $\mu(t)$ at the point $t=a$ has the power singularity:

$$\mu(t) = \frac{\mu_1(t)}{(t-a)^\gamma}, \qquad \gamma = \alpha + j\beta, \qquad 0 \le \alpha \prec 1.$$

The function $\mu_1(t)$ satisfies a Hoelder condition.
Writing the integral:

$$\Phi(z) = \frac{1}{2\pi j}\int_a^b \frac{\mu_1(t)\,dt}{(t-a)^\gamma(t-z)},$$

as a sum of two integrals:

$$\Phi(z) = \frac{1}{2\pi j}\int_a^b \frac{\mu_1(t)-\mu_1(a)}{(t-a)^\gamma(t-z)}dt + \frac{\mu_1(a)}{2\pi j}\int_a^b \frac{dt}{(t-a)^\gamma(t-z)}. \tag{3.12}$$

The function $\mu_1(t)$ satisfies a Hoelder condition and the behavior of $\Phi(z)$ at $z \sim a$ is determining by the second integral.
Here we have:

$$\Omega(z) = \frac{1}{2\pi j}\int_a^b \frac{(t-a)^{-\gamma}}{t-z}dt. \tag{3.13}$$

From the formula (3.4), it follows that:

$$\Omega^+(t)-\Omega^-(t) = (t-a)^{-\gamma} \quad \text{when } t \in L. \tag{3.14}$$

We now introduce the function:

$$\omega(z) = (z-a)^{-\gamma}$$

with the branch point $z = a$. The cut running from the point $z = a$ into infinity is drawn along the contour L (Fig.3.3) and the single–valued function $\omega(z)$ branch is chosen, so that on the left cut side (going along the contour L from the point a) $\omega^+(t) = (t-a)^{-\gamma}$.

Then in the opposite point of the cut

$$\omega^-(t) = (t-a)^{-\gamma}\left(e^{2\pi j}\right)^{-\gamma} = (t-a)^{-\gamma}\,e^{-2\pi j\gamma}$$

and we move around the branch point counterclockwise.
Therefore:

$$\omega^+(t) - \omega^-(t) = (t-a)^{-\gamma}\left(1 - e^{-2\pi j\gamma}\right).$$

So, for the function:

$$\omega_1(z) = \frac{\omega(z)}{1 - e^{-2\pi j\gamma}} = \frac{e^{\pi j\gamma}(z-a)^{-\gamma}}{2j\sin\gamma\pi}$$

the difference is

$$\omega_1^+(t) - \omega_1^-(t) = (t-a)^{-\gamma},$$

that coincides with the value (3.14). Consequently, the functions $\Omega(z)$ and $\omega_1(z)$ can be different and this difference is an analytical function (presented near the point a as a sum of the positive powers of $(z-a)$). So, the expression (3.12) can be written as:

$$\Phi(z) \approx \mu_1(a)\frac{e^{\pi j\gamma}}{2j\sin\gamma\pi}(z-a)^{-\gamma}\,, \qquad z \sim a\,, \tag{3.15}$$

which follows from:

$$\Phi(t) = \frac{1}{2}\left(\Phi^+(t) + \Phi^-(t)\right) = \frac{\mu_1(a)}{2j}(t-a)^{-\gamma}\,ctg\,\gamma\pi\,, \quad t \sim a\,. \tag{3.16}$$

If the density has the singularity near the point b, then:

$$\mu(t) = \mu_1(t)(t-b)^{-\gamma}\,, \quad \gamma = \alpha + i\beta\,, \qquad 0 \le \alpha \prec 1,$$

with $\mu_1(t)$ satisfying a Hoelder condition. Then the formula (3.8) can be written as:

$$\Phi(z) = \frac{1}{2\pi j}\int_a^b \frac{\mu_1(t) - \mu_1(b)}{(t-b)^\gamma(t-z)}dt + \frac{\mu_1(b)}{2\pi j}\int_a^b \frac{dt}{(t-b)^\gamma(t-z)}\,, \tag{3.17}$$

separating the value $\mu_1(b)$ in the point b. At $z \to b$ the singularity of the first integral is weaker than that of the second integral. Therefore, it is enough to consider the behavior of the function:

$$\Omega(z) = \frac{1}{2\pi j} \int_a^b \frac{(t-b)^{-\gamma}}{t-z}\, dt$$

near the point $z = b$. According to the formula (3.4) we have:

$$\Omega^+(z) - \Omega^-(z) = (t-b)^{-\gamma}, \qquad t \in L. \tag{3.18}$$

We now introduce the function:

$$\omega(z) = (z-b)^{-\gamma}$$

with the branching point $z = b$. We draw the cut running from the point $z = b$ into infinity along the contour L, as is shown in Fig.3.4. We choose the single-valued function in the plane with the cut branch of this function $\omega(z)$, so that on the left side of this cut (moving along the contour L from the point a) the function is:

$$\omega^+(t) = (t-b)^{-\gamma}.$$

The value of the function on the opposite cut side differs by the factor $\left(e^{-2\pi j}\right)^{-\gamma}$, because we move round the branching point clockwise, whereas

$$\omega^-(t) = (t-b)^{-\gamma}\, e^{2\pi j\gamma}.$$

Therefore:

$$\omega^+(t) - \omega^-(t) = (t-b)^{-\gamma}\left(1 - e^{2\pi j\gamma}\right).$$

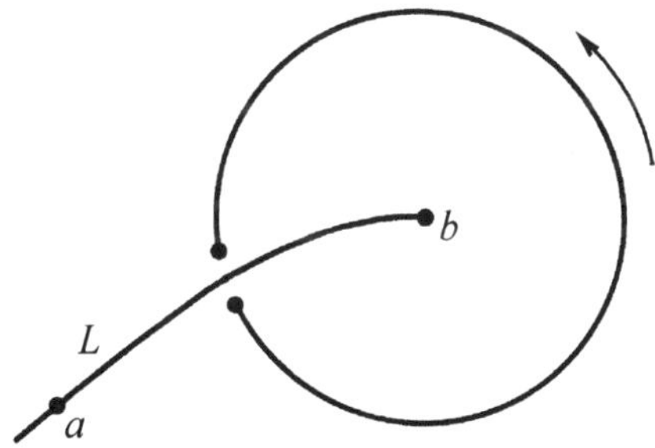

Fig. 3.4. *A view of the plane cut running from the point b into infinity* ∞.

Hence the function:

$$\omega_1(z) = \frac{\omega(z)}{1 - e^{2\pi j\gamma}} = \frac{-e^{-\pi j\gamma}}{2j\sin\gamma\pi}(z-b)^{-\gamma}$$

and the difference is:

$$\omega_1^+(t) - \omega_1^-(t) = (t-b)^{-\gamma},$$

this coincides with the formula (3.18). Consequently, the functions $\Omega(z)$ and $\omega_1(z)$ difference is the analytic function, which has no singularities for finite z. So the expression (3.17) can be presented as:

$$\Phi(z) \approx -\mu_1(b)\frac{e^{-\pi j\gamma}}{2j\sin\gamma\pi}(z-b)^{-\gamma}, \qquad z \sim b. \tag{3.19}$$

So here:

$$\Phi(t) = \frac{1}{2}\left(\Phi^+(t)+\Phi^-(t)\right) \approx \frac{\mu_1(b)}{2j}(t-b)^{-\gamma} ctg\,\gamma\pi, \qquad t \sim b. \tag{3.20}$$

Let the density of the integrand have singularity of power kind in the point $t=c$, which does not coincide with the ends of the *arc L* and then:

$$\mu(t) = \mu_1(t)/(t-c)^{-\gamma}, \qquad t \in L_1,$$

$$\mu(t) = \mu_2(t)/(t-c)^{-\gamma}, \qquad t \in L_2 \qquad \text{and} \qquad L_1 + L_2 = L. \tag{3.21}$$

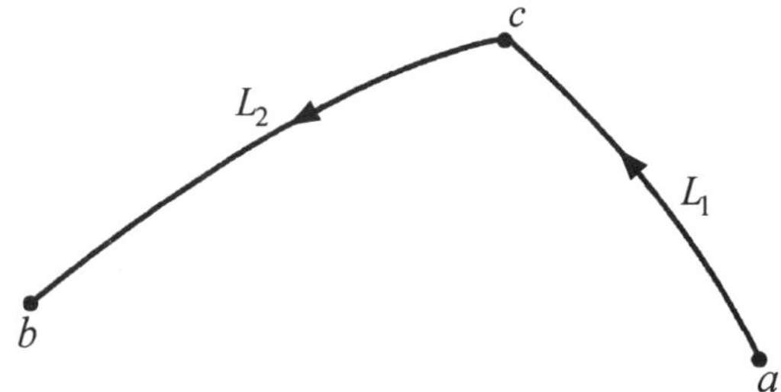

***Fig. 3.5.** An explanation of Cauchy's type integral behavior at the ends of the integration contour.*

In Fig.3.5 the arrows indicate the direction in which contour L is moving. The *arcs* L_1 and L_2 join at the corner point c. The functions $\mu_1(t)$ and $\mu_2(t)$ are not equal at the corner points $\mu_1(c) \neq \mu_2(c)$. The functions $\mu_1(t)$ and $\mu_2(t)$ as we wrote before satisfies a Hoelder condition. Here we write the expression (3.8) as the sum of two functions:

$$\Phi(z) = \Phi_1(z) + \Phi_2(z), \tag{3.22}$$

where

$$\Phi_1(z) = \frac{1}{2\pi j}\int_a^c \frac{\mu_1(t)\,dt}{(t-c)^{\gamma}(t-z)} \tag{3.23}$$

and

$$\Phi_2(z) = \frac{1}{2\pi j}\int_c^b \frac{\mu_2(t)\,dt}{(t-c)^{\gamma}(t-z)}, \qquad \gamma = \alpha + j\beta. \tag{3.24}$$

The behavior of the second integral $\Phi_2(z)$ near point c is described by the formulae (3.15) and (3.16), in which one has to change a into c. Remembering that this function:

$$f_1(z) = (z-c)^{-\gamma} \tag{3.25}$$

is the single–valued function in the variable z plane, with the cut running from the branching point $z = c$ along the *arc* L_2 and further into infinity. The function branch is chosen so that

$$f_1^+(z) = (t - c)^{-\gamma}, \qquad\qquad t \in l_{,2},$$

$$f_1^-(z) = (t - c)^{-\gamma} e^{-2\pi j\gamma}, \qquad\qquad t \in L_2. \qquad\qquad (3.26)$$

The behavior of $\Phi_1(z)$ near the point c is described by the expressions (3.19) and (3.20), where we replace c for b. We understand that the function $(z - b)^{-\gamma}$ in formula (3.19) is expressed like $f_2(z) = (z - c)^{-\gamma}$, which is the single–valued function in the plane with the cut running from the branching point $z = c$ and along the *arc* L_1 and on further into infinity. We chose the single–valued function $f_2(z)$ when the point z moves to the point t of the contour L_1 from the left side:

$$f_2^+(z) = (t - c)^{-\gamma} \equiv f_1(t), \qquad t \in L_1. \qquad\qquad (3.27)$$

The function $f_2(t)$ value coincides with the value of the function $f_1(t)$ calculated at the same point (the function $f_1(z)$ has no discontinuity in the points of the *arc* L_1). The values of $f_2(z)$ on the opposite side of the cut is:

$$f_2^-(t) = (t - c)^{-\gamma} e^{2\pi j\gamma} = f_1(t) e^{2\pi j\gamma}, \qquad t \in L_1. \qquad\qquad (3.28)$$

Formulae (3.27) and (3.28) show that $f_2(z) = f_1(z)$ for the point z is near *arc* L_1 but from the left side of *arc* L_1. Formulae (3.27) and (3.28) show that $f_2(z) = f_1(z)e^{2\pi j\gamma}$ for the point z is near *arc* L_1 but to the right side of it.

So according to the formula (3.15) we have:

$$\Phi_2(z) \approx \mu_2(c) \cdot \frac{e^{\pi j\gamma}}{2j \sin \gamma\pi} (z - c)^{-\gamma}, \qquad z \sim c. \qquad\qquad (3.29)$$

The function (3.25) is meant for $(z \sim c)^{-\gamma}$ and the point z may be both to the left and to the right side of the contour L. From formula (3.19), we write:

$$\Phi_1(z) \approx -\mu_1(c) \cdot \frac{e^{-\pi j\gamma}}{2j \sin \gamma\pi} (z - c)^{-\gamma}, \qquad z \sim c, \qquad\qquad (3.30)$$

when the point z is from the left side of the *arc* L and

$$\Phi_1(z) \approx -\mu_1(c) \cdot \frac{e^{\pi j\gamma}}{2j \sin \gamma\pi} (z - c)^{-\gamma}, \qquad z \sim c, \qquad\qquad (3.31)$$

when the point z is from the right side of the contour L. Consequently, for function (3.22) we obtain:

$$\Phi(z) \approx \frac{\mu_2(c) e^{\pi j\gamma} - \mu_1(c) e^{-\pi j\gamma}}{2j \sin \gamma\pi} (z - c)^{-\gamma}, \qquad z \sim c, \qquad\qquad (3.32)$$

when the point z is from the left side of the contour L.

For function (3.22) we obtain:

$$\Phi(z) \approx \frac{\mu_2(c)e^{\pi j\gamma} - \mu_1(c)e^{\pi j\gamma}}{2j\sin\gamma\pi}(z-c)^{-\gamma}, \qquad z \sim c, \tag{3.33}$$

when the points z is from the right side of the contour L. If $\mu_1(c) = \mu_2(c)$ then the expression (3.33) vanishes and the singularity of the function $\Phi(z)$ near the point c is determined by the terms that we discard. We see in [3.3] that in this case:

$$\Phi(z) \approx \frac{\Phi_0(z)}{|z-c|^{\alpha_0}}, \qquad \alpha_0 \prec \alpha.$$

Here the function $\Phi_0(z)$ is extendable into L in the vicinity of the point c and satisfies a Hoelder condition.

The values of the integral (3.22) in the points of the contour $L = L_1 + L_2$ can be found by the formula:

$$\Phi(t) = \left(\Phi^+(t) + \Phi^-(t)\right)/2.$$

Remembering that the factor $(z-c)^{-\gamma}$ in the formulae (3.32) and (3.33) is the function (3.25), where we obtain:

$$\Phi(t) \approx \left(\frac{\mu_2(c)e^{\pi j\gamma}}{2j\sin\pi\gamma} - \frac{\mu_1(c)ctg\,\pi\gamma}{2j}\right)(t-c)^{-\gamma}, \qquad t \sim c,\, t \in L_1$$

and

$$\Phi(t) \approx \left(\frac{\mu_2(c)ctg\,\pi\gamma}{2j} - \frac{\mu_1(c)e^{-\pi j\gamma}}{2j\sin\pi\gamma}\right)(t-c)^{-\gamma}, \qquad t \sim c,\, t \in L_2. \tag{3.34}$$

The behavior of Cauchy type integral near the point where several *arcs* join will be the same as we saw in earlier formulae (3.22)-(3.34).

Thus, the peculiarities of the function $\Phi(z)$, which are defined according to formula (3.8) are fully described by the behavior of the density $\mu(t)$ at each point of the contour L.

When we write the components of electric and magnetic fields in integral representations, which give the solutions of Laplace's or Maxwell's equations then these representations will consist of the Cauchy type integral (3.8). After satisfying certain boundary conditions (that was formulated in electrodynamical problems) we arrived at a SIEs system. The density $\mu(t)$ is the solution of the SIEs system:

$$f(t_0) = \mu(t_0) + \frac{1}{\pi j}\int_L \frac{\mu(t)\,dt}{t-t_0} + \frac{1}{\pi j}\int_L \mu(t)K(t,t_0)\,dt. \tag{3.35}$$

The second term of the system (3.35) has a nonsingular kernel $K(t,t_0)$, which is obtained as a result of imposing certain boundary conditions on the solution of differential equations. In that, the electromagnetic field behavior peculiarities are determined by these boundary conditions.

3.3. Solution of Singular Integral Equations

When we solve electrodynamical problems a SIE, which contain the Cauchy type integral will occur. These SIEs can be solved in quadratures. In this book, the SIEs are solved numerically according to the following scheme. In the equation:

$$f(t_0) = \mu(t_0) + \int_L \mu(t)K(t,t_0)\,dt \tag{3.36}$$

the integral

$$\int_L \mu(t)K(t,t_0)\,dt = \sum_{i=1}^{n} \int_{t_i-\Delta/2}^{t_i+\Delta/2} \mu(t)K(t,t_0)\,dt$$

is written as the sum of integrals over the sections of a small length Δ. Every such integral, according to the Bogoliubov–Krylov method [3.5], is written and calculated by this formula:

$$\int_{t_i-\Delta/2}^{t_i+\Delta/2} \mu(t)K(t,t_0)\,dt \approx \mu(t_i) \int_{t_i-\Delta/2}^{t_i+\Delta/2} K(t,t_0)\,dt \,.$$

We determined in the equation (3.36) that the magnitudes $t_0 = t_1$, $t_0 = t_2$, $t_0 = t_3$ $\dots t_0 = t_n$. As a result we arrive at a system:

$$f(t_i) = \mu(t_i) + \sum_{j=1}^{n} \mu(t_j) \int_{t_j-\Delta/2}^{t_j+\Delta/2} K(t,t_i)\,dt \,, \qquad i = 1,2,\dots n \,, \tag{3.37}$$

of linear integral equations for determining the values of $\mu(t_i)$ at separate points t_i of the contour L. The integrals in (3.37) are calculated in the sense of the principal value. Sokhotsky–Plemelj formulae are applied to the Cauchy type integrals (3.1), when $i = j$ the integral in (3.37) is omitted.

· In solving an electrodynamical problem we need to compute a system of linear algebraic equations (3.37). It is more convenient to write this system in the matrix form:

$$Ax = B \,, \tag{3.38}$$

where A is the matrix of the coefficients:

$$a_{ij} = \delta_{ij} + \int_{t_j-\Delta/2}^{t_j+\Delta/2} K(t,t_i)\,dt \,.$$

The magnitude B is a column of free members $b_i = f(t_i)$ and x is a column of the unknowns $x_i \equiv \mu(t_i)$. One of the simplest methods to solution the system of linear algebraic equations (3.38) is the Gauss method, where the equations is reduced to a system with the matrix A, having only zeros on one side of the main diagonal. When the order n of the system (3.38) increases, then the number of operations needed to solve the system also must increases as in n^3. The computation time is proportional to n^2 per iteration [3.6] and in the case of a well converging process, it can be by one or two orders less than in the direct method. In our computations, we used the direct method and sometimes the Seidel and Jacobi iteration methods [3.7].

In the Seidel method, the successive approximation computation is carried out by the formula:

$$x_i^{(k)} = \frac{1}{a_{ii}}\left(b_i - \sum_{j=1}^{i-1} a_{ij}x_j^{(k)} - \sum_{j=i+1}^{n} a_{ij}x_j^{(k=1)}\right), \qquad i=1,2,\cdots,n, \tag{3.39}$$

where the superscript (k) denotes the number of iterations. The iteration process continues until the inequality $\max\left|x_i^{(k)} - x_j^{(k-1)}\right| \prec \delta_0$ is satisfied. The quantity δ_0 characterizes the computation accuracy. In the Jacobi method numerical realization, the column of unknowns is singled out in the equation: (3.38)

$$x = Cx + B, \quad C = E - A, \tag{3.40}$$

where E is the unit matrix. The formula (3.40) is used in the iteration process:

$$x^{(k)} = Cx^{(k-1)} + B, \qquad k = 1,2,\ldots.$$

The elements of the column B are used as the starting values:

$$x^{(0)} = B.$$

The simplest and the most sufficient sign of the iteration method convergence is the condition [3.6] and [3.7]:

$$\max_i \sum_{j=1}^{n}\left|C_{ij}\right| \prec 1 \quad \text{or} \quad \max_j \sum_{i=1}^{n}\left|C_{ij}\right| \prec 1.$$

The system order n depends on the necessary accuracy for certain electrodynamical problems. The matrix A of the system (3.38) coefficients is broken into N^2 blocks. In the computation process we used zero marks for the blocks with zero arrays [3.8]. This essentially shortens the time for calculating electrodynamical characteristics of waveguide structures that consist of gyrotropic media. And so to receive a certain accuracy of calculations for gyrotropic structures we must take a larger value of the system order n. The main parameters of gyrotropic media will be considered in the next chapter.

4. ANISOTROPIC MEDIA

Up until now we have only considered striplines with isotropic substrates. The material of the substrate can also be anisotropic. For example, substrates of waveguide structures could be created of anisotropic crystals, magnetized ferrite or semiconductor materials that have anisotropic properties. Certain ferrite and semiconductor materials, which are placed into external constant magnetic fields, are also called gyrotropic media.

When solving certain electrodynamical problem no new elements will occur (as compared to striplines with isotropic substrates) for striplines with anisotropic substrates, where the anisotropy is described by a symmetric permittivity tensor $\varepsilon_{ik} = \varepsilon_{ki}$. Here it is convenient to work in the coordinate system in which the tensor ε_{ik} has the diagonal form. The divergence of the electric induction:

$$divD = 0 ,$$

where $D_i = \sum_k \varepsilon_{ik} E_k$ and $\vec{E} = -grad\ V$.

We can write this equation through the electric potential V like this:

$$\varepsilon_{11}\frac{\partial^2 V}{\partial x^2} + \varepsilon_{22}\frac{\partial^2 V}{\partial y^2} + \varepsilon_{33}\frac{\partial^2 V}{\partial z^2} = 0 .$$

After the scale transformation:

$$x'\sqrt{\varepsilon_{11}} = x , \quad y'\sqrt{\varepsilon_{22}} = y , \quad z'\sqrt{\varepsilon_{33}} = z$$

the last differential equation turns into Laplace's equation:

$$\frac{\partial^2 V}{\partial x'^2} + \frac{\partial^2 V}{\partial y'^2} + \frac{\partial^2 V}{\partial z'^2} = 0 .$$

In this way we arrive at the harmonic function.

It is quite a complex matter when the substrates are gyrotropic materials with electromagnetic properties that are described by antisymmetric tensors having components ε_{ik} or μ_{ik}. The values of these components depend on the constant magnetic field.

In this chapter, we consider super high frequency (SHF) tensor of ferrite permeability and SHF tensor of semiconductor permittivity. These materials are widely used in electronics where the applied constant magnetic field can change their properties within certain limits.

4.1. The Main Parameters of Ferrite Media

In Maxwell's equations (1.1), the connection between the magnetic induction vector $\vec{B}$ and magnetic field strength vector $\vec{H}$ for anisotropic media described:

$$B_i = \sum_j \mu_{ij} H_j \quad \text{or} \quad \vec{B} = \ddot{\mu}\vec{H}, \tag{4.1}$$

by the permeability tensor $\ddot{\mu}$ of rank two in the Cartesian coordinates.

Ferrites are nonmagnetic materials with low conductivity. Therefore, the ferromagnetic properties of ferrite media are mainly stipulated by spin magnetic moments of electrons. Real ferrite devices work when the magnetic field is strong enough to saturate the ferrite (i.e. when there is no domain structure in it). The magnetization vector $\vec{M}$ (magnetic moment per volume unit) of the system is equal to the vector sum of magnetic moments of electrons in the volume unit. The magnetic moment $\vec{M}$ placed in the magnetic field $\vec{B}$ is acted upon by the force moment:

$$\vec{T} = \left[\vec{M} \times \vec{B}\right], \tag{4.2}$$

so that a change in the moment of momentum $\vec{I}$:

$$\frac{d\vec{I}}{dt} = \left[\vec{M} \times \vec{B}\right]. \tag{4.3}$$

Besides having its own magnetic moment $\vec{m}$, every electron has the mechanic moment (spin) $\vec{s}$, which are interconnected by the relation:

$$\vec{m} = -\gamma_0 \vec{s}, \tag{4.4}$$

where $\gamma_0 = g\dfrac{[e]}{2m_e}$ is the gyromechanic ratio. Here e is the electron charge, m_e is the rest mass of electron and $m_e = 9.107 \cdot 10^{-31}$ kg. Where g is the Lande factor, for the electron $g = 2$ and $\gamma_0 = 1.76 \cdot 10^{11}$ c/kg.

After substituting the relation (4.4) into the formula (4.3) we will arrive at the Landau–Lifshits' equation of motion without losses:

$$\frac{d\vec{M}}{dt} = -\gamma_0 \left[\vec{M} \times \vec{B}\right]. \tag{4.5}$$

The vector $\vec{B}$ in the equation (4.5) means the electric induction, without taking into account the magnetic moments, i.e. $\vec{B} = \mu_0 \vec{H}$ (μ_0 is the magnetic constant of a vacuum).

When magnetized ferrite medium is at saturation, the outside field and the boundary conditions determine the field, which consists of a constant component $\vec{B}_{oi} = \mu_0 \vec{H}_{0i}$ and an alternating component (of SHF field) $\vec{B}_{\sim} = \mu_0 \vec{H}_{\sim}$. In a similar way, we can determine the magnetization vector, which consists of a constant component $\vec{M}_0$ and an alternating $\vec{M}_\sim$. The vectors $\vec{M}_0$ are parallel to $\vec{B}_{0i}$ for isotropic medium, which is magnetized to saturation (in the absence of an alternating magnetic field).

Let the alternating components depend on time as $e^{i\omega t}$ and the amplitudes of alternating components are small when compared to the constant components. Here we assume that the vectors $\vec{M}_0$ and $\vec{B}_{oi}$ are directed along the axis z in Cartesian coordinates.

Writing the equation (4.5) into the projections and neglecting the small terms, we obtain:

$$i\omega M_{\sim x} = -\gamma_0 \left(B_{0i} M_{\sim y} - M_0 \mu_0 H_{\sim y} \right),$$

$$i\omega M_{\sim y} = -\gamma_0 \left(M_0 \mu_0 H_{\sim x} - B_{0i} M_{\sim x} \right),$$

$$i\omega M_{\sim z} = 0. \tag{4.6}$$

The solution of the system (4.6) is:

$$M_{\sim x} = \chi H_{\sim x} + iv H_{\sim y},$$

$$M_{\sim y} = -iv H_{\sim x} + \chi H_{\sim y}. \tag{4.7}$$

Where χ and v are defined through relations:

$$\chi = \frac{\Omega_H}{\Omega_H^2 - \Omega^2}, \qquad v = \frac{\Omega}{\Omega_H^2 - \Omega^2},$$

$$\Omega_H = \frac{H_{0i}}{M_0}, \qquad \Omega = \frac{\omega}{\mu_0 \gamma_0 M_0}. \tag{4.7a}$$

The quantities x, v, Ω_H, Ω are dimensionless magnitudes.

In these expressions, the inner constant magnetic field $\vec{H}_{0i}$ is determined by the outside constant magnetic field but is different from it.

The relations (4.7) in the form $\vec{M}_\sim = \ddot{\chi} \vec{H}_\sim$ give the components of the susceptibility tensor $\ddot{\chi}$ in explicit form connecting the alternating component of magnetization $\vec{M}_\sim$ with a SHF magnetic field $\vec{H}_\sim$ in the ferrite medium (in linear approximation). Considering the definition of magnetic inductance: $\vec{B} = \mu_0 \left(\vec{H} + \vec{M} \right)$, we find:

$$\vec{B}_\sim = \mu_0 \left(1 + \ddot{\chi} \right) \vec{H}_\sim.$$

So that the tensor of the relative permeability has the form:

$$\ddot{\mu} = \begin{vmatrix} 1+\chi & iv & 0 \\ -iv & 1+\chi & 0 \\ 0 & 0 & 1 \end{vmatrix}. \tag{4.8}$$

When we changed the direction of the constant magnetic field to the opposite direction then z–projections of the magnetic fields $\vec{H}_{0i}$, as well as the magnetic moment $\vec{M}_0$ change their signs. Therefore, after these changes, the diagonal elements of the tensor (4.8) stay unchanged and the nondiagonal elements change signs.

The tensor character of the relations (4.7) means that the alternating field applied along the axis x creates magnetization components in both directions of axis x and axis y. Another peculiarity of these relations is the resonance character of vibrations of the magnetization vector at the frequency:

$$\omega = \gamma_0 B_{0i}, \tag{4.9}$$

when components of the alternating magnetization take up the infinite values. We note that if the values of $\vec{M}_\sim$ is large then derivation given above are not correct. So the expression "the infinite value" should be understood only conditionally. When magnetization components have infinity values, this means we did not take into account the damping processes (we considered lossless media). It is also necessary to stress that the resonance condition (4.9) is only valid for boundless media. If we were to consider finite ferrite samples, their shape would have some essential influence on resonance conditions. We will consider resonance conditions depending on the crystallographic anisotropy later.

To correctly evaluate the amplitudes of the alternating fields and magnetization it is necessary to take into account the magnetic losses, which are always present in the real ferrite. Consideration of these losses leads to the appearance of finite value components of the magnetic susceptibility and permeability tensors, even at the ferromagnetic resonance.

When we take into account losses in the equation (4.5), it is necessary to add a dissipative term. We designate the dissipative term by $\vec{R}$ and we arrive at:

$$\frac{d\vec{M}}{dt} = -\gamma_0 \mu_0 \left[\vec{M} \times \vec{H} \right] + \vec{R}. \tag{4.10}$$

There are several forms of the dissipative term $\vec{R}$. We now consider the most widespread form, which was suggested by T. Hilbert, for the dissipative term [4.1]:

$$\vec{R} = \frac{\alpha_0}{M} \left[\vec{M} \times \frac{d\vec{M}}{dt} \right]. \tag{4.11}$$

Here the dimensionless parameter α_0 characterize the losses.

Written in the form of (4.11) we presuppose the presence of an effective field of "friction force", when it is antiparallel to the field $\vec{H}$ and proportional to the speed of the magnetization $\vec{M}$ change.

If the losses were low we substituted the derivative $d\vec{M}/dt$ from the equation (4.5) in the expression (4.11) and we obtained the dissipative term in the form of [4.2]:

$$\vec{R} = \eta\left\{\vec{H} - \frac{\vec{M}}{M^2}\left(\vec{M}\vec{H}\right)\right\},\qquad(4.12)$$

where the coefficient $\eta = \alpha_0 \mu_0 \gamma_0 M$ (second^{-1}).

Here the alternating components of the magnetization and field are less in comparison to the constant components of the magnetization and field, so that $\left(\vec{H}\cdot\vec{M}\right)/HM \approx 1$. Then the expression (4.12) (out of parentheses the ratio $\vec{H}_0/\vec{M}_0$) transforms into the dissipative term of Bloch equation [4.3]:

$$\vec{R} = \omega_r\left\{\chi_0\vec{H} - \vec{M}\right\}.\qquad(4.13)$$

Here the dimensionless value $\chi_0 = M_0/H_0$ is the static susceptibility and the value $\omega_r = \eta/\chi_0$ is the relaxation frequency. The quantity inverse to the relaxation frequency is called the relaxation time $\tau_r = 1/\omega_r$.

In the formulae above the dissipative term (which describe these losses) have only one parameter but there are dissipative term forms having two parameters. The last forms having two parameters are of no use in electrodynamical analysis. In practice it is enough to use a dissipative term with one parameter. There are two forms for dissipative terms having two parameters T_1 and T_2, which are longitudinal and transverse relaxation times respectively.

The dissipative term proposed in [4.4] has the form in Cartesian coordinates:

$$R = -i_x\frac{M_x}{T_2} - i_y\frac{M_y}{T_2} - i_z\frac{M_z - M_0}{T_1},$$

and the expression:

$$R = \frac{i_x}{T_2}\left(\chi_0 H_x - M_x\right) + \frac{i_y}{T_2}\left(\chi_0 H_y - M_y\right) + \frac{i_z}{T_1}\left(\chi_0 H_z - M_z\right),$$

corresponds to Bloch–Wangsness form [4.5].

A formula for dissipative term allows us to write expressions of elements of magnetic susceptibility or permeability tensors taking into account losses. Here we decided to take the dissipative term in the form (4.13) but any other form would give the equivalent results if losses are low.

Writing the equation (4.10) into the projections in Cartesian (rectangular) axes and taking into account (4.13) we solve the system of equations with respect to alternating components of magnetization. Then we arrive at:

$$M_{\sim x} = \chi H_{\sim x} + iv H_{\sim y},$$

$$M_{\sim y} = -iv H_{\sim x} + \chi H_{\sim y},$$

$$M_{\sim z} = \chi_{\parallel} H_{\sim z},\tag{4.14}$$

where the complex components of the magnetic susceptibility tensor:

$$\chi = \frac{\Omega_H + i\Omega_r \left(\Omega - i\Omega_r\right)/\Omega_H}{\Omega_H^2 - \left(\Omega - i\Omega_r\right)^2},$$

$$\nu = \frac{\Omega}{\Omega_H^2 - \left(\Omega - i\Omega_r\right)^2},$$

$$\chi_{\parallel} = \frac{-i\Omega_r}{\Omega_H \left(\Omega - i\Omega_r\right)},$$

$$\Omega_r = \frac{\omega_r}{\gamma_0 \mu_0 M_0},\tag{4.15}$$

are more general than the formulae (4.7a). Separating the real and imaginary parts:

$$\dot{\chi} = \chi' - i\chi'', \quad \dot{\nu} = \nu' - i\nu'', \quad \dot{\chi}_{\parallel} = \chi_{\parallel}' - i\chi_{\parallel}'',$$

one can write with accuracy:

$$\chi' = \frac{\Omega_H \left(\Omega_H^2 - \Omega^2 + \Omega_r^2\right) + 2\Omega^2 \Omega_r^2 / \Omega_H}{\left(\Omega_H^2 - \Omega^2 + \Omega_r^2\right)^2 + 4\Omega^2 \Omega_r^2},$$

$$\chi'' = \frac{\Omega_r \Omega \left(\Omega_H^2 + \Omega^2\right)}{\Omega_H \left[\left(\Omega_H^2 - \Omega^2 + \Omega_r^2\right)^2 + 4\Omega^2 \Omega_r^2\right]},$$

$$\nu' = \frac{\Omega \left(\Omega_H^2 - \Omega^2 + \Omega_r^2\right)}{\left(\Omega_H^2 - \Omega^2 + \Omega_r^2\right)^2 + 4\Omega^2 \Omega_r^2},$$

$$\nu'' = \frac{2\Omega^2 \Omega_r}{\left(\Omega_H^2 - \Omega^2 + \Omega_r^2\right)^2 + 4\Omega^2 \Omega_r^2},$$

$$\chi_{\parallel}' = \frac{\Omega_r^2}{\Omega_H \left(\Omega_r^2 + \Omega^2\right)}, \qquad \chi_{\parallel}'' = \frac{\Omega \Omega_r}{\Omega_H \left(\Omega_r^2 + \Omega^2\right)}.\tag{4.16}$$

It is obvious that at the resonance frequency $\Omega^2 = \Omega_H^2 + \Omega_r^2 \approx \Omega_H^2$ the tensor (4.16) components do not turn into infinity but become the values:

$$\chi_{res} = \frac{1}{2\Omega_H} - \frac{i}{2\Omega_r}, \qquad \nu_{res} = -\frac{i}{2\Omega_r}.$$

The relations (4.16) for the magnetic susceptibility tensor components connect the high frequency magnetic field inside the ferrite media with alternating magnetization components.

4.2. Magnetic Susceptibility Tensor Taking into Account Crystalographic Anisotropy

The magnetization vector $\vec{M}$ in the ferrite crystal is acted upon by the outside magnetic field $\vec{H}_0$, the demagnetizing field (which describe the influence of the sample boundaries) and the inner field. This inner field has a crystal lattice symmetry. The equilibrium orientation of the magnetization vector $\vec{M}$ is determined by the summary action of these fields. It is assumed that there are no inner mechanic tensions in the sample and that the magnetization is space uniform. The equilibrium orientation of the magnetization vector $\vec{M}$ is found from the minimum of the free energy of the anisotropy per crystal volume unit:

$$\vec{F} = -\mu_0 \left(\vec{M} \vec{H}^{ef} \right) + \vec{F}_{an}. \tag{4.17}$$

Here the constant field $\vec{H}_{ef}$ is determined by taking into account the sample boundary shape and $\vec{F}_{an}$ is free energy of the anisotropy depending on the magnetization vector $\vec{M}$ orientation with respect to crystallographic axes. The function $\vec{F}_{an}$ must be invariant under all transformations transferring the lattice into itself as well as the time sign change.

The last demand leads to the expansion of the function $\vec{F}_{an}$ in vector magnetization $\vec{M}$ on the even direction cosine powers [4.6], because the magnetic moment $\vec{M}(-t) = -\vec{M}(t)$ is proportional to the mechanic moment.

In the case of cubic crystal, where the invariance under reflection in coordinate planes in the expansion of the function $\vec{F}_{an}$ the odd powers of cosines α_x α_y, α_z are excluded. Here α_x, α_y, α_z are angles between the magnetization vector $\vec{M}$ and the cube edges, for example, product $\alpha_x \alpha_y$. For the lowest even power (quadratic dependence), the only possibility left is the combination of α_x^2, α_y^2, α_z^2. The invariance under transformations $x \Leftrightarrow y$, $x \Leftrightarrow z$, $y \Leftrightarrow z$ (reflection in planes passing through the opposite edges), which arrive at the linear combination:

$$\alpha_x^2 + \alpha_y^2 + \alpha_z^2 = 1, \tag{4.18}$$

with equal coefficients. However, the expression (4.18) is equal to one, so that the function $\vec{F}_{an}$ expansion begins with the fourth power.

It is convenient to write the function $\vec{F}_{an}$ as a serial in terms of uniform harmonic polynomials f_n [4.7] that satisfies the equation:

$$\left(\frac{\partial^2}{\partial \alpha_x^2} + \frac{\partial^2}{\partial \alpha_y^2} + \frac{\partial^2}{\partial \alpha_z^2} \right) f_n\left(\alpha_x, \alpha_y, \alpha_z\right) = 0 .$$

In this case the expansion coefficients (called the anisotropy constants), depend on temperature in the usual way. For the crystals of cubic symmetry, we have [4.7] and [4.8]:

$$F_{an} = k_2\left[\alpha_x^4 + \alpha_y^4 + \alpha_z^4 - 3\left(\alpha_x^2\alpha_y^2 + \alpha_x^2\alpha_z^2 + \alpha_y^2\alpha_z^2 \right) \right] + k_3\left[\alpha_x^6 + \alpha_y^6 + \alpha_z^6 - \right.$$

$$\left. - \frac{15}{2}\left(\alpha_x^4\alpha_y^2 + \alpha_x^2\alpha_y^4 + \alpha_x^4\alpha_z^2 + \alpha_x^2\alpha_z^4 + \alpha_y^4\alpha_z^2 + \alpha_y^2\alpha_z^4 \right) + 90\alpha_x^2\alpha_y^2\alpha_z^2 \right] + \ldots .$$

Using equality (4.18) for the last expression we arrive at another formula:

$$F_{an} = \frac{1}{2}K_2\left(\alpha_x^4 + \alpha_y^4 + \alpha_z^4 \right) + \frac{1}{2}K_3\alpha_x^2\alpha_y^2\alpha_z^2 + \ldots , \tag{4.19}$$

where

$$K_2 = 5k_2 + \frac{21}{2}k_3 , \qquad K_3 = 231k_3 .$$

Here we assumed that the other anisotropy coefficients $K_4,\ldots$ are zeros.

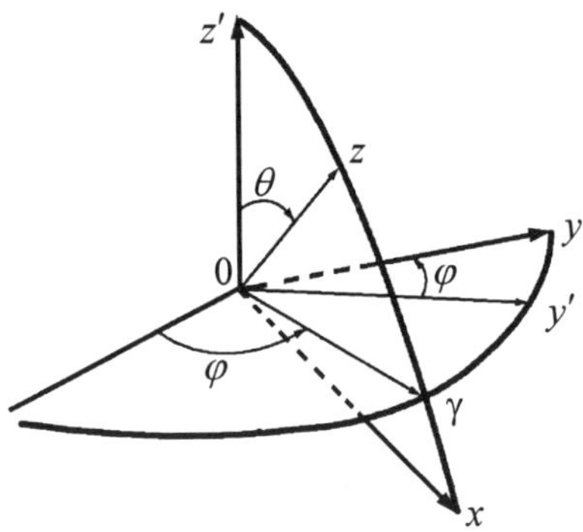

Fig. 4.1. *Transformation of the coordinates system* x', y', z' *to the system* x, y, z .

We directed the axes x', y', z' along the cube edges. The angles θ and φ (Fig.4.1) determine the orientation of the magnetization vector $\vec{M}$ with respect to these axes x', y', z'. This coordinate system x, y, z (shown in Fig.4.1) is obtained from the system x', y', z'. At first we rotate around the axis z' by the angle φ and then we rotate around the new axis y by the angle θ.

Now we introduced the unit vectors $\vec{i}$, $\vec{j}$, $\vec{k}$ and $\vec{i}'$, $\vec{j}'$, $\vec{k}'$ in the systems under consideration. Bearing in mind the formulae:

$$\vec{i} = \vec{i}'\cos\theta\cos\varphi + \vec{j}'\cos\theta\sin\varphi - \vec{k}'\sin\theta,$$

$$\vec{j} = -\vec{i}'\sin\varphi + \vec{j}'\cos\varphi,$$

$$\vec{k} = \vec{i}'\sin\theta\cos\varphi + \vec{j}'\sin\theta\sin\varphi + \vec{k}'\cos\theta \tag{4.20}$$

we get the expression for the free energy of the anisotropy

$$F_{an} = \frac{1}{2}K_2\left[\sin^4\theta\left(\cos^4\varphi + \sin^4\varphi\right) + \cos^4\theta\right] +$$

$$+\frac{1}{2}K_3\sin^4\theta\cos^2\theta\sin^2\varphi\cos^2\varphi + \dots. \tag{4.21}$$

In the case of magneto uniaxial crystals, when the symmetry axis coincides with the axis z, one gets the expression:

$$F_{an} = k_1\left(\alpha_x^2 + \alpha_y^2 - 2\alpha_z^2\right) + k_2\left[\left(\alpha_x^2 + \alpha_y^2\right)^2 - 8\left(\alpha_x^2 + \alpha_y^2\right)\alpha_z^2 + \frac{8}{3}\alpha_z^4\right] + \dots,$$

which can be taken in the form:

$$F_{an} = -K_1\alpha_z^2 + K_2\alpha_z^4. \tag{4.22}$$

Here the magnitudes are:

$$K_1 = 3k_1 + 10k_2 \quad \text{and} \quad K_2 = \frac{35}{3}k_2.$$

According to the formula (4.22), we have:

$$F_{an} = -K_1\cos^2\theta + K_2\cos^4\theta. \tag{4.23}$$

The law of the angular momentum changing leads to the equation:

$$\frac{d\vec{M}}{dt} = -\gamma_0\vec{T}, \tag{4.24}$$

which describes the change of the magnetization vector $\vec{M}$ in time. Here vector $\vec{M}$ is oriented along axis z. The magnitude $\vec{T}$ is the moment of force acting upon $\vec{M}$ and γ_0 is a magnetomechanic ratio. Projection of $\vec{T}$ (for example, on the axis z') is equal to the derivative of the free energy (4.17) with respect to the angle of the rotation around the axis z'. This derivative of the free energy of the anisotropy is taken with the opposite sign. For the first term in the expression (4.17), the orientation of vector $\vec{M}$ is defined by the angles θ and φ. And the vector $\vec{H}^{ef}$ is defined by the angles θ' and φ' so that:

$$F_n = -\mu_0 \left(\vec{M}\vec{H}^{ef} \right) = -\mu_0 M H^{ef} \left[\sin\theta \sin\theta' \cos(\varphi - \varphi') + \cos\theta \cos\theta' \right].$$

Therefore

$$T_{Hz'} = -\partial F_n / \partial \varphi = -\mu_0 M H^{ef} \sin\theta \sin\theta' (\varphi - \varphi').$$

That corresponds to the expression in Cartesian coordinates:

$$\vec{T}_{Hz'} = \mu_0 \left[\vec{M}\vec{H}^{ef} \right]_{z'}.$$

In the equilibrium state:

$$\vec{T} = \mu_0 \left[\vec{M}_0 \vec{H}^{ef} \right] + \vec{T}_{an,0} = 0 , \tag{4.25}$$

where $\vec{M}_0$ is the equilibrium magnitude of the magnetization vector $\vec{M}$ oriented along the axis z and $\vec{T}_{an,0}$ is the moment of the anisotropy force. If $\vec{T}_{an,0} \neq 0$, then the orientation of the magnetization vector $\vec{M}$ does not coincide with the direction of $\vec{H}^{ef}$. The condition (4.25) determines the angles θ and φ of the vector $\vec{M}_0$ equilibrium orientation at the given field $\vec{H}^{ef}$.

If we apply the constant magnetic field $\vec{H}^{ef}$ and the alternating field $\vec{H}_\sim$ then the acting field becomes $\vec{H} = \vec{H}^{ef} + \vec{H}_\sim , \vec{H}_\sim \ll \vec{H}^{ef}$ and the magnetization vector $\vec{M} = \vec{M}_0 + \vec{M}_\sim$ (when $\vec{M}_\sim \ll \vec{M}_0$) becomes dependent on time. Because the magnitude of the moment of force:

$$\vec{T} = \mu_0 \left[\vec{M}_\sim \vec{H}^{ef} \right] + \mu_0 \left[\vec{M}_0 H_\sim \right] + \vec{t}_{an} \neq 0 ,$$

where $\vec{t}_{an} \neq 0$ and $\vec{T}_{an} = \vec{T}_{an,0} + \vec{t}_{an}$.

Here the magnitude $\vec{t}_{an}$ is the additional moment in which the anisotropy field acts upon the vector $\vec{M}_0$, when it deviates from the equilibrium.

We determined the projections $t_{an,x}$, $t_{an,y}$ in the coordinate system in which the axis z is directed along the equilibrium vector $\vec{M}_0$. And the axis y is normal to the plane passed through axes z' and z and lying in the plane $x'y'$. The axis x is in the plane $z'z$ as shown in Fig.4.1.

Near the equilibrium, which is supposed to be stable, the free energy expansion has the form:

$$F_{an} = F_0 + \frac{1}{2}\left[\frac{\partial^2 F}{\partial\theta^2}(\Delta\theta)^2 + 2\frac{\partial^2 F}{\partial\theta\,\partial\alpha}\Delta\theta \cdot \Delta\alpha + \frac{\partial^2 F}{\partial\alpha^2}(\Delta\alpha)^2 \right],$$

where α is the angle of rotation around the axis x. Here the first derivatives in the equilibrium $\partial F_{an}/\partial\theta = 0$ and $\partial F_{an}/\partial\alpha = 0$, that in the vector form is written by the formula (4.25). The additional moment of the anisotropy field:

$$t_{an,y} = -\frac{\partial F_{an}}{\partial(\Delta\theta)} = -\frac{\partial^2 F_{an}}{\partial\theta^2}\Delta\theta - \frac{\partial^2 F_{an}}{\partial\theta\,\partial\alpha}\Delta\alpha = -\frac{\partial^2 F_{an}}{\partial\theta^2}\cdot\frac{M_{\sim x}}{M_0} + \frac{\partial^2 F_{an}}{\partial\theta\,\partial\alpha}\cdot\frac{M_{\sim y}}{M_0},$$

because the vector is not changing its magnitude (the anisotropic media is magnetized up to the saturation of the media).

Similarly:

$$t_{an,x} = -\frac{\partial F_{an}}{\partial(\Delta\alpha)} = -\frac{\partial^2 F_{an}}{\partial\theta\,\partial x}\Delta\theta - \frac{\partial^2 F_{an}}{\partial\alpha^2}\Delta\alpha = -\frac{\partial^2 F_{an}}{\partial\theta\,\partial\alpha}\cdot\frac{M_{\sim x}}{M_0} + \frac{\partial^2 F_{an}}{\partial\alpha^2}\cdot\frac{M_{\sim y}}{M_0}.$$

Naturally, $\vec{t}_{an,z} = 0$ because the rotation around the axis z does not bring the vector $\vec{M}_0$ out of equilibrium.

We now introduce the dimensionless designations:

$$\Omega = \frac{\omega}{\gamma_0\mu_0 M_0} - \frac{i}{\mu_0 M_0^2}\cdot\frac{\partial^2 F_{an}}{\partial\theta\,\partial\alpha},$$

$$\Omega_1 = \frac{H_z^{ef}}{M_0} + \frac{1}{2\mu_0 M_0^2}\left(\frac{\partial^2 F_{an}}{\partial\theta^2} + \frac{\partial^2 F_{an}}{\partial\alpha^2}\right),$$

$$\delta = \frac{1}{2\mu_0 M_0^2}\left(\frac{\partial^2 F_{an}}{\partial\alpha^2} - \frac{\partial^2 F_{an}}{\partial\theta^2}\right). \tag{4.26}$$

Then the motion equations take the form:

$$i\Omega^* M_{\sim x} + (\Omega_1 + \delta) M_{\sim y} = H_{\sim y},$$

$$(\Omega_1 - \delta) M_{\sim x} - i\Omega M_{\sim y} = H_{\sim x}, \tag{4.27}$$

where we find:

$$M_{\sim x} = \frac{\Omega_1 + \delta}{\Omega_1^2 - \delta^2 - \Omega\Omega^*} H_{\sim x} + \frac{i\Omega}{\Omega_1^2 - \delta^2 - \Omega\Omega^*} H_{\sim y},$$

$$M_{\sim y} = \frac{-i\Omega^*}{\Omega_1^2 - \delta^2 - \Omega\Omega^*} H_{\sim x} + \frac{\Omega_1 - \delta}{\Omega_1^2 - \delta^2 - \Omega\Omega^*} H_{\sim y}. \tag{4.28}$$

In other words, the dynamic magnetic susceptibility tensor can be written as the Hermitian matrix:

$$\tilde{\chi} = \begin{vmatrix} \chi + \Delta & iv & 0 \\ -iv* & \chi - \Delta & 0 \\ 0 & 0 & 0 \end{vmatrix}, \tag{4.29}$$

where

$$\chi = \frac{\Omega_1}{\Omega_1^2 - \delta^2 - \Omega\Omega^*},$$

$$v = \frac{\Omega}{\Omega_1^2 - \delta^2 - \Omega\Omega^*},$$

$$\Delta = \frac{\delta}{\Omega_1^2 - \delta^2 - \Omega\Omega^*}, \tag{4.30}$$

the mark "$*$" means the complex conjugate.

The tensor (4.29) is a sum of symmetric and antisymmetric tensors. The symmetric tensor has the only nonzero nondiagonal element $-v''$ where $v = v' + iv''$. It can be turned into zero in the new coordinate system obtained by the rotation of the starting coordinate system around the axis z by the angle β. The element:

$$\chi_{1'2'} = iv' - \tfrac{1}{2}\left(\chi_{11} - \chi_{22}\right)\sin 2\beta - v''\cos 2\beta,$$

is expressed through the elements of the tensor (4.29). Where from the rotation angle sought it is:

$$tg\, 2\beta = \frac{2v''}{\chi_{22} - \chi_{11}}.$$

In this coordinate system, the tensor (4.29) has the nonzero elements:

$$\chi_{1'1'} = \frac{\chi_{11} + \chi_{22}}{2} - \frac{0.5\left(\chi_{11} - \chi_{22}\right)^2 + 2v''^2}{\sqrt{\left(\chi_{11} - \chi_{22}\right)^2 + 4v''^2}},$$

$$\chi_{2'2'} = \frac{\chi_{11} + \chi_{22}}{2} + \frac{0.5\left(\chi_{11} - \chi_{22}\right)^2 + 2v''^2}{\sqrt{\left(\chi_{11} + \chi_{22}\right)^2 + 4v''^2}},$$

$$\chi_{1'2'} = -\chi_{2'1'} = iv'. \tag{4.31}$$

When we derived at the expression (4.29), we never used the explicit crystal symmetry. Here it means the formula (4.29) represents the general susceptibility tensor form. The tensor in the form (4.31) for cubic crystals is given in [4.9]. One can see from (4.26), (4.29) and (4.30) that $\dfrac{\partial^2 F_{an}}{\partial \alpha^2} = \dfrac{\partial^2 F_{an}}{\partial \theta^2}$, which means $\Delta = 0$. Then the tensor (4.29) coincides with the tensor in the isotropic case (see the equations (4.7)).

We now consider in more detail the calculation of the derivatives with respect to the rotation angle α. While rotating around the axis x by the angle α the vector $\vec{M}_0$ rotates in the plane yz and takes the position $\vec{M}_0'/M_0 = \vec{k}\cos\alpha - \vec{j}\sin\alpha$, where $\vec{k}$ and $\vec{j}$ are unit vectors of z and y axes. Let $\vec{\gamma}$ be the unit vector directed along the line of the intersection of the zz' and $x'y'$ planes. Evidently, $\vec{k} = \vec{k}'\cos\theta + \vec{\gamma}\sin\theta$, so that

$$\vec{M}_0'/M_0 = \vec{k}'\cos\theta\cos\alpha + \vec{\gamma}\sin\theta\cos\alpha - \vec{j}\sin\alpha.$$

Now we find the new angles θ' and φ' defining the vector $\vec{M}_0'$ orientation:

$$\cos\theta' = \cos\theta\cos\alpha \approx \cos\theta\cdot\left(1-\alpha^2/2\right),$$

$$\sin\theta' = \sqrt{\sin^2\theta\cos^2\alpha + \sin^2\theta} \approx \sin\theta\cdot\left(1+ctg^2\theta\cdot\alpha^2/2\right),$$

$$\cos\varphi' = \frac{\cos\varphi\sin\theta\cos\alpha + \sin\varphi\cdot\sin\alpha}{\sqrt{\sin^2\theta + \cos^2\theta\sin^2\alpha}} \approx \cos\varphi\left(1-\frac{\alpha^2}{2\sin^2\theta}\right) + \alpha\frac{\sin\varphi}{\sin\theta},$$

$$\sin\varphi' = \frac{\sin\varphi\sin\theta\cos\alpha - \sin\alpha\cos\varphi}{\sqrt{\sin^2\theta + \cos^2\theta\sin^2\alpha}} \approx \sin\varphi\left(1-\frac{\alpha^2}{2\sin^2\theta}\right) - \alpha\frac{\cos\varphi}{\sin\theta}. \tag{4.32}$$

The magnitudes (4.32) are enough to find with the accuracy up to the second order of smallness. Formulae (4.32) enable one to calculate the derivatives $\dfrac{\partial F_{an}}{\partial\alpha}$, $\dfrac{\partial^2 F_{an}}{\partial\alpha\,\partial\theta}$ and $\dfrac{\partial^2 F_{an}}{\partial\alpha^2}$ in the equilibrium state when of $\alpha = 0$ [4.10].

Taking into account losses in the form of (4.13) the magnetic susceptibility tensor is described by the matrix:

$$\ddot{\chi} = \begin{vmatrix} \chi+\Delta+i\Delta_r & i\left(v-i\left(v_r+\delta v_r\right)\right) & 0 \\ -i\left(v^* -i\left(v_r-\delta v_r\right)\right) & \chi-\Delta+i\Delta_r^* & 0 \\ 0 & 0 & \chi_{\|} \end{vmatrix}, \tag{4.33}$$

here the magnitudes are expressed like this:

$$\chi = \frac{1}{\alpha}\left(\Omega_1 + \frac{\Omega^2_r}{\Omega_H}\right), \qquad \Delta = \frac{\delta}{\alpha}, \qquad \Delta_r = \frac{\Omega_r\Omega}{\Omega_H\alpha},$$

$$v = \frac{\Omega}{\alpha}, \qquad v_r = \frac{1}{\alpha}\left(\Omega_r - \frac{\Omega_r\Omega_1}{\Omega_H}\right), \qquad \delta v_r = \frac{\Omega_r\delta}{\Omega_H\alpha},$$

where

$$\alpha = \Omega_1^2 - \delta^2 - \left(\Omega - i\Omega_r\right)\left(\Omega^* - i\Omega_r\right),$$

$$\chi_{\|} = \frac{\Omega_r}{\Omega_H\left[\Omega_r + \dfrac{i}{2}\left(\Omega+\Omega^*\right)\right]}.$$

are written in designations of Ω, Ω_1 and δ defined by the formula (4.26). The magnitude Ω_r is defined by the formula (4.15) and the magnitude Ω_H is defined by the formula $\Omega_H = H_z^{ef}/M_0$. The tensor (4.33) is not Hermitian.

4.3. The Permittivity Tensor of Semiconductor Media

The permittivity of semiconductor media in a constant magnetic field $\vec{B}_0$ expresses itself through conductivity $\vec{\sigma}$ in Ohm's law for the current density:

$$\vec{j} = \vec{\sigma}\vec{E}. \tag{4.34}$$

Here the value $\vec{\sigma}$ connecting vectors $\vec{j}$ and $\vec{E}$ is usually a tensor. Substituting the expression (4.34) in Maxwell's equation we will obtain this:

$$rot\vec{H} = i\omega\vec{\varepsilon}_k\varepsilon_0\vec{E} + \vec{j}.$$

The last equation we will write in this form:

$$rot\vec{H} = i\omega\varepsilon_0\left(\vec{\varepsilon}_k - i\frac{\vec{\sigma}}{\omega\varepsilon_0}\right)\vec{E} \equiv i\omega\vec{\varepsilon}\vec{E},$$

where the permittivity tensor [4.11] has this form:

$$\vec{\varepsilon} = \vec{\varepsilon}_k - i\frac{\vec{\sigma}}{\omega\varepsilon_0}. \tag{4.35}$$

Here the permittivity tensor $\vec{\varepsilon}$ consists of ohmic losses and the tensor $\vec{\varepsilon}_k$ has imaginary components.

Anyone could obtain a conductivity tensor $\vec{\sigma}$ when they consider the motion of a charge e having the effective mass m^*, which reflects the crystal lattice influence on the particle motion in the electromagnetic field. The motion equation:

$$\frac{d}{dt}\left(m^*\vec{v}'\right) = e\left\{\vec{E} + \left[\vec{v}\vec{B}\right]\right\} - \frac{m^*\vec{v}}{\tau}, \tag{4.36}$$

consist the phenomenological "friction force", which is characterized by the relaxation time τ. Here the magnitude τ is allowed to be constant. Vectors of electric and magnetic fields are sums of uniform and alternating fields of small amplitudes. Thus, one has to change the formula (4.36):

$$\vec{v} \to \vec{v}_0 + \vec{v}, \quad \vec{E} \to \vec{E}_0 + \vec{E} \quad \text{and} \quad \vec{B} \to \vec{B}_0 + \vec{B}.$$

Bearing this in mind that the quantities $\vec{v}_0$, $\vec{E}_0$, $\vec{B}_0$ correspond to the stationary motion and preserving the quantities of the first order of smallness we obtained the algebraic equation:

$$i\omega m^*\vec{v} = e\left\{\vec{E} + \left[\vec{v}\vec{B}_0\right]\right\} - \frac{m^*\vec{v}}{\tau},$$

(instead of (4.36)) in the case of harmonic time dependence $e^{i\omega t}$. It is more convenient to rewrite the last equation as:

$$\vec{v} + \frac{\tau^* e}{m^*}\left[\vec{B}_0\vec{v}\right] = \frac{e\tau^*}{m^*}\vec{E}, \tag{4.37}$$

where $\tau^* = \tau/(1+i\omega\tau)$.

Taking the scalar product of the equation (4.37) and the vector $\vec{B}_0$ we see that:

$$\left(\vec{B}_0\vec{v}\right) = \frac{e\tau^*}{m^*}\left(\vec{B}_0\vec{E}\right).$$

$$(4.38)$$

Now using the vector product of (4.37) and the vector $\vec{B}_0$ we find:

$$\left[\vec{B}_0\vec{v}\right] + \frac{e\tau^*}{m^*}\left\{\vec{B}_0\left(\vec{B}_0\vec{v}\right) - B_0^2\vec{v}\right\} = \frac{e\tau^*}{m^*}\left[\vec{B}_0\vec{E}\right],$$

and for the vector product we obtained this:

$$\left[\vec{B}_0\vec{v}\right] = \frac{e\tau^*}{m^*}\left\{\left[\vec{B}_0\vec{E}\right] + B_0^2\vec{v} - \frac{e\tau^*}{m^*}\left(\vec{B}_0\vec{E}\right)\vec{B}_0\right\}.$$

Substituting the last formula into (4.37), we finally get the expression for the speed:

$$\vec{v} = \frac{e\dfrac{\tau^*}{m^*}}{1+\left(\dfrac{e\tau^*}{m^*}B_0\right)^2}\left\{\vec{E} - \frac{e\tau^*}{m^*}\left[\vec{B}_0\vec{E}\right] + \left(\frac{e\tau^*}{m^*}\right)^2\left(\vec{B}_0\vec{E}\right)B_0\right\}.$$

$$(4.39)$$

The current density is defined by the expression:

$$\vec{j} = ne\vec{v} ,$$

where n is the concentration of the charges. Then we substitute in this last equation the formula (4.39) and comparing this with the formula (4.34) we find the conductivity tensor:

$$\sigma_{rs} = \frac{ne^2\dfrac{\tau^*}{m^*}}{1+\left(\dfrac{e\tau^*}{m^*}B_0\right)^2}\left\{\delta_{rs} + e_{rsk}\frac{e\tau^*}{m^*}B_{0k} + \left(\frac{e\tau^*}{m^*}\right)^2 B_{0r}B_{0s}\right\}.$$

$$(4.40)$$

Here the summation made over the repeating subscript k, e_{rsk} is the antisymmetric with respect to all subscripts tensor; its nonzero elements are obtained from $e_{123} = 1$ and also by changing the subscripts (for example $e_{213} = -1$). The formulae (4.35) and (4.40) define the permittivity tensor.

For clarity sake, we write the conductivity tensor elements (4.40) in the coordinate system with z axis directed along the vector $\vec{B}_0$, so that $B_{0x} = B_{0y} = 0$ and $B_{0z} = B_0$. In this coordinate system the components of the permittivity tensor are:

$$\varepsilon_{xx} = \varepsilon_{k,xx} - i\frac{\omega_r\tau^*}{1+\left(\omega_c\tau^*\right)^2}\frac{\omega_r}{\omega} ,$$

$$\varepsilon_{yy} = \varepsilon_{k,yy} - i\frac{\omega_r \tau^*}{1 + \left(\omega_c \tau^*\right)^2}\frac{\omega_r}{\omega},$$

$$\varepsilon_{xy} = \varepsilon_{k,xy} - i\frac{\omega_r^2 \tau^{*2}}{1 + \left(\omega_c \tau^*\right)^2}\frac{\omega_c}{\omega},$$

$$\varepsilon_{yx} = \varepsilon_{k,yx} + i\frac{\omega_r^2 \tau^{*2}}{1 + \left(\omega_c \tau^*\right)^2}\frac{\omega_c}{\omega},$$

$$\varepsilon_{xz} = \varepsilon_{k,xz}, \qquad \varepsilon_{yz} = \varepsilon_{k,yz},$$

$$\varepsilon_{zx} = \varepsilon_{k,zx}, \qquad \varepsilon_{zy} = \varepsilon_{k,zy},$$

$$\varepsilon_{zz} = \varepsilon_{k,zz} - i\frac{\tau^* \omega_r^2}{\omega}, \tag{4.41}$$

where we used designations:

$$\omega_r^2 = \frac{ne^2}{m^* \varepsilon_0},$$

$$\omega_c = \frac{e}{m^*} B_0.$$

Anyone can obtain the lossless case by substituting the volume $\tau \to \infty$ in the formulae (4.41) and $\tau^* = 1/i\omega$. Then one can see from these formulae that the elements ε_{xx}, ε_{yy}, ε_{xy} and ε_{yx} at $\omega \to \omega_c$ have a resonance behavior.

In the formula (4.40) we considered that the current was formed by the one-type charge carriers, which are described by the concentration n, charge e, effective mass m^* and the relaxation time τ. When there are charge carriers of several types, we assume that one type charge motion is independent from the other type charge motion. The formula (4.40) can be written in a general form when we determine the conductivity for all types of charges and then add these conductivities together.

Up to now the effective mass m^* was considered to be a scalar quantity. However, the quantity m^* reflects phenomenologycally the influence of the crystal lattice on the charge motion. The effective mass, generally speaking, is described by a tensor $\ddot{m}^*$ so that the momentum $\ddot{m}^* \vec{v}$ direction does not coincide with direction of speed $\vec{v}$. The motion equation for the situation above is written like this:

$$\frac{d}{dt}\ddot{m}^* \vec{v} = e\left\{\vec{E} + \left[\vec{v}\vec{B}\right]\right\} - \frac{\ddot{m}^* \vec{v}}{\tau},$$

where the "friction forces" are characterized by one (constant) parameter τ. Assuming harmonic dependence on time, we obtain the algebraic equation:

$$\ddot{m}^{*}\vec{v} + e\tau^{*}\left[\vec{B}_{0}\vec{v}\right] = e\tau^{*}\vec{E} ,$$

where $\tau^{*} = \tau/(1+i\omega\tau)$.   (4.42)

A further solution to this problem is carried out in the coordinate system in which the effective mass tensor has a diagonal form with elements m_{1}^{*}, m_{2}^{*} and m_{3}^{*} on the main diagonal. Writing the relations (4.42) in projections and solving this system of three equations with respect to v_{x}, v_{y} and v_{z} we present the result in vector form:

$$\vec{v} = \frac{\tau^{*}e}{1+\dfrac{\tau^{*2}e^{2}}{m_{1}^{*}m_{2}^{*}m_{3}^{*}}\left(\vec{B}_{0},\ddot{m}^{*}\vec{B}_{0}\right)}\left\{\ddot{m}^{-1}\vec{E} + \frac{\tau^{*}e}{m_{1}^{*}m_{2}^{*}m_{3}^{*}}\left[\vec{E},\ddot{m}^{*}\vec{B}_{0}\right] + \right.$$

$$\left. + \frac{e^{2}\tau^{*2}}{m_{1}^{*}m_{2}^{*}m_{3}^{*}}\left(\vec{E}\vec{B}_{0}\right)\vec{B}_{0}\right\} .   (4.43)$$

The formula 4.43 is valid in any coordinate system. Therefore, we have the formula for the conductivity tensor:

$$\sigma_{rs} = \frac{ne^{2}\tau^{*}}{1+\dfrac{\tau^{*2}e^{2}}{m_{1}^{*}m_{2}^{*}m_{3}^{*}}\left(\vec{B}_{0},\ddot{m}^{*}\vec{B}_{0}\right)}\left\{\left(\ddot{m}^{-1}\right)_{rs} + \frac{e\tau^{*}}{m_{1}^{*}m_{2}^{*}m_{3}^{*}}e_{rsk}m_{kp}\vec{B}_{0p} + \right.$$

$$\left. + \frac{e^{2}\tau^{*2}}{m_{1}^{*}m_{2}^{*}m_{3}^{*}}\vec{B}_{0r}\vec{B}_{0s}\right\} ,   (4.44)$$

where it is assumed summation over the repeating subscripts. Evidently, it is more convenient to use this formula in the coordinate system in which the mass tensor $\ddot{m}^{*}$ and the inverse tensor $\ddot{m}^{-1}$ has the diagonal form: $m_{rs}^{*} = \ddot{m}_{r}^{*}\delta_{rs}$ and

$$m_{rs}^{-1} = \frac{1}{\ddot{m}_{r}^{*}}\delta_{rs} .$$

We have now finishing describing anisotropic media and in the next chapter for this media we will analyze the reciprocity theorem, boundary conditions and representations for electric inductions, magnetic fields, electric charges, as the relations between matrices of capacitances and inductances.

5. COMMON DEPENDENCIES FOR TRANSMITION LINES IN TEM – APPROXIMATION

In chapter 2 we considered simple striplines with uniform fillings, but in this chapter we are going to analyze striplines with piece-wise uniform fillings. For most striplines with piece-wise uniform fillings it is impossible to obtain a solution in visible formulae (as in chapter 2). So we are going to use our methods of SIE to solve these kinds of problems in TEM-approximation.

Mathematical investigation of striplines in TEM–approximation in case of piece-wise fillings (Fig.5.1) means we must solve Laplace's equation with certain boundary conditions on contours dividing different media. The idea method is to write the integral representation of the solution in each region, which satisfy the differential equation (2.2) and which has the unknown function found from the boundary conditions. As soon as one succeeds to do this problem is solved.

To use Laplace's equation to solution a problem means we must make an assumption that the second term in Helmholtz's equation (1.4) is small when compared to the first term. The rough evaluation of this assumption validity can be done in the following way: let L be the characteristic size of the transmission line cross-section, then we replace the evaluation $\partial^2 E_z / \partial x^2$ with E_z / L^2. If the magnitude E_z / L^2 is a lot larger than $\left[(2\pi\varepsilon\mu/\lambda)^2 - (2\pi/\Lambda)^2 \right] E_z$ (where λ is a wavelength in a vacuum and Λ is a wavelength in a waveguide) then the last term will be discarded. And so, we arrive at Laplace's equation instead Helmholtz's equation. This approximation is called the transverse wave approximation (or TEM–approximation), because the longitudinal electromagnetic field components are not taken into account when using this approach. Generally speaking, it is possible to take into consideration the longitudinal components but the results in approximation of harmonic functions (see sec. 6.4) would differ little from that of TEM-approximation.

5.1. Integral Representation

For the integral representation, we use the solution of Poisson's equation with the point-source of intensity a placed in the origin:

$$\frac{\partial^2 \varphi}{\partial x^2} + \frac{\partial^2 \varphi}{\partial y^2} = a\delta(x)\delta(y), \tag{5.1}$$

where $\delta(x)$ is the Dirac function having properties:

$$\delta(x) = 0, \quad x \neq 0,$$

$$\int_{-b}^{+b} f(x)\delta(x)dx = f(0). \tag{5.2}$$

It is known [5.1] that the solution of the equation (5.1):

$$\varphi \sim \ln r, \quad r^2 = x^2 + y^2,$$

satisfies Laplace's equation everywhere except at the point $x = y = 0$.

Therefore, the function:

$$\varphi(x, y) = \int_L \mu(x', y') \ln \sqrt{(x-x')^2 + (y-y')^2}\, ds, \tag{5.3}$$

satisfies Laplace's equation everywhere except the points of the contour $L = L_0 + L_1 + L_2 + \dots$. In the formula (5.3) ds is the element of the contour arc, where the density $\mu(x', y')$ is determined from the boundary conditions. Boundary conditions for an electric field lead to the uniform integral equations with respect to the function $\mu(x', y')$.

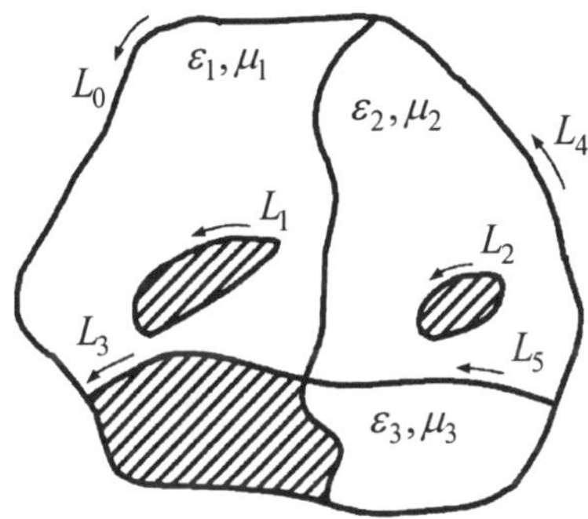

Fig. 5.1. Cross-section of a piece - wise uniform regular transmission line.

Application of the Bogoliubov–Krylov method, as it was described in sec.3.3, transforms the integral equation into a system of linear algebraic equations for certain values of $\mu(x_i, y_i)$, $i = 1,2,\dots,\ n$, at n points on the contour. This system has a nontrivial solution, if its determinant is equal to zero. There is no free parameter, which could turn the determinant into zero in the previous system. Therefore the representation (5.3) does not solve this problem. We supplement the representation (5.3) with terms like this:

$$A_1 \ln \sqrt{(x - x_1)^2 + (y - y_1)^2}\,.$$

Here A_1 is an unknown coefficient and the point (x_1, y_1) is inside the contour bounder of the metal. The number of such terms is equal to the number of metal conductors in the stripline. Here in the solution of Laplace's equation:

$$\varphi(x, y) = \int_L \mu(x', y') \ln \sqrt{(x-x')^2 + (y-y')^2}\, ds \;+\; \sum_{k=1}^{m} A_k \ln \sqrt{(x-x_k)^2 + (y-y_k)^2} \qquad (5.4)$$

one has to determine the constants A_k. Where we find the constants A_k setting the potential values φ_k on each conductor (or charges Q_k *). This formula (5.4) we will use to express the electric field. When we calculate a magnetic field we must set the value of the longitudinal currents, which run along each conductor.

When we apply Bogoliubov–Krylov's method, we obtain the nonuniform system of $n+m$ linear equations to find n values of $\mu(x_k, y_k)$, $k = 1,2,...n$ and m of the constants A_k, $k = 1,2,...m$, where m is the number of conductors. This is the only solution, because the system determinant is not equal to zero.

Here it is worthy to remember that the solution (5.4) of the static problem is only an approximation of the corresponding boundary electrodynamical problem in SHF and EHF ranges. Maxwell's equations (1.1) are usually solved using the complex amplitudes, when the dependence on time is described by the factor $e^{i\omega t}$. Then in the formula (5.4), the quantities φ, μ and A_k are to be understood as complex. An example the coefficients A_k are replaced by the complex quantity:

$$\dot{A}_k = A_k' + iA_k''. \qquad (5.5)$$

The introduction of complex amplitudes, which allows us to take into account the losses in materials of stripline substrates in a simple way.

It was shown in chapter 2 that it is convenient to describe the plane fields by means of complex functions. Because of this we can write the formula (5.4) in this form:

$$\dot{\Phi}(z) = \int_L \dot{\mu}(t) \ln(t - z)\, ds + \sum_{k=1}^{m} \dot{A}_k \ln(z - z_k) \qquad (5.6)$$

where $z = x + jy$ is the complex coordinate of the point which the potential determines in $t = x' + jy'$ is the coordinate of the point of the contour L.

The magnitude $z_k = x_k + jy_k$ is the coordinate of the point inside the contour bordering the kth conductor. Bearing in mind, that the function $\dot{\mu}$ and constants A_k are real with respect to coordinates z and t, we see that the potential (5.4) is the real part of the complex potential (5.6) and the derivative $\dot{\Phi}'(z)$ gives the complex field (2.4).

In principle, using the SIE method, allows us to calculate a multiconductor regular line with any topology and dissipative filling.

Since the equations for the electric and magnetic fields in TEM-approximation are not coupled, the solution for each field is carried separately.

---

* The rigorous mathematical proof of the boundary problem solution uniqueness is given, for example in [3.3].

To describe the magnetic field, we use the vector potential with the only longitudinal component $A_z \neq 0$ and the transversal field:

$$\vec{B} = rot\vec{A}. \tag{5.7}$$

In formulae (1.3) where $E_z \sim A_z$ and the component A_z also can be represented in this integral form (5.6).

5.2. Derivation of Singular Integral Equations

Here certain SIEs are derived when satisfying boundary conditions. The following boundary conditions in points of contour surrounding different magnetodielectric media are: an equality of the normal components of the electric inductance:

$$D_{n+} = D_{n-}, \tag{5.8}$$

then the tangential components of electric field strength:

$$E_{\tau+} = E_{\tau-}. \tag{5.9}$$

And boundary conditions in points of contour surrounding perfect metal are an equality of the tangential component of the electric field strength to:

$$E_\tau = 0. \tag{5.10}$$

Subscripts " + " and " − " in the formulae (5.8) and (5.9) indicate that the points in which the quantities are determined and approach the contour from the left or right correspondingly.

For the magnetostatic version of the problem, we require on the media division line, the equality of tangential components of magnetic field strength:

$$H_{\tau+} = H_{\tau-} \tag{5.11}$$

and the equality of normal components of magnetic inductance (magnetic flux density):

$$B_{n+} = B_{n-}. \tag{5.12}$$

In the points of contour surrounding *perfect* metal the normal component of magnetic inductance is:

$$B_n = 0. \tag{5.13}$$

Now we will demonstrate the solution to an electrostatic problem. Here we write the permittivity of the media taking into account its losses in this form:

$$\dot{\varepsilon} = \varepsilon' - i\varepsilon''. \tag{5.14}$$

The sign of the permittivity is supplied with subscripts for different media. In each region the amplitude of the electric inductance:

$$\dot{D}(z) = -\dot{D}_x(x, y) + j\dot{D}_y(x, y)$$

and the electric field:

$$\dot{E}(z) = -\dot{E}_x(x,y) + j\dot{E}_y(x,y) \tag{5.15}$$

are connected through the constitutive relation:

$$\dot{D}(z) = \dot{\varepsilon E}(z). \tag{5.16}$$

Here all values are complex, an example would be:

$$\dot{E}_x(x,y) = E_x'(x,y) + iE_x''(x,y),$$

$$\dot{E}_y(x,y) = E_y'(x,y) + iE_y''(x,y),$$

$$\dot{D}_x(x,y) = D_x'(x,y) + iD_x''(x,y) \text{ and so on.}$$

When we calculate the tangential components:

$$E_\tau^\pm(t_0) = E_x^\pm(t_0)\tau_x(t_0) + E_y^\pm(t_0)\tau_y(t_0) \tag{5.17}$$

and the normal components:

$$E_n^\pm(t_0) = E_x^\pm(t_0)n_x(t_0) + E_y^\pm(t_0)n_y(t_0) \tag{5.18}$$

of an electric field in the point t_0 on the contour L, when point z approaches t_0 from the regions S_+ and S_- (Fig.5.2). The region S_+ is on the left side of the contour as we move along it. The unit vector $\vec{\tau}$ is directed along the contour tangent. The unit normal vector $\vec{n}$ is in the direction of region S_+. Therefore, the projections:

$$\tau_x = n_y = \cos\theta(t_0), \ \tau_y = -n_x = \sin\theta(t_0), \tag{5.19}$$

where $\theta(t_0)$ is the angle between the contour tangent at the point t_0 and the positive direction of the axis x, can be written:

$$dt = ds \cdot e^{j\theta(t)} \tag{5.20}$$

for any point t on the contour L.

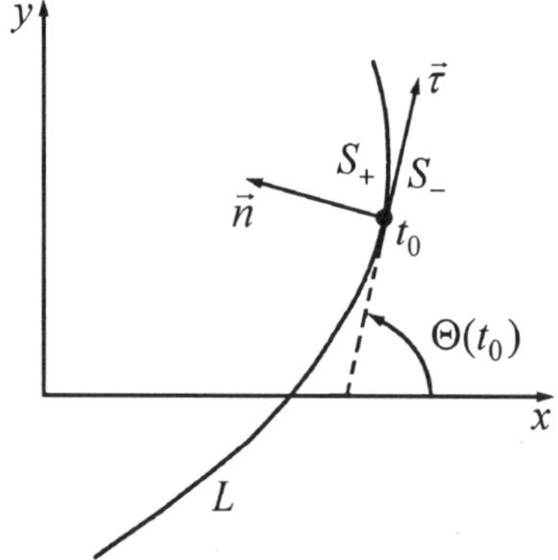

Fig. 5.2. *The designations when contour L is smooth.*

Differentiating the expression (5.6) we find the electric field:

$$\dot{E}(z) = -\int_L \frac{\dot{\mu}(t)ds}{t-z} - \sum_k \frac{\dot{A}_k}{z_k - z}. \tag{5.21}$$

For the points $z \to t_0$ from either side of the contour L when taking into account formulae (3.4) and (5.20) we obtain this:

$$\dot{E}^{\pm}(t_0) = -2\pi j \left\{ \pm \frac{1}{2} \dot{\mu}(t_0) e^{-j\theta(t_0)} + \frac{1}{2\pi j} \int_L \frac{\dot{\mu}(t)ds}{t-t_0} \right\} - \sum_k \frac{\dot{A}_k}{z_k - t_0}.$$

According to the definition (5.15) we obtained:

$$\dot{E}_x^{\pm}(t_0) = \pm \pi \dot{\mu}(t_0) \sin \theta(t_0) + \mathrm{Re}_j \left[\int_L \frac{\dot{\mu}(t)ds}{t-t_0} + \sum_k \frac{\dot{A}_k}{z_k - t_0} \right],$$

$$\dot{E}_y^{\pm}(t_0) = \mp \pi \dot{\mu}(t_0) \cos \theta(t_0) - \mathrm{Im}_j \left[\int_L \frac{\dot{\mu}(t)ds}{t-t_0} + \sum_k \frac{\dot{A}_k}{z_k - t_0} \right]. \tag{5.22}$$

Following the formulae (5.17)–(5.19) we find:

$$\dot{E}_{\tau}^{\pm}(t_0) = \mathrm{Re}_j \left\{ e^{j\theta(t_0)} \left[\int_L \frac{\dot{\mu}(t)ds}{t-t_0} + \sum_k \frac{\dot{A}_k}{z_k - t_0} \right] \right\}, \tag{5.23}$$

$$\dot{E}_n^{\pm}(t_0) = \mp \pi \dot{\mu}(t_0) - \mathrm{Im}_j \left\{ e^{j\theta(t_0)} \left[\int_L \frac{\dot{\mu}(t)ds}{t-t_0} + \sum_k \frac{\dot{A}_k}{z_k - t_0} \right] \right\}. \tag{5.24}$$

One can see from the expression (5.23) that the tangential component does not have discontinuity on the contour L, so the condition (5.9) is satisfied because the electric field representation is (5.21). The normal component (5.24) has discontinuity in the contour points. Thus the boundary condition (5.8) leads to the SIE:

$$(\varepsilon^+ + \varepsilon^-)\pi \dot{\mu}(t_0) + (\varepsilon^+ - \varepsilon^-) \mathrm{Im}_j \left\{ e^{j\theta(t_0)} \left[\int_L \frac{\dot{\mu}(t)ds}{t-t_0} + \sum_k \frac{\dot{A}_k}{z_k - t_0} \right] \right\} = 0, \tag{5.25}$$

for the determination of densities $\dot{\mu}(t)$ and constants $\dot{A}_k$. For the points t_0, which is on the contour surrounding *perfect* metal and according to the condition (5.10) we can write the expression:

$$\mathrm{Re}_j \left\{ e^{j\theta(t_0)} \left[\int_L \frac{\dot{\mu}(t)ds}{t-t_0} + \sum_k \frac{\dot{A}_k}{z_k - t_0} \right] \right\} = 0. \tag{5.26}$$

Finally, we require for the mth conductor to have the potential (5.4) equal to the constant $\dot{\varphi}_m$:

$$\dot{\varphi}_m = \int_L \dot{\mu}(x', y') \ln \sqrt{(x_m - x')^2 + (y_m - y')^2}\, ds + \sum_{k=1}^{M} \dot{A}_k \ln \sqrt{(x_m - x_k)^2 + (y_m - y_k)^2} ,$$

where $m = 1, 2, ..., M$. $\qquad\qquad\qquad\qquad\qquad\qquad\qquad\qquad$ (5.27)

The points (x_m, y_m) can be chosen at any place on the conductor surface, which is convenient for the calculation integrals. The equations (5.25)–(5.27) are the full system of SIEs, which allows us to find the solution to the formulated electrostatic problem above.

One can also solve a magnetostatic problem in the same way as for the electrostatic problem. When we would like to calculate the magnetostatic field we have to use the expression for magnetic inductance $\vec{B} = rot\ \vec{k}A$ where:

$$B_x = \frac{\partial A}{\partial y}, \quad B_y = -\frac{\partial A}{\partial x} . \qquad\qquad\qquad\qquad\qquad (5.28)$$

Here we introduce the complex potential (5.6):

$$\Phi(z) = u(x, y) + jv(x, y) ,$$

with the real part coinciding with the function A . Here the derivative of the complex potential:

$$\Phi'(z) = \frac{\partial u}{\partial x} + j\frac{\partial v}{\partial x} = \frac{\partial u}{\partial x} - j\frac{\partial u}{\partial y} ,$$

gives the complex representation of the magnetic field:

$$\Phi'(z) \equiv \hat{B}(z) = -B_y - jB_x .$$

Repeating the calculations when approaching $z \to t_0$ from the right or left contour side, we have the same designations as for an electrostatic problem:

$$B_x^{\pm}(t_0) = \pm\pi\dot{\mu}(t_0)\cos\theta(t_0) + \mathrm{Im}_j\left\{\left[\int_L \frac{\dot{\mu}(t)ds}{t - t_0} + \sum_k \frac{\dot{A}_k}{z_k - t_0}\right]\right\},$$

$$B_y^{\pm}(t_0) = \pm\pi\dot{\mu}(t_0)\sin\theta(t_0) + \mathrm{Re}_j\left\{\left[\int_L \frac{\dot{\mu}(t)ds}{t - t_0} + \sum_k \frac{\dot{A}_k}{z_k - t_0}\right]\right\},$$

$$B_\tau^{\pm}(t_0) = \pm\pi\dot{\mu}(t_0) + \mathrm{Im}_j\left\{e^{j\theta(t_0)}\left[\int_L \frac{\dot{\mu}(t)ds}{t - t_0} + \sum_k \frac{\dot{A}_k}{z_k - t_0}\right]\right\},$$

$$B_n^{\pm}(t_0) = \mathrm{Re}_j\left\{e^{j\theta(t_0)}\left[\int_L \frac{\dot{\mu}(t)ds}{t - t_0} + \sum_k \frac{\dot{A}_k}{z_k - t_0}\right]\right\}. \qquad (5.29)$$

So the boundary condition (5.12) is satisfied because the representation of the complex potential is expressed by the formula (5.6). Here we write the equation (5.11) explicitly:

$$\left(\frac{1}{\mu^+}+\frac{1}{\mu^-}\right)\pi\mu(t_0)+\left(\frac{1}{\mu^+}-\frac{1}{\mu^-}\right)\times$$

$$\times \mathrm{Im}_j\left\{e^{j\theta(t_0)}\left[\int_L\frac{\mu(t)\,ds}{t-t_0}+\sum_k\frac{\dot A_k}{z_k-t_0}\right]\right\}=0\,. \tag{5.30}$$

In points of contour t_0 surrounding *perfect* metal, the boundary condition (5.13) can be written as:

$$\mathrm{Re}_j\left\{e^{j\theta(t_0)}\left[\int_L\frac{\mu(t)\,ds}{t-t_0}+\sum_k\frac{\dot A_k}{z_k-t_0}\right]\right\}=0\,. \tag{5.31}$$

The remaining equations were obtained calculating the magnetic induction (per unit length of the transmission line) through the surface connecting any two conductors:

$$\int_1^2(B\,ds)=\int_1^2\{B_x\,dy-B_y\,dx\}=\int_1^2 dA=A(2)-A(1)\,,$$

where we used the expressions (2.5a) at $\ell=1$. The point (x_m,y_m) can be placed at any point on the surface of *perfect* metal because the vector lines of the magnetic induction do not cross the metal as we see from the formula (5.13). So every metal surface has its definite value of potential A. Thus, we get an equation similar to formula (5.27).

Now comparing the equations for electrostatic and magnetostatic problems one can see that the second equation can be obtained from the first equation changing the permittivities ε into inverse permeabilities μ^{-1}.

5.3. Reciprocity Theorem

The solution of SIE systems (5.25)–(5.27) defines the values of densities $\dot\mu(t)$ on all points of contours, dividing different media, and constants $\dot A_k$. This enables us to calculate by formula (5.21) the electric field components at any point of the transmission line cross-section for any given excitation.

Usually we characterize certain transmission lines through the capacitance induction coefficients C_{mr} according to the definition:

$$Q_m=\sum_r C_{mr}\varphi_r\,. \tag{5.32}$$

Here Q_m is the charge of mth conductor, φ_m is its potential. Charges on conductors which are under given potentials φ_m are determined by the normal component of the induction D_n on *perfect* metal surfaces. Similarly, in the case of magnetostatic problems, we introduce coefficients of induction L_{mn}:

$$A_m=\sum_n L_{mn}I_n\,. \tag{5.33}$$

Here A_m is the value of the vector potential on the mth conductor coinciding with the magnetic inductance $\bar{B}$ through the surface, which connects the mth conductor with our chosen conductor (such as the ground metal plate of a stripline). The value I_n is the current flowing through the conductor. Currents under given potentials A_m are determined through the tangential components H_τ of the magnetic field at the *perfect* metal surface.

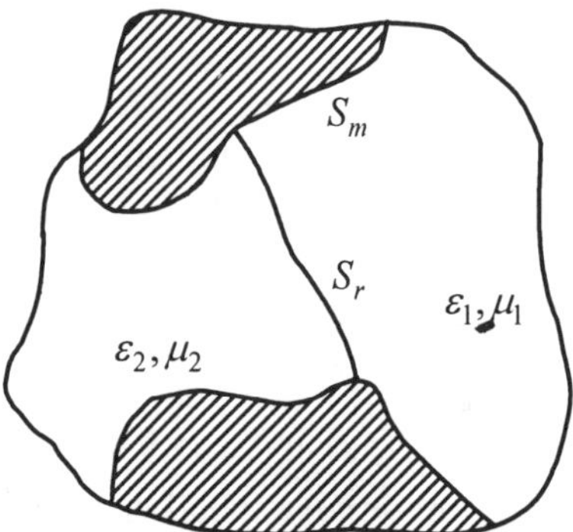

Fig. 5.3. *Cross-section of a piece-wise uniform multiconductor transmission line.*

Now we consider the piece-wise uniform multiconductor line shown in Fig.5.3. Here we use Gauss's formula:

$$\oint_S (\bar{a}d\bar{s}) = \int_V div\bar{a} \ dV \, , \tag{5.34}$$

written for each separate volume filled with a uniform media having a certain permittivity and permeability. The surface S surrounding volume V can consist of surfaces S_r dividing nonmetal media and metal surfaces S_m. First, we put in volume:

$$\bar{a} = \varphi^{(1)}(\ddot{\varepsilon} \ grad\varphi^{(2)}) \, . \tag{5.35}$$

Where $\varphi^{(1)}$ is the potential in the first experiment, when the conductors have charges of $Q_m^{(1)}$. The magnitude $\varphi^{(2)}$ is the potential in the second experiment, when conductors have charges of $Q_m^{(2)}$. Now adding all the equalities of type (5.34) we find:

$$\sum_m \int_{S_m} \varphi^{(1)} \left(\ddot{\varepsilon} \, grad\varphi^{(2)}, ds \right) + \sum_r \int_{S_r} \left(\varphi_+^{(1)} \left(\ddot{\varepsilon} \, grad\varphi^{(2)} \right)_+ - \right.$$

$$\left. -\varphi_-^{(1)}(\ddot{\varepsilon} \ grad\varphi^{(2)})_-, ds \right) = \sum_i \int_{V_i} \left(grad\varphi^{(1)}, \ddot{\varepsilon} \ grad\varphi^{(2)} \right) dV \tag{5.36}$$

because

$$div(\ddot{\varepsilon} \ grad\varphi^{(2)}) = div \ \vec{D}^{(2)} = 0 \, ,$$

since it is assumed that there are no free charges. In integrals of the first sum of the equality (5.36) the factor $\varphi^{(1)}$ is constant at every point of the *perfect* metal surface and the factor $\varphi^{(1)}$ can be taken out of the integral sign. The integral that remains:

$$\int_{S_m} (\ddot{\varepsilon}\ \text{grad}\,\varphi^{(2)}, ds) = Q_m^{(2)}$$

gives the charge on the mth conductor in the second experiment. In integrals of the second sum (5.36), potentials $\varphi_+^{(1)} = \varphi_-^{(1)}$ are equal on different sides of the surface dividing two dielectric media due to the condition (5.9) and

$$\left(\left(\ddot{\varepsilon}\ \text{grad}\,\varphi^{(2)} \right)_+ , ds \right) = \left(\left(\ddot{\varepsilon}\ \text{grad}\,\varphi^{(2)} \right)_- , ds \right)$$

due to the equality (5.8), so that the second sum vanishes and we obtain:

$$\sum_m \varphi_m^{(1)} Q_m^{(2)} = \sum_i \int_{V_i} (\text{grad}\ \varphi^{(1)}, \ddot{\varepsilon}\ \text{grad}\ \varphi^{(2)}) dv. \tag{5.37}$$

We now will consider these three particular cases:

1. Let all the media be isotropic and nonconducting. Then permittivity ε_i is a scalar real factor. Changing the formula (5.37) by replacing the subscripts $1 \Leftrightarrow 2$, we will receive the same integral and therefore:

$$\sum_m \varphi_m^{(1)} Q_m^{(2)} = \sum_m \varphi_m^{(2)} Q_m^{(1)}. \tag{5.38}$$

This equality (5.38) expresses the reciprocity theorem. Putting here the formula (5.32), we find that the matrix $\|C_{mr}\|$:

$$C_{mr} = C_{rm} \tag{5.39}$$

is symmetric. The property (5.39) does not take place for the transmission line made of lossy materials.

2. If the media is described by the symmetric tensor $\ddot{\varepsilon}_{ik} = \ddot{\varepsilon}_{ki}$ with real elements $\ddot{\varepsilon}_{ik}$, then we get the relation (5.38) and the property (5.39) in the same way.

3. Media are being described by Hermitian tensor:

$$\ddot{\varepsilon}_{ik} = \ddot{\varepsilon}_{ki}^*, \tag{5.40}$$

therefore the transmission line does not have losses. In this case, the potentials φ_m and Q_m charges may have different phases, so that the coefficients C_{mr} maybe a complex quantities.

When we replaced the magnitude $\varphi^{(1)}$ by its complex conjugated magnitude in the formula (5.37) we obtained:

$$\sum_m \varphi_m^{(1)*} Q_m^{(2)} = \sum_i \int_{V_i} (\text{grad}\,\varphi^{(1)*}, \ddot{\varepsilon}\ \text{grad}\,\varphi^{(2)}) dv. \tag{5.41}$$

In expression (5.41) we change the subscripts $1 \Leftrightarrow 2$ and then we take the complex conjugated expression. In the right hand side, we get the same integral as in the formula (5.41) due to Hermitian property (5.40), i.e.

$$\sum_m \varphi_m^{(1)*} Q_m^{(2)} = \sum_m \varphi_m^{(2)} Q_m^{(1)*},$$

Taking into account the definition (5.32), it follows that the matrix:

$$C_{mr} = C_{rm}^* \tag{5.42}$$

is Hermitian. At the end of the previous section, in the formula (5.33) the inductance coefficients L_{mr} have similar properties as in (5.42). The matrix L_{mr} is Hermitian.

The following calculations can be carried out in the following way. First, we place in the formula (5.34) this expression:

$$\vec{a} = \left[\vec{A}^{(1)}, (\ddot{\mu})^{-1} rot \vec{A}^{(2)} \right]. \tag{5.43}$$

Here $\vec{A}^{(1)} = \vec{k} A^{(1)}$ is the vector potential of a magnetic field in the first experiment, when the conductors carry currents $\vec{I}_m^{(1)}$. And magnitudes $\vec{A}^{(2)}$, $\vec{I}_m^{(2)}$ are the corresponding quantities in the second experiment. Writing the formula (5.34) with vector (5.43) for each region, taken it as uniform magnetic, and adding it we arrive at:

$$\sum_m \int_{S_m} \left(\left[A^{(1)}, H^{(2)} \right], ds \right) + \sum_r \int_{S_r} \left(\left[A_+^{(1)}, H_+^{(2)} \right] - \left[A_-^{(1)}, H_-^{(2)} \right], ds \right) =$$

$$= \sum_i \int_{V_i} div \left[A^{(1)}, H^{(2)} \right] dv = \sum_i \int_{V_i} \left(rot A^{(1)}, (\ddot{\mu})^{-1} rot A^{(2)} \right) dv \tag{5.44}$$

because $rot \vec{H}^{(2)} = 0$ due to the absence of outside currents in the media.

In the first sum (5.44) where the summation is being carried out over the conductor contours:

$$\left(\left[A^{(1)}, H^{(2)} \right], n \right) = A^{(1)} \left([n, k] H^{(2)} \right) = A^{(1)} H_\tau^2 \text{, the value of the potential } A^{(1)} \text{ does}$$

not depend on the position of the point on the *perfect* metal surface and $A^{(1)}$ can be taken out of the integral sign. The tangential component of the magnetic field $H_\tau^{(2)}$ gives the density of the longitudinal current and so:

$$\int_{S_m} \left(\left[A^{(1)}, H^{(2)} \right], ds \right) = A_m^{(1)} I_m^{(2)}.$$

Further from the formula (5.28), it follows the equality

$$B_n = (\tau \, grad A) = \partial A / \partial \tau$$

from the condition (5.12), consequently, $A_+ = A_-$, so that the second sum in the formula (5.44) vanishes due to (5.11). Finally, we obtain the expression:

$$\sum_m A_m^{(1)} I_m^{(2)} = \sum_i \int_{V_i} rotA^{(1)} \left(\ddot{\mu}\right)^{-1} rotA^{(2)}) \ , \tag{5.45}$$

which leads to formulae similar to (5.39) or (5.40) to determine inductances L_{mr} depending on permeability $\ddot{\mu}$ properties.

5.4. Connection Between Matrices of Capacitances and Inductances

The solution to electrostatic and magnetostatic problems described in section 5.2, is one of many examples, where different physical problems (from the point of view of a mathematical solution) appears to be identical. But in this case, there is rigid dependence between matrices of capacitances and inductances defined by the formulae (5.32) and (5.33). And so we formulate the dependences for anisotropic media:

$$D_x = \varepsilon_{xx} E_x + \varepsilon_{xy} E_y \ ,$$

$$H_x = \left(\mu^{-1}\right)_{xx} B_x + \left(\mu^{-1}\right)_{xy} B_y \ . \tag{5.46}$$

Identity of mathematical formulation is verified trough direct comparison of the formulae. The boundary conditions on *perfect* metal for electrostatic and magnetostatic problems (5.10) and (5.13) are correspondently expressed through the equations (5.26) and (5.31). The boundary conditions (5.9) and (5.12) though our representations of tangential components of electric field strength and normal components of magnetic inductance (as we can see in the formulae (5.23) and (5.29)) are identically satisfied. The definitions of potentials φ and A are given by formula (5.4). In the boundary conditions of (5.8) and (5.11) appear tensors $\ddot{\varepsilon}$ and $\ddot{\mu}$, which characterize anisotropic media. We obtained from the formulae (5.22) in accordance with the definition (5.46) these expressions:

$$D_n^{\pm} = \mp \pi \mu(t_0)(\varepsilon_{xx}^{\pm} \sin^2 \theta(t_0) + \varepsilon_{yy}^{\pm} \cos^2 \theta(t_0)) -$$

$$-\varepsilon_{xx}^{\pm} \sin \theta(t_0) \operatorname{Re} \Phi' - \varepsilon_{yy}^{\pm} \cos \theta(t_0) \operatorname{Im} \Phi' \pm \frac{1}{2} \pi \mu(t_0) \left(\varepsilon_{xy}^{\pm} + \varepsilon_{yx}^{\pm} \right) \sin 2\theta(t_0) +$$

$$+ \varepsilon_{xy}^{\pm} \sin \theta(t_0) \operatorname{Im} \Phi' + \varepsilon_{yx}^{\pm} \cos \theta(t_0) \operatorname{Re} \Phi' \ . \tag{5.47}$$

Here we introduced the designation:

$$\Phi' = \int_L \frac{\mu(t)ds}{t - t_0} + \sum_k \frac{A_k}{z_k - t_0} \ . \tag{5.48}$$

Similarly, from the first two formulae (5.29) and the definition (5.46), we receive:

$$H_\tau^\pm = \pm \pi \mu(t_0) \left[\left(\ddot{\mu}^{-1} \right)_{xx}^\pm \cos^2 \theta(t_0) + \left(\ddot{\mu}^{-1} \right)_{yy}^\pm \sin \theta(t_0) \right] +$$

$$+ \left(\ddot{\mu}^{-1} \right)_{xx}^\pm \cos \theta(t_0) \operatorname{Im} \Phi' + \left(\ddot{\mu}^{-1} \right)_{yy}^\pm \sin \theta(t_0) \operatorname{Re} \Phi' \pm$$

$$\pm \frac{1}{2} \pi \mu(t_0) \left[\left(\ddot{\mu}^{-1} \right)_{xy}^\pm + \left(\ddot{\mu}^{-1} \right)_{yx}^\pm \right] \cdot \sin 2\theta(t_0) +$$

$$+ \left(\ddot{\mu}^{-1} \right)_{xy}^\pm \cos \theta(t_0) \operatorname{Re} \Phi' + \left(\ddot{\mu}^{-1} \right)_{yx}^\pm \sin \theta(t_0) \operatorname{Im} \Phi' \ . \tag{5.49}$$

Here, the boundary condition (5.8) takes this form:

$$\pi \mu(t_0) \left[\left(\varepsilon_{xx}^+ + \varepsilon_{xx}^- \right) \sin^2 \theta(t_0) + \left(\varepsilon_{yy}^+ + \varepsilon_{yy}^- \right) \cos^2 \theta(t_0) \right] -$$

$$- \frac{1}{2} \pi \mu(t_0) \left(\varepsilon_{xy}^+ + \varepsilon_{xy}^- + \varepsilon_{yx}^+ + \varepsilon_{yx}^- \right) \sin 2\theta(t_0) +$$

$$+ \left(\varepsilon_{xx}^+ - \varepsilon_{xx}^- \right) \sin \theta(t_0) \operatorname{Re} \Phi' + \left(\varepsilon_{yy}^+ - \varepsilon_{yy}^- \right) \cos \theta(t_0) \operatorname{Im} \Phi' -$$

$$- \left(\varepsilon_{xy}^+ - \varepsilon_{xy}^- \right) \sin \theta(t_0) \operatorname{Im} \Phi' - \left(\varepsilon_{yx}^+ - \varepsilon_{yx}^- \right) \cos \theta(t_0) \operatorname{Re} \Phi' = 0 , \tag{5.50}$$

which is more general then the equation (5.25).
In the same way, we write the integral equation:

$$\pi \mu(t_0) \left[\left(\left(\ddot{\mu}^{-1} \right)_{xx}^+ + \left(\ddot{\mu}^{-1} \right)_{xx}^- \right) \cos^2 \theta(t_0) + \left(\left(\ddot{\mu}^{-1} \right)_{yy}^+ + \left(\ddot{\mu}^{-1} \right)_{yy}^- \right) \sin^2 \theta(t_0) \right] +$$

$$+ \left(\left(\ddot{\mu}^{-1} \right)_{xx}^+ - \left(\ddot{\mu}^{-1} \right)_{xx}^- \right) \cos \theta(t_0) \operatorname{Im} \Phi' + \left(\left(\ddot{\mu}^{-1} \right)_{yy}^+ - \left(\ddot{\mu}^{-1} \right)_{yy}^- \right) \sin \theta(t_0) \operatorname{Re} \Phi'_+ +$$

$$+ \frac{1}{2} \pi \mu(t_0) \left(\left(\ddot{\mu}^{-1} \right)_{xy}^+ + \left(\ddot{\mu}^{-1} \right)_{xy}^- + \left(\ddot{\mu}^{-1} \right)_{yx}^+ + \left(\ddot{\mu}^{-1} \right)_{yx}^- \right) \sin 2\theta(t_0) +$$

$$+ \left(\left(\ddot{\mu}^{-1} \right)_{xy}^+ - \left(\ddot{\mu}^{-1} \right)_{xy}^- \right) \cos \theta(t_0) \operatorname{Re} \Phi' -$$

$$\left(\left(\ddot{\mu}^{-1} \right)_{yx}^+ + \left(\ddot{\mu}^{-1} \right)_{yx}^- \right) \sin \theta(t_0) \operatorname{Im} \Phi' = 0 , \tag{5.51}$$

which expresses the boundary condition (5.11) and generalizes the equation (5.30) for anisotropic media. When we compare the formulae (5.50) and (5.51) we see that the first equation (5.50)is transferred into the second (5.51) after this change:

$$\varepsilon_{xx} \to \left(\ddot{\mu}^{-1} \right)_{yy} , \qquad \varepsilon_{yy} \to \left(\ddot{\mu}^{-1} \right)_{xx} ,$$

$$\ddot{\varepsilon}_{xy} \to - \left(\ddot{\mu}^{-1} \right)_{yx} \quad \text{and} \quad \varepsilon_{yx} \to - \left(\ddot{\mu}^{-1} \right)_{xy} . \tag{5.52}$$

This means that the solution of a magnetostatic problem coincides with the solution of an electrostatic problem when we use the relations (5.52). In case of an electrostatic problem the sources of electric fields are charges and in the case of magnetostatic problem the sources of magnetic fields are currents. We can verify this approach by direct calculations. Let the normal vector be directed into *perfect* metal (Fig.5.2). and then the surface density of electric charges is described in this form:

$$\sigma = -D_n^- = -\pi\mu(t_0)(\varepsilon_{xx}^- \sin^2\theta(t_0) + \varepsilon_{yy}^- \cos^2\theta(t_0)) +$$

$$+ \varepsilon_{xx}^- \sin\theta(t_0)\operatorname{Re}\Phi' + \varepsilon_{yy}^- \cos\theta(t_0)\operatorname{Im}\Phi' +$$

$$+ \frac{1}{2}\pi\mu(t_0)\left(\varepsilon_{xy}^- + \varepsilon_{yx}^-\right)\sin 2\theta(t_0) - \varepsilon_{xy}^- \sin\theta(t_0)\operatorname{Im}\Phi' - \varepsilon_{yx}^- \cos\theta(t_0)\operatorname{Re}\Phi'. \qquad (5.53)$$

The density of the surface currents are described in this form:

$$j = H_\tau^- = -\pi\mu(t_0)\left(\left(\ddot{\mu}^{-1}\right)_{xx}^- \cos^2\theta(t_0) + \left(\ddot{\mu}^{-1}\right)_{yy}^- \sin^2\theta(t_0)\right) +$$

$$+ \left(\ddot{\mu}^{-1}\right)_{xx}^- \cos\theta(t_0)\operatorname{Im}\Phi' + \left(\ddot{\mu}^{-1}\right)_{yy}^- \sin\theta(t_0)\operatorname{Re}\Phi' -$$

$$- \frac{1}{2}\pi\mu(t_0)\left(\left(\ddot{\mu}^{-1}\right)_{xy}^- + \left(\ddot{\mu}^{-1}\right)_{yx}^-\right)\sin 2\theta(t_0) + \left(\ddot{\mu}^{-1}\right)_{xy}^- \cos\theta(t_0)\times$$

$$\times\operatorname{Re}\Phi' + \left(\ddot{\mu}^{-1}\right)_{yx}^- \sin\theta(t_0)\operatorname{Im}\Phi'. \qquad (5.54)$$

Here we see that the change (5.52) transforms the expression (5.53) into (5.54).Since the connection between the charges and the potentials is expressed through the matrix of capacitances $\ddot{C}$ (5.32), the currents and vector potential values is expressed through the matrix inverse to the matrix of inductances $\ddot{L}$ (5.33), one could write:

$$L_{\ddot{\mu}} = C_{\ddot{\mu}^{-1}}^{-1}. \qquad (5.55)$$

Here the interchange of the subscripts $\ddot{\varepsilon}$ and $\ddot{\mu}^{-1}$ means the relation (5.52). In isotropic media, the relation (5.52) becomes very simple and the magnitude ε replaced by μ^{-1} [5.2].

5.5. Eigenwaves of a Lossless Multiconductor Transmission Line

In TEM-approximation the multiconductor transmission line is characterized by matrices of capacitance $\ddot{C}$ and inductance $\ddot{L}$. Therefore, we will use the theory of long transmission lines because it is convenient to do it in matrix notations. Here we introduce the vector-column of voltage:

$$V = \begin{pmatrix} V_1 \\ V_2 \\ \vdots \\ V_n \end{pmatrix}$$

and the vector-column of current:

$$I = \begin{pmatrix} I_1 \\ I_2 \\ \vdots \\ I_n \end{pmatrix}. \tag{5.56}$$

Where V_j is the voltage of the jth conductor with respect to the singled out conductor of a multiconductor line and I_k is the current carried on the corresponding conductor. The equations of long lines have the form:

$$-\frac{\partial}{\partial z}V = \frac{\partial}{\partial t}L \cdot I \,,$$

$$-\frac{\partial}{\partial z}I = \frac{\partial}{\partial t}C \cdot V \,, \tag{5.57}$$

where z is the point coordinate that is counted along the line. Assuming that the dependence of the wave process on time t and coordinate z is described by the exponent factor $e^{i\omega t - ihz}$, we can rewrite the system of differential equations (5.57) in the algebraic form:

$$hV = \omega LI \,,$$

$$hI = \omega \cdot C \cdot V \,. \tag{5.58}$$

It is more convenient to solve the system when excluding either the currents I :

$$L \cdot C \cdot V = \frac{h^2}{\omega^2} \cdot V \,, \tag{5.59}$$

or voltages V :

$$C \cdot L \cdot I = \frac{h^2}{\omega^2} I \,. \tag{5.60}$$

The equations (5.59) and (5.60) present the problem:

$$Ax = \lambda x \,, \qquad\qquad \lambda = h^2 / \omega^2 \tag{5.61}$$

of the determination of eigenvalues λ_i and the corresponding eigenvectors $x_j \neq 0$, $j = 1, \dots, n$. The number of vectors is equal to the order of the matrix A, $A = LC$ or $A = CL$. The eigenvalues are roots of the nth order equation:

$$\det \left| A - \lambda \hat{E} \right| = 0 \tag{5.62}$$

where $\hat{E}$ is the identity matrix. The formula (5.62) expresses the condition of the existence of the nontrivial solution of the uniform linear equation system (5.61). we define the scalar product of two vectors x and y as:

$$(x, y) = X^+ Y = \sum_i x_i^* y_i \,.$$

Here $A_{ik}^+ = A_{ki}^*$ where matrix A^+ is Hermitian conjugated to the matrix A. When the transmission line fillings are the materials having no losses, then the matrices L and C are Hermitian:

$$L^+ = L, \quad C^+ = C. \tag{5.63}$$

Let the vectors I_k and V_k be the eigenvectors of the problem (5.59) and (5.60). Multiply the equation (5.59), which is written for the vector V_k, from the left side by the vector I_k then we obtain:

$$I_k^+ LCV_i = \lambda_i I_k^+ V_i ,$$

and it follows from formula (5.60) that:

$$I_k^+ LC = \lambda_k^* I_k^+ .$$

Then comparing the last two equations we obtain:

$$(\lambda_k^* - \lambda_i)I_k^+ V_i = 0, \qquad \text{where} \quad \lambda_k^* = \lambda_k .$$

Since the transmission line fillings are the materials having no losses, the propagation constant is real number. Vectors I_k are orthogonal to the vectors V_i then we can write:

$$I_k^+ V_i = 0 \quad \text{for} \quad k \neq i, \qquad \lambda_k \neq \lambda_i . \tag{5.64}$$

That means that the vectors V_i make a nonorthogonal basis of vectors in n dimensional space and the vectors I_k make the additional system of vectors. As an example, the vector I_1 is orthogonal to all vectors V_i, where $i = 2,3...,n$. The exception is the vector V_1 (presentation of a vector by covariant and contravariant components). Schematically, the situation is shown in the Fig.5.4 in the case of a two-dimensional space. Any vector can be written as decomposition either in one unit vectors or in others.

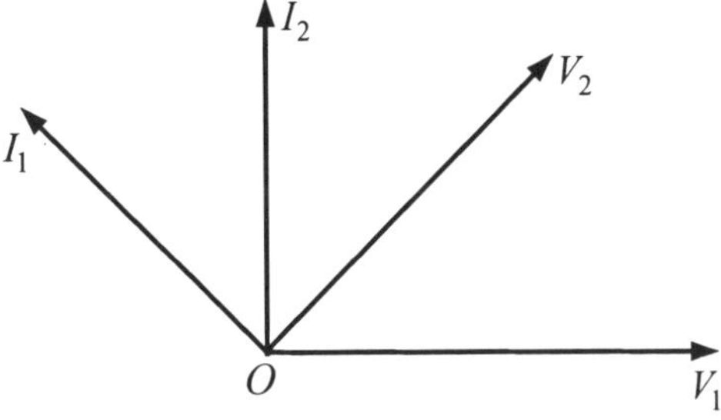

Fig. 5.4. Helps explain a nonorthogonal basis of vectors in a two-dimensional space.

When subscripts in (5.64) coincide then the product:

$$\mathrm{Re}(I_k^+ V_k / 2) = P_k , \tag{5.65}$$

describes the average power carried by the EM wave along the transmission line. In every cross-section $(z = const)$ of the transmission line the power is the same. And the orthogonality property (5.64) show the total power carried on the transmission line is equal to the sum of powers carried by separate particular eigenwaves.

The average energy of the magnetic field per unit length of the transmission line and the energy of electric field for each eigenwave is given by the expression:

$$E = (I^+L\,I + V^+C\,V)/4 \tag{5.66}$$

and has the dimension J/m. The power carried by the transmission line is defined by a formula equivalent to (5.65):

$$P = (V^+I + I^+V)/4 , \tag{5.67}$$

and has the dimension $J/\sec$. The ratio P/E has the dimension of velocity and coincides with the phase velocity of this particular eigenwave, which follows explicitly from the equations (5.58):

$$\upsilon = \frac{P}{E} = \frac{V^+I + I^+V}{I^+LI + V^+CV} = \frac{\omega}{h} . \tag{5.68}$$

Let the signal propagate along the transmission line and this signal at the cross-section (such as $z = 0$) will be represent by vector V. Here we can write the decomposition of the signal voltage:

$$V = \sum_k a_k V_k ,$$

in eigenvectors V_k. Then the vector of currents of the signal:

$$I = \sum_k a_k I_k ,$$

is written as a decomposition in eigenvectors I_k with the same coefficients. It is convenient to normalize the eigenvectors to the unit (unit power): $I_k^+V_i = \delta_{ik}$. Here we define the energy velocity [5.3] as:

$$\upsilon = \frac{I^+V}{(I^+LI + V^+CV)/2} = \omega \frac{\sum a_k^* a_k}{\sum h_k a_k^* a_k} . \tag{5.69}$$

Since the phase velocities of eigenwaves are different, the signal propagating along the transmission line changes its shape. The energy velocity which is determined by (5.69) was used in the paper [5.4] for investigating MSLs.

5.6. Eigenwaves of a Lossy Transmission Line

The long line equations (5.58) with the added terms iRI and igV which take into account the losses of the transmission line take this form:

$$hV = \omega LI - iRI , \qquad hI = \omega CV - igV .$$

Here R is the impedance matrix and g is the conductivity matrix (per unit length of the transmission line). Here $i^2 = -1$ where i is the imaginary unit. When solving Maxwell's equations the losses usually are included in the permittivity of the media, which fills the transmission lines. Similarly, in the transmission long line theory, the matrices R and g are united with L and C. When solving the equations (5.58) the matrices L and C are not Hermitian conjugated but are: $L^+ \neq L$, $C^+ \neq C$. We will now consider the problem (5.61) together with this new problem:

$$A^+ Y = \mu Y , \tag{5.70}$$

for the determination of eigenvectors y_k and eigenvalues μ_k of the matrix A^+, which is Hermitian conjugated to A.

The eigenvectors of problem (5.61) (similar to the problem (5.70)), make the system of linearly independent vectors, if the vectors correspond in pairs to the different eigenvalues. Here we let:

$$Ax_i = \lambda_i x_i , \quad x_i \neq 0 \quad \lambda_i \neq \lambda_k \quad i \text{ and } k = 1,2 \cdots, m. \tag{5.71}$$

We suppose that the vectors are linearly dependent and so we can write it like this:

$$\sum_{i=1}^{m} C_i x_i = 0 . \tag{5.72}$$

Apply the operator $A - \lambda_1 E$ from the left side of the formula (5.72), the results can be expressed like this:

$$\sum_{i=2}^{m} C_i (\lambda_i - \lambda_1) \, x_i = 0 . \tag{5.73}$$

We now apply the operator $A - \lambda_2 E$ to the formula (5.73) and we obtain the sum:

$$\sum_{i=3}^{m} C_i (\lambda_i - \lambda_1) \, (\lambda_i - \lambda_2) x_i = 0 , \tag{5.74}$$

which has one term less. Continuing the process we come to the equality:

$$C_m (\lambda_m - \lambda_1) \cdot (\lambda_m - \lambda_2) \dots (\lambda_m - \lambda_{m-1}) = 0 ,$$

from that $C_m = 0$. Thus, one can prove that all the coefficients $C_i = 0$, $i = 1,2,...,m$ in the formula (5.72), i.e. the vectors x_i are linearly independent.

If several eigenvectors belong to the same eigenvalues then any linear combination of the eigenvectors belong to this eigenvalues as well. We assume here that all the eigenvectors x_i are linearly independent. The system of vectors y_i is also linearly independent.

The eigenvalues of the problem (5.70) are the roots of the equation:

$$\det \left| A^+ - \mu E \right| = 0 . \tag{5.75}$$

Changing the formula (5.75) rows by columns and taking then the complex conjugated, we have this result:

$$\det\left|A - \mu^{*}E\right| = 0. \tag{5.76}$$

Comparing (5.76) with (5.62) one can see that the problems (5.61) and (5.70) have complex conjugated eigenvalues:

$$\mu_i = \lambda_i^{*}. \tag{5.77}$$

Now multiplying the equality $Ax_i = \lambda_i x_i$ from the left side by y_k^{+} and bearing in mind the relation (5.77) we obtain:

$$\left(\lambda_i - \lambda_k\right) y_k^{+} x_i = 0.$$

We normalize the vectors y_k and then we will receive this:

$$y_k^{+} x = \delta_{k,i}. \tag{5.78}$$

When one eigenvalue has several vectors, we make the linear combinations of these vectors, which then satisfies the relation (5.78).

The system of additional to x_i vectors y_k can be built up without solving the problem (5.70). In this case one needs Gramm's determinant:

$$\Gamma(x_1, x_2, ..., x_n) = \begin{vmatrix} x_1^{+} x_1 & x_1^{+} x_2 & \cdots & x_1^{+} x_n \\ x_2^{+} x_1 & x_2^{+} x_2 & \cdots & x_2^{+} x_n \\ \vdots & \vdots & & \vdots \\ x_n^{+} x_1 & x_n^{+} x_2 & \cdots & x_n^{+} x_n \end{vmatrix}. \tag{5.79}$$

As is known, the determinant is not equal to zero for linearly independent vectors. The direct verification shows that the vector[*]:

$$y^{(1)} = \frac{1}{\Gamma} \begin{vmatrix} x_1 & x_1^{+} x_2 & \cdots & x_1^{+} x_n \\ x_2 & x_2^{+} x_2 & \cdots & x_2^{+} x_n \\ \vdots & \vdots & & \vdots \\ x_n & x_n^{+} x_2 & \cdots & x_n^{+} x_n \end{vmatrix},$$

is orthogonal to all vectors x_i, $i = 2,3,...,n$, except x_1, and $y^{(1)+} x_1 = 1$.

The vector $y^{(i)}$ is obtained from the formula (5.79) changing the ith column by the vector-column x_i, that is:

$$y^{(i)} = \frac{1}{\Gamma} \begin{vmatrix} x_1^{+} x_1 & \cdots & x_1^{+} x_{i-1} & x_1 & \cdots & x_1^{+} x_n \\ x_2^{+} x_1 & \cdots & x_2^{+} x_{i-1} & x_2 & & x_2^{+} x_n \\ \vdots & & \vdots & \vdots & & \vdots \\ x_n^{+} x_1 & \cdots & x_n^{+} x_{i-1} & x_n & \cdots & x_n^{+} x_n \end{vmatrix}, \tag{5.80}$$

---

[*] It is meant the decomposition in elements of the first column.

then

$$y^{(i)+}x_k = \delta_{i,k}. \tag{5.81}$$

Bearing in mind what we said previously about "right" eigenvectors V_k, I_k in the equations (5.58)and (5.59) or (5.60), it is then necessary to introduce the "left" eigenvectors V^{k+}, I^{k+}, which satisfies the equations:

$$I^+L = \frac{h}{\omega}V^+, \quad V^+C = \frac{h}{\omega}I^+, \tag{5.82}$$

Here we determine:

$$V^+CL = \frac{h^2}{\omega^2}V^+, \tag{5.83}$$

$$I^+LC = \frac{h^2}{\omega^2}I^+. \tag{5.84}$$

Vectors $V^{(k)}$, $I^{(k)}$ belonging to eigenvalues h_k^2/ω in equations (5.82)–(5.84), now do not coincide with the eigenvectors V_k, I_k of the equations (5.58)–(5.60), but:

$$V^{(k)+}I_i = \delta_{ki},$$

$$I^{(k)+}V_i = \delta_{ki}. \tag{5.85}$$

Let the signal, described by the voltage–vector in the transmission line cross-section $z = 0$. Decomposing the voltage–vector in the eigenvectors of voltage:

$$V = \sum_k a^{(k)}V_k = \sum_k a_k V^{(k)}.$$

We also obtained the decomposition of the current:

$$I = \sum_k a^{(k)}I_k = \sum_k a_k I^{(k)}.$$

The power is expressed by this formula:

$$V^+I = I^+V = \sum_k a_k^* a^{(k)}.$$

For magnetic and electric energies we get the expressions:

$$\frac{1}{2}I^+LI = \frac{1}{2}\sum_k \frac{h_k}{\omega} a_k^* a^{(k)},$$

$$\frac{1}{2}V^+CV = \frac{1}{2}\sum_k \frac{h_k}{\omega} a_k^* a^{(k)}.$$

Therefore, for group (energy) velocity, we have:

$$\upsilon = \omega \frac{\mathrm{Re}\sum a_k^* a^{(k)}}{\mathrm{Re}\sum h_k a_k^* a^{(k)}}. \tag{5.86}$$

The velocity depends on coordinate z because of different phase shifts of the eigenwaves and their different attenuations.

It is worth to noting, that as soon as we have the matrix C of capacitances (taking into account the losses) then writing it in this form:

$$C = \frac{1}{2}(C + C^+) - \frac{i}{2}(iC - iC^+) \equiv H_1 - iH_2,$$

where matrices $H_1^+ = H_1$ and $H_2^+ = H_2$ are Hermitian conjugated matrices. We see that H_2 directly defines the admittance matrix. These formulae can be used in calculating semiconductor elements such as transistors.

5.7. Long Line Circuits

When calculating electric circuits made of long line segments, one must be guided by Ohm's law and current conservation law. Also it does not matter which side you approach the potential the circuit is the same of any point.

Here we have an example of a simple electric circuit shown in Fig.5.5. In the segment before the cross-section AB and after the cross-section CD there is only one-conductor line. Between the cross-sections AB and CD the line has two-conductors loaded at the points B and C by the impedances Z_1 and Z_2 correspondently.

Let the electromagnetic wave propagate to the point A from the left side, having a voltage vector V_+ with the unit amplitude $a^{(+)} = 1$. Then the wave V_+ excites an electromagnetic wave V_- at the point A. Here the V_- wave having the unknown amplitude $a^{(-)}$ is going in the opposite direction. The V_- wave determines the eigenwaves of two-conductor segment. These eigenwaves having arrived at the cross-section CD, excite the waves, which are returning and the wave V_3. The wave V_3 is propagating to the right from the point D and having the amplitude $a^{(3)}$. We now have a stationary regime, when there are direct waves V_i, I_i and back waves V_{-i}, I_{-i} on each section, (except the one to the right of the point D).

We write equations to determine amplitudes of all eigenwaves and we normalized them. According to the first Kirchhoff's low the sum of all currents going to point A turns into zero:

$$I_+ + a^{(-)}I_- - (a^{(1)}I_{11} + a^{(-1)}I_{-11} + a^{(2)}I_{21} + a^{(-2)}I_{-21}) = 0, \qquad (5.87)$$

where I_{11} is the current of the first eigenwave carried by the first conductor. Similarly at point B (taking into account the current I_1, which is carried by the impedance Z_1), we have:

$$I_1 + a^{(1)}I_{12} + a^{(-1)}I_{-12} + a^{(2)}I_{22} + a^{(-2)}I_{-22} = 0. \qquad (5.88)$$

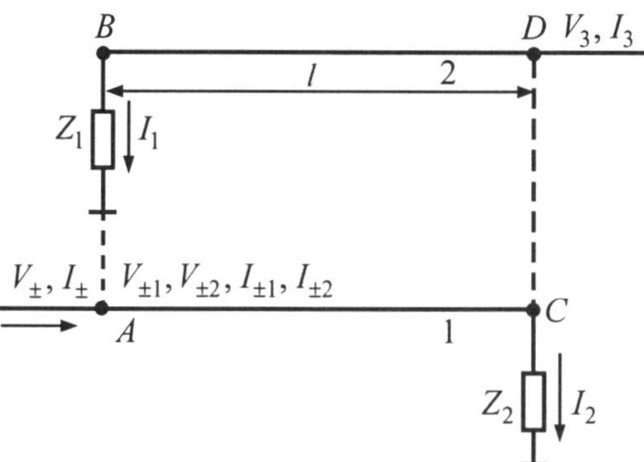

Fig. 5.5. *An electric circuit of line segments with distributed parameters loaded at points B and C by impedances Z_1 and Z_2.*

In the same way, taking into account the dependence $e^{\pm ihz}$ we write:

$$-a^{(3)}I_3 + a^{(1)}I_{12}e^{-ih_1 l} + a^{(-1)}I_{-12}e^{ih_1 l} + a^{(2)}I_{22}e^{-ih_2 l} + a^{(-2)}I_{-22}e^{ih_2 l} = 0, \qquad (5.89)$$

the current conservation law for point D. The current conservation law:

$$-I_2 + a^{(1)}I_{11}e^{-ih_1 l} + a^{(-1)}I_{-11}e^{ih_1 l} + a^{(2)}I_{21}e^{-ih_2 l} + a^{(-2)}I_{-21}e^{ih_2 l} = 0, \qquad (5.90)$$

C for point. Here I_2 is the current that is carried by the impedance Z_2 and l is the length of the two-conductor lines. Further, we calculate the voltage of point A when approaching it from the left and the right side:

$$V_+ + a^{(-)}V_- = a^{(1)}V_{11} + a^{(-1)}V_{-11} + a^{(2)}V_{21} + a^{(-2)}V_{-21}. \qquad (5.91)$$

Now we calculate the voltage of point D and we obtain the equation:

$$a^{(3)}V_3 = a^{(1)}V_{12}e^{-ih_1 l} + a^{(-1)}V_{-12}e^{ih_1 l} + a^{(2)}V_{22}e^{-ih_2 l} + a^{(-2)}V_{-22}e^{ih_2 l}. \qquad (5.92)$$

According to Ohm's law, we have two equations:

$$I_1 Z_1 = a^{(1)}V_{12} + a^{(-1)}V_{-12} + a^{(2)}V_{22} + a^{(-2)}V_{-22}, \qquad (5.93)$$

$$I_2 Z_2 = a^{(1)}V_{11}e^{-ih_1 l} + a^{(-1)}V_{-11}e^{ih_1 l} + a^{(2)}V_{21}e^{-ih_2 l} + a^{(-2)}V_{-21}e^{-ih_2 l}, \qquad (5.94)$$

connecting currents in loads Z_1 and Z_2 with voltages at points B and C correspondently. Now we have eight equations (5.87)–(5.94) to determine the unknown amplitudes $a^{(-)}$, $a^{(\pm 1)}$, $a^{(\pm 2)}$, $a^{(3)}$ and currents I_1, I_2 in impedances.

Similarly, we can write equations for more complex eclectic circuits (Fig.5.6). In Fig.5.5 for an example, we connected the points A with B and C with D by the line of length l_1 with distributed parameters. Here we the eigenwaves in the transmission line are : $V_{\pm 4}$, $V_{\pm 5}$, $I_{\pm 4}$, $I_{\pm 5}$ as shown in Fig.5.6. When comparing the previous circuits (Fig.5.5) with (Fig.5.6) we see that for the last circuit we have an additional four unknown amplitudes of waves propagating along AB and CD.

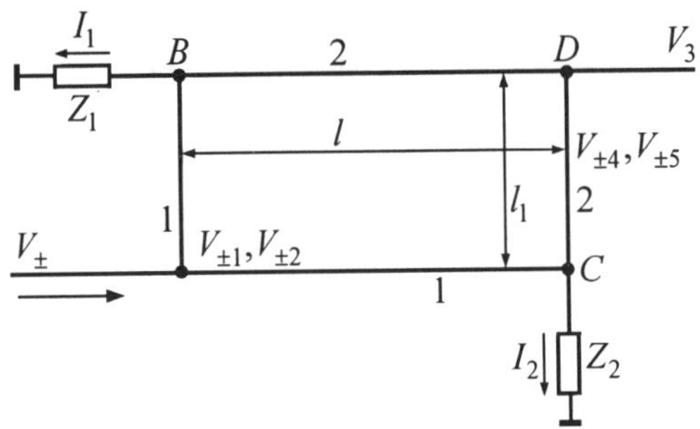

Fig. 5.6. *An electric circuit of line segments with distributed parameters loaded by impedances* Z_1 *and* Z_2.

Finally we obtained twelve equations. Four of them are:

$$I_+ + a^{(-)}I_- - (a^{(1)}I_{11} + a^{(-1)}I_{-11} + a^{(2)}I_{21} + a^{-2}I_{-21}) -$$

$$- (a^{(4)}I_{41} + a^{(-4)}I_{-41} + a^{(5)}I_{51} + a^{(-5)}I_{-51}) = 0,$$

$$-I_1 + a^{(4)}I_{41}e^{-ih_4l_1} + a^{(-4)}I_{-41}e^{ih_4l_1} + a^{(5)}I_{51}e^{-ih_5l_1} + a^{(-5)}I_{-51}e^{ih_5l_1} -$$

$$-(a^{(1)}I_{12} + a^{(-1)}I_{-12} + a^{(2)}I_{22} + a^{(-2)}I_{-22}) = 0,$$

$$-I_2 + a^{(1)}I_{11}e^{-ih_1l} + a^{(-1)}I_{-11}e^{ih_1 l} + a^{(2)}I_{21}e^{-ih_2 l} + a^{(-2)}I_{-21}e^{ih_2 l} -$$

$$- (a^{(4)}I_{42} + a^{(-4)}I_{-42} + a^{(5)}I_{52} + a^{(-5)}I_{-52}) = 0,$$

$$-a^{(3)}I_3 + a^{(1)}I_{12}e^{-ih_1l} + a^{(-1)}I_{-12}e^{ih_1 l} + a^{(2)}I_{22}e^{-ih_2 l} + a^{(-2)}I_{-22}e^{ih_2 l} +$$

$$+ a^{(4)}I_{42}e^{-ih_4l_1} + a^{(-4)}I_{-42}e^{ih_4l_1} + a^{(5)}I_{52}e^{-ih_5l_1} + a^{(-5)}I_{-52}e^{ih_5l_1} = 0,$$

which express the current conservation law at points A, B, C and D. Here we write the equalities of voltages at the same points A, B, C and D approaching them along different lines (Fig.5.6):

$$V_+ + a^{(-)}V_- = a^{(1)}V_{11} + a^{(-1)}V_{-11} + a^{(2)}V_{21} + a^{-2}V_{-21} =$$

$$= (a^{(4)}V_{41} + a^{(-4)}V_{-41} + a^{(5)}V_{51} + a^{(-5)}V_{-51}),$$

$$I_1Z_1 = a^{(1)}V_{12} + a^{(-1)}V_{-12} + a^{(2)}V_{22} + a^{(-2)}V_{-22} =$$

$$= a^{(4)}V_{41}e^{-ih_4l_1} + a^{(-4)}V_{-41}e^{ih_4l_1} + a^{(5)}V_{51}e^{-ih_5l_1} + a^{(-5)}V_{-51}e^{ih_5l_1},$$

$$I_2Z_2 = a^{(1)}V_{11}e^{-ih_1l} + a^{(-1)}V_{-11}e^{ih_1l} + a^{(2)}V_{21}e^{-ih_2l} + a^{(-2)}V_{-21}e^{ih_2l} =$$

$$= a^{(4)}V_{42} + a^{(-4)}V_{-42} + a^{(5)}V_{52} + a^{(-5)}V_{-52},$$

$$a^{(3)}V_3 = a^{(1)}V_{12}e^{-ih_1l} + a^{(-1)}V_{-12}e^{ih_1l} + a^{(2)}V_{22}e^{-ih_2l} + a^{(-2)}V_{-22}e^{ih_2l} =$$

$$= a^{(4)}V_{42}e^{-ih_4l_1} + a^{(-4)}V_{-42}e^{ih_4l_1} + a^{(5)}V_{52}e^{-ih_5l_1} + a^{(-5)}V_{-52}e^{ih_5l_1}.$$

5.8. Realization of the Bogoliubov - Krylov Method

When solving the system of the SIEs (5.25)–(5.27) and in a general case (5.50) every integral is replaced by a sum of integrals over small segments Δ of the contour so that we write:

$$\int\limits_{t_k-\Delta_k/2}^{t_k+\Delta_k/2} \frac{\mu(t)ds}{t-t_0} = \int\limits_{t_k-\Delta_k/2}^{t_k+\Delta_k/2} \frac{\mu(t)e^{-j\theta(t)}dt}{t-t_0} \approx$$

$$\approx \mu(t_k)e^{-j\theta(t_k)} \; \ln\frac{t_k+\Delta_k/2-t_0}{t_k-\Delta_k/2-t_0} \equiv \mu(t_k)\cdot\Phi'(t_0,t_k), \tag{5.95}$$

where $\theta(t)$ is the angle between the contour tangent at the point t and the x axis. So, the system of integral equations is replaced by a system of linear equations in the form:

$$\sum_k a_{ik}x_k = b_i. \tag{5.96}$$

Here the unknowns x_k is the values of the density $\mu(t)$ (in the contour point $t=t_k$) or of the constants A_m. Let x_k at $k=1,2,...,N_1+N_2$ are the quantities $\mu(t_k)$ for the points t_k ($k=1,2,...,N_1$), which are on the contours, dividing dielectric media and for $k=N_1+1,...,N_1+N_2$ the point t_k belongs to the contour bordering a metal conductor. Here $x_{N_1+N_2+m}=A_m$ and M is the number of conductors. In these designations:

$$b_i = 0, \qquad i=1,...N_1.+N_2,$$

$$b_{N_1+N_2+m} = \varphi_m, \tag{5.97}$$

where φ_m is the set potential of the mth conductor. In practical calculations it is convenient to choose the potentials equal to 1 or 0. The ith equation of system (5.96) (at $i\le N_1+N_2$) is obtained by substituting the value of the point coordinate $t_0=t_i$ into the equations (5.26) and (5.50). The remaining M equations of the system (5.96) are defined by the formula (5.27). First we write the diagonal elements a_{ii} of the matrix $\|a_{ik}\|$ of the system (5.96). The point $t_0=t_i$ belongs to the contour dividing dielectrics when $i\le N_1$. From the equation (5.50) we see that:

$$a_{ii} = \pi\left[\left(\varepsilon_{xx}^+ + \varepsilon_{xx}^-\right)\sin^2\theta(t_i)+\left(\varepsilon_{yy}^+ + \varepsilon_{yy}^-\right)\cos^2\theta(t_i)\right] -$$

$$-\frac{\pi}{2}(\varepsilon_{xy}^+ + \varepsilon_{xy}^- + \varepsilon_{yx}^+ + \varepsilon_{yx}^-)\sin 2\theta(t_i), \quad i=1,...,N_1 . \tag{5.98}$$

The point $t_0=t_i$ belongs to the contour bordering the metal region, when $N_1 \prec i\le N_1+N_2$ and we obtained from the equation (5.26):

$$a_{ii} = 0, \quad i = N_1 + 1,..., N_1 + N_2 . \tag{5.99}$$

Finally from the equations (5.27), we find:

$$a_{ii} = \ln \sqrt{(x_{(i)} - x_{[i]})^2 + (y_{(i)} - y_{[i]})^2} ,$$

$$i = N_1 + N_2 + 1,..., N_1 + N_2 + M \tag{5.100}$$

where $x_{(i)}$, $y_{(i)}$ are coordinates of a certain point on the metal surface and $x_{[i]}$, $y_{[i]}$ are the coordinates of the point inside the metal.

The nondiagonal elements of the equations' system are obtained from the integral terms of the same formulae and from the sum of the formula (5.27). For $i \le N_1$ we obtained from the equation (5.50):

$$a_{ik} = \left(\varepsilon_{xx}^+ - \varepsilon_{xx}^-\right) \sin \theta(t_i) \, \mathrm{Re}\, \Phi'(t_i, t_k) + \left(\varepsilon_{yy}^+ - \varepsilon_{yy}^-\right) \cos \theta(t_i) \, \mathrm{Im}\, \Phi'(t_i, t_k) -$$

$$- \left(\varepsilon_{xy}^+ - \varepsilon_{xy}^-\right) \sin \theta(t_i) \, \mathrm{Im}\, \Phi'(t_i, t_k) - \left(\varepsilon_{yx}^+ - \varepsilon_{yx}^-\right) \cos \theta(t_i) \, \mathrm{Re}\, \Phi'(t_i, t_k) ,$$

$$i \ne k, \quad i = 1,..., N_1, \quad k = 1,..., N_1 + N_2 , \tag{5.101}$$

$$a_{i,N_1+N_2+m} = \left(\varepsilon_{xx}^+ - \varepsilon_{xx}^-\right) \sin \theta(t_i) \, \mathrm{Re}\, \frac{1}{z_m - t_i} + \left(\varepsilon_{yy}^+ - \varepsilon_{yy}^-\right) \cos \theta(t_i) \, \mathrm{Im}\, \frac{1}{z_m - t_i} -$$

$$- \left(\varepsilon_{xy}^+ - \varepsilon_{xy}^-\right) \sin \theta(t_i) \, \mathrm{Im}\, \frac{1}{z_{[m]} - t_i} - \left(\varepsilon_{yx}^+ - \varepsilon_{yx}^-\right) \cos \theta(t_i) \, \mathrm{Re}\, \frac{1}{z_{[m]} - t_i} ,$$

$$i = 1,2,..., N_1 , \qquad m = 1,2,..., M . \tag{5.102}$$

And when $i = N_1 + 1, \cdots, N_1 + N_2$ then it follows from the equation (5.26) that:

$$a_{ik} = \mathrm{Re}\left\{ e^{j\theta(t_i)} \Phi'(t_i, t_k) \right\},$$

$$i = N_1 + 1,..., N_1 + N_2 , \qquad k = 1,..., N_1 + N_2 , \tag{5.103}$$

$$a_{i,N_1+N_2+m} = \mathrm{Re}\left\{ e^{j\theta(t_i)} \frac{1}{z_{(m)} - t_i} \right\}, \qquad m = 1,2,..., M . \tag{5.104}$$

Similarly, from equations (5.27) we find:

$$a_{N_1+N_2+m,k} = \int\limits_{t_k - \Delta/2}^{t_k + \Delta/2} \ln \sqrt{(x_{(m)} - x)^2 + (y_{(m)} - y)^2} \; ds ,$$

$$m = 1,2,..., M , \qquad k = 1,2,..., N_1 + N_2 , \tag{5.105}$$

$$a_{N_1+N_2+m, \, N_1+N_2+m'} = \ln \sqrt{(x_{(m)} - x_{[m']})^2 + (y_{(m)} - y_{[m']})^2} ,$$

$$m \text{ and } m' = 1,2,..., M . \tag{5.106}$$

The integral in the formula (5.105) always reduces to the integral:

$$\int \ln(x^2 + a^2)\,dx = x \ln(x^2 + a^2) - 2x + 2a \; arctg\, \frac{x}{a} .$$

5.9. Consideration of Geometric Symmetry

The presence of any symmetry in the geometry of the cross-section allows us to diminish the number of unknowns x_k in the system (5.96) or to increase the accuracy of computations for a given number of points in the counter. This can be achieved by setting the potentials φ_m, which are characteristic for the line having a certain symmetry. We will now illustrate an example of symmetry plane.

Fig.5.7 shows the contour L_1 bordering a metal conductor, the contour L_2 dividing gyrotropic media and the symmetric contours L_1' and L_2' corresponding to them. When our contours have geometrical symmetry the potentials of the metal surfaces satisfy the condition:

$$I_0\varphi(B_1) = \varphi(B_1') = \sigma\varphi(B_1) \tag{5.107}$$

where I_0 is the operator of the reflection in the symmetry plane. Since the double application of the operator I_0 corresponds to the identity transformation when each point stays in its place, then $\sigma^2 = 1$ and therefore:

$$\sigma = \pm 1. \tag{5.108}$$

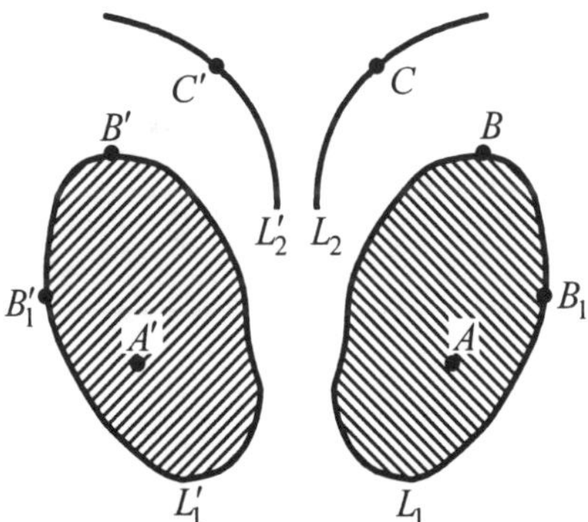

Fig. 5.7. Geometrically symmetric contours bordering metal and gyrotropic media.

Now we will write the equation (5.27) in designations corresponding to Fig.5.7:

$$\varphi(B_1) = 1 = \int\limits_{L_1}\mu(B)\ln R_{BB_1}\,ds_B + \int\limits_{L_1'}\mu(B')\ln R_{B'B_1}\,ds_{B'} + A_1\ln R_{AB_1} +$$

$$+A_1'\ln R_{A'B_1} + \int\limits_{L_2}\mu(C)\ln R_{CB_1}\,ds_C + \int\limits_{L_2'}\mu(C')\ln R_{C'B_1}\,ds_{C'}, \tag{5.109}$$

$$\varphi(B_1') = \pm 1 = \int\limits_{L_1}\mu(B)\ln R_{BB_1'}\,ds_B + \int\limits_{L_1'}\mu(B')\ln R_{B'B_1}\,ds_{B'} + A_1\ln R_{AB_1'} +$$

$$+A_1' \ln R_{A'B_1'} + \int_{L_2} \mu(C) \ln R_{CB_1'} ds_C + \int_{L_2'} \mu(C') \ln R_{C'B_1} ds_{C'} . \tag{5.110}$$

Since the points A, B, B_1 and C are symmetrical to the points A', B', B_1' and C', then the distances are $R_{BB_1} = R_{B'B_1'}$, $R_{B'B_1} = R_{BB_1'}$ and so on. Therefore, the equation (5.110) is satisfied if the density values are:

$$\mu(B') = \pm\mu(B),$$

$$\mu(C') = \pm\mu(C), \tag{5.111}$$

and the constants are: $A_1' = \pm A_1$.

Also the calculations of the coefficients which connect two symmetric points α_{ik} are simplified. We now demonstrate an example of gyrotropic media described by the tensor with the elements:

$$\varepsilon_{xx} = \varepsilon_{yy} \equiv \varepsilon_d ,$$

$$\varepsilon_{xy} = -\varepsilon_{yx} \equiv \varepsilon_{nd} . \tag{5.112}$$

Let the ith and k^{th} points transfer to each other after reflection in the symmetry plane as shown in Fig.5.8.

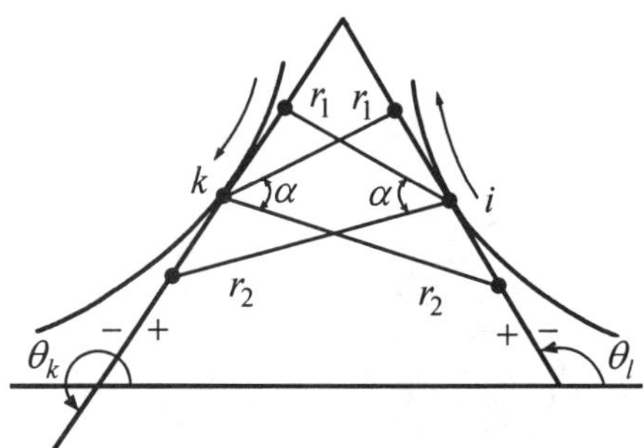

Fig. 5.8. *When calculating matrix elements A_{ik} of geometric symmetry contours.*

Now we consider the coefficients a_{ik}, when i, $k = 1,2,...,N_1$, are the points belonging to the contours dividing two gyrotropic media. According to the formulae (5.101), (5.112) and (5.95) of the function $\Phi'(t_i,t_k)$, we have:

$$a_{ik} = \left(\varepsilon_d^+ - \varepsilon_d^-\right) \; \mathrm{Im}\left\{ e^{j(\theta_i-\theta_k)} \; \ln \frac{t_k + \Delta_k/2 - t_i}{t_k - \Delta_k/2 - t_i} \right\} +$$

$$+\left(\varepsilon_{nd}^+ - \varepsilon_{nd}^-\right) \; \mathrm{Re}\left\{ e^{j(\theta_i-\theta_k)} \; \ln \frac{t_k + \Delta_k/2 - t_i}{t_k - \Delta_k/2 - t_i} \right\} . \tag{5.113}$$

One can see from Fig.5.8 that the angle is:

$$\Theta_k = 2\pi - \Theta_i ,$$

$$\frac{t_k + \Delta_k/2 - t_i}{t_k - \Delta_k/2 - t_i} = \frac{r_2}{r_1} e^{i\alpha}.$$

Where r_1 and r_2 are modules of the nominator and denominator on the left side and the angle α is the difference of arguments of corresponding complex numbers. In this same way in Fig.5.8 we find:

$$\frac{t_i + \Delta_i/2 - t_k}{t_i - \Delta_i/2 - t_k} = \frac{r_1}{r_2} e^{i\alpha}.$$

Therefore, instead of (5.113) we have:

$$a_{ik} = \left(\varepsilon_d^+ - \varepsilon_d^-\right)\left[\cos 2\theta_i\alpha + \sin 2\theta_i \ln(r_2/r_1)\right] +$$
$$+ \left(\varepsilon_{nd}^+ - \varepsilon_{nd}^-\right)\left[\cos 2\theta_i \ln(r_2/r_1) - \sin 2\theta_i\,\alpha\right].$$

And we will write:

$$a_{ki} = \left(\varepsilon_d^+ - \varepsilon_d^-\right)\left[\cos 2\theta_i\alpha + \sin 2\theta_i \ln(r_2/r_1)\right] -$$
$$- \left(\varepsilon_{nd}^+ - \varepsilon_{nd}^-\right)\left[\cos 2\theta_i \ln(r_2/r_1) - \sin 2\theta_i\,\alpha\right], \qquad i \text{ and } k = 1,2,...,N_1.$$

Here the coefficients a_{ik}, which are proportional to the difference of diagonal elements are symmetric with respect to the change of subscripts $i \Leftrightarrow k$. Also the coefficients, which are proportional to the difference of nondiagonal elements is antisymmetric. When the ith and k^{th} points are on the metal, i and $k = N_1 + 1,..., N_1 + N_2$ the formula (5.103) is in the explicit form:

$$a_{ik} = \mathrm{Re}\left\{ e^{j(\theta_i - \theta_k)} \; \ln\frac{t_k + \Delta_k/2 - t_i}{t_k - \Delta_k/2 - t_i} \right\},$$

which shows that:

$$a_{ik} = -a_{ki}, \qquad i \text{ and } k = N_1 + 1,..., N_1 + N_2.$$

The formula (5.106) show that:

$$a_{ik} = a_{ki}, \qquad i \text{ and } k = N_1 + N_2 + 1,..., N_1 + N_2 + M.$$

Here we are reminded that these last relations are valid for the coefficients connecting symmetric points. Also that the symmetry points in geometry of the transmission line cross–section can be used in other cases [5.5]. These formulae allowed us decrease the order of our linear algebraic system.

Finally, the formulae which we arrived at in this chapter we will use in chapter six for the numerical analysis of MSLs with lossless isotropic substrates in TEM–approximation.

6. ANALYSIS OF MICROSTRIP LINES WITH ISOTROPIC AND ANISOTROPIC SUBSTRATES IN TEM–APPROXIMATION BY THE SIE METHOD

6.1. Microstrip Lines with Isotropic Substrates

Microstrip lines (MSLs) with isotropic substrates (Fig.6.1) are simple objects to study by using the SIE method [6.1]–[6.3]. Some similar MSLs have been calculated by other methods. Here in this chapter, we compared known data from different references and saw that these examples only gave an idea about the potential of our SIE method.

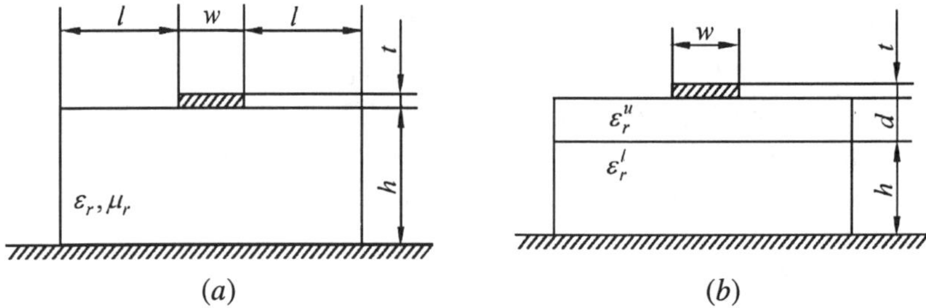

Fig. 6.1. Cross–section of a one–strip MSL: (a) with a magnetodielectric substrate; (b) with a layer dielectric substrate.

Here we see MSLs with one layer and two layers substrates Fig.6.1a,b. The sizes of microstrips are: w and t are the width and thickness, l is the shoulder, d is thickness of upper layer and h is the height of lower layer substrate.

Table 6.1 shows a comparison of results calculated by our SIE method and were given in [6.4]. In Ref. [6.4] there was no indication of the relative MSL sizes: relative distance l/h between the metal strip and the substrate lateral edges as well as the metal strip thickness t/h. But in our calculations, we used the relative MSL sizes $l/h = 2$ and $t/h = 0.005$.

Table 6.2 gives wave impedance for two values t/h. We see here that if you ignore the thickness of the metal strip it could possibly lead to some errors. The ratios l/h and t/h are not given in [6.5]. Here ε_r^u, ε_r^l is the relative permittivity of the upper and lower layers respectively.

Table 6.1. *The effective permittivity values of one–strip MSLs with dielectric substrates.*

MSL parameters		$\sqrt{\varepsilon_{MSL}^{ef}}$	
ε_r	w/h	Ref. [6.4]	Our calculations
1.5	2.0	1.16	1.16
2.0	2.0	1.34	1.33
6.0	0.2	1.93	1.90
6.0	2.0	2.08	2.11
8.0	2.0	2.48	2.53
15.0	2.0	3.50	3.44
20.0	1.2	3.62	3.61
20.0	2.0	3.80	3.95

Table 6.2. *The wave impedance values of one–strip MSLs* ($l/h = 2$, $\varepsilon_r^l = 12$)

MSL parameters			$Z(\Omega)$		
ε_r^u	d/h	w/h	t/h	Ref. [6.5]	Our calculations
—	0	0.1	0.005	92	95
—	0	0.4	0.02	69	72
—	0	1.0	0.02	45	46
—	0	2.0	0.02	31	31
—	0	2.0	0.005	31	28
—	0	4.0	0.02	19	18
4.5	0.01	0.4	0.02	70	84
4.5	0.01	1.0	0.02	47	52

Table 6.2 gives wave impedance for two values t/h. We see here that if you ignore the thickness of the metal strip it could possibly lead to some errors. The ratios l/h and t/h are not given in [6.5]. Here ε_r^u, ε_r^l is the relative permittivity of the upper and lower layers respectively.

Table 6.3. *Values of the relative wavelength in one–strip MSLs* ($\varepsilon_r^l = 2.55$, $\varepsilon_r^u = 1$)

MSL parameters			λ/λ_0	
d/h	w/h	t/h	Ref. [6.5]	Our calculations
0.4	0.43	0.0164	0.84	0.87
0.8	0.86	0.033	0.83	0.83
1.6	1.72	0.066	0.77	0.80

Table 6.3 presents the experimental values of the relative wavelengths λ/λ_0 (λ_0 is a wavelength in a vacuum) results that were taken from [6.5] and were calculated by using the SIE method.

***Table 6.4.** The effective permeability values of one–strip MSLs with magnetic substrates.*

μ_r	w/h	μ_{MSL}^{ef}	
		Ref. [6.6]	Our calculations
0.4	0.5	0.53	0.53
0.5	0.1	0.64	0.69
0.5	0.4	0.63	0.64
0.5	2.0	0.59	0.59
0.5	6.0	0.55	0.54
0.8	0.5	0.87	0.87

Table 6.4 contains the effective permeability values μ_{MSL}^{ef} of the MSL with the geometry of Fig.6.1a. This MSL has the magnetic substrate with the relative permeability μ_r, which is given in Table 6.4. We used the ratios $t/h = 0.005$ and $l/h = 1.6$ in our calculations. Ref. [6.6] does not show these ratios although they admit that l/h is infinite.

***Table 6.5.** The wave impedance values of two–strip MSLs ($\varepsilon_r = 16$)*

w/h	s/h	$Z_e(\Omega)$		$Z_0(\Omega)$	
		Ref. [6.7]	Our calculations	Ref. [6.7]	Our calculations
0.1	0.1	130	128	42	38
0.1	5.0	89	94	89	92
0.2	2.0	75	78	68	70
0.2	5.0	73	75	73	73
0.3	5.0	65	65	64	63
0.4	2.0	60	60	53	53
0.5	0.1	75	73	26	23
0.5	5.0	53	52	53	51
0.6	5.0	50	47.5	50	46.5
0.6	7.0	50	48	50	47
1.0	1.0	43	41	34	30

Table 6.5 presents our calculations using the SIE method and are given in [6.7]. The wave impedance values for the even Z_e and the odd Z_0 eigenwaves of two–strip MSLs with the dielectric substrate and size ratios $l/h = 2$ and $t/h = 0.005$. Ref. [6.7] does not mention these size ratios.

It is obvious when we analyse two–strip MSLs that the strips do not interact with each other, if the distance between the metal strips increase (such as $s/h = 10$). So the characteristics of a two–strip MSL become similar to a one–strip MSL simply because they do not interact with each other. We verified our computer programs by the previous criteria. So then by increasing the distance between the metal strips we transformed a two–strip MSL into a one–strip MSL with the dielectric, magnetized ferrite or magnetized semiconductor substrate. Our calculations show that the results deviated by less than 2%.

Here we show the calculated results for directional couplers, which were created on the bases of two MSLs (Fig.6.2a,b). Our goal was to determine the MSL sizes from given electric parameters. The MSLs were calculated using the standard characteristic impedance equal to 50 Ω.

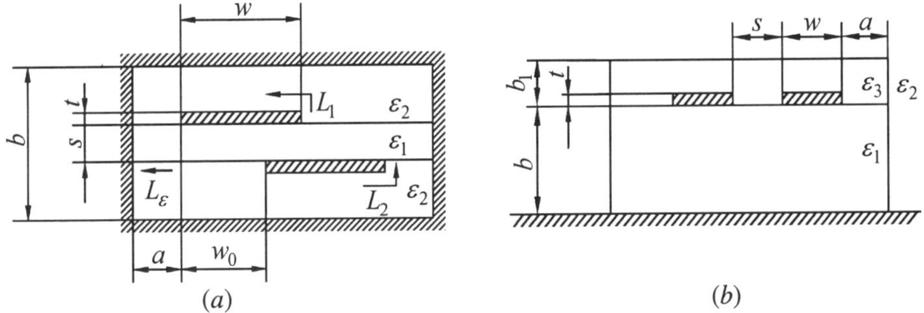

Fig. 6.2. *(a) Cross–section of a screened two–strip MSL; (b) cross–section of an open two–strip MSL.*

The MSL sizes derived from our calculations always showed some scattering due to the finite accuracy of the numerical computation. To smooth out the results we used the least square method [6.8]. The sizes of the MSL shown in Fig.6.2a are given in Table 6.6 as a function of the transition attenuation constant α and were measured in decibels [6.3]. Also the MSL shown in Fig.6.2b was calculated for the fixed values $\varepsilon_2 = \varepsilon_3 = 1$, $\varepsilon_1 = 10$, $a/b = 2$, and $t/h = 0.001$. As our calculations show the electric parameters of the MSL do not depend on the ratio a/b when $a/b \geq 2$. In Table 6.6 ε_e^{ef} and ε_o^{ef} are effective permittivities for an even or odd excitation of an MSL respectively.

Table 6.7 show the calculated results for the MSL (Fig.6.2b) with the substrate of three dielectrics. Our calculations show that when changing b_1 and ε_3 one can achieve equality of propagation constants on both eigenwaves. Here we have an example of such MSL parameters: $Z = 50\ \Omega$, $\varepsilon_1 = 10$, $\varepsilon_2 = 1$, $\varepsilon_3 = 25$, $\alpha = 9.12\,\text{dB}$, $b = 1\,\text{cm}$, $s/b = 0.3$, $w/b = 0.64$, $t/b = 0.0001$, $a/b = 2$ and $b_1/b = 0.09$.

Table 6.6. *Dependence of MSL sizes (Fig.6.2a) on transition attenuation constant $\alpha\,(dB)$ for fixed parameters $Z = 50\ \Omega$, $a/b = 1$ and $t/b = 0.0075$.*

Formula number	Calculation Formula	Limits of variation of α, dB	The fixed stripline parameters
1	$w = 0.5 + 0.127\alpha$ $s = -0.0187 + 0.047\alpha$	$2.6 \le \alpha \le 3.2$	$w_0 = 0$ $\varepsilon_1 = \varepsilon_2 = 1$
2	$w = 0.139 + 0.12\alpha$ $s = -0.0387 + 0.0468\alpha$	$2.7 \le \alpha \le 3.2$	$w_0 = 0$ $\varepsilon_1 = \varepsilon_2 = 2$
3	$w = 0.0934 + 0.104\alpha$ $s = -0.036 + 0.043\alpha$	$2.7 \le \alpha \le 3.2$	$w_0 = 0$ $\varepsilon_1 = \varepsilon_2 = 2.5$
4	$w = 0.0884 + 0.1109\alpha$ $s = -0.0353 + 0.0409\alpha$	$2.7 \le \alpha \le 3$	$w_0 = 0$ $\varepsilon_1 = 2$ $\varepsilon_2 = 2.5$ $\varepsilon_0^{ef} = 2.083$ $\varepsilon_e^{ef} = 2.448$
5	$w = 0.027 + 0.134\alpha$ $s = -0.04 + 0.0443\alpha$	$2.7 \le \alpha \le 3.2$	$w_0 = 0$ $\varepsilon_1 = 2.3$ $\varepsilon_2 = 2.5$ $\varepsilon_0^{ef} = 2.334$ $\varepsilon_e^{ef} = 2.478$
6	$w = 0.3885 + 0.03763\alpha - 0.00084\alpha^2$ $3 \le \alpha \le 23$ $\varepsilon_0^{ef} = 2.411 + 0.003\alpha - 0.258/\alpha$ $\varepsilon_0^{ef} = 2.475$ $at\quad \alpha \succ 23dB;\ \ w = 0.8; \varepsilon_e^{ef} \cong \varepsilon_o^{ef} = 2$ $w_0 = 0.5016 + 0.0305\alpha - 0.00005\alpha^2$ $3 \le \alpha \le 40$	$3 \le \alpha \le 23$ $3 \le \alpha \le 40$	$s = 0.0929$ $\varepsilon_1 = 2.3$ $\varepsilon_2 = 2.5$
7	$w = 0.7463 + 0.004\alpha - 1.01/\alpha$ $\varepsilon_o^{ef} = 2.353 + 0.0041\alpha - 0.844/\alpha$ $\varepsilon_e^{ef} = 2.43$ $(at\ \ \alpha \succ 25\ dB: w = 0.8;$ $\varepsilon_e^{ef} \approx \varepsilon_o^{ef} = 2.42)$ $w_0 = 0.442 + 0.03392\alpha - 1.6248/\alpha$	$3 \le \alpha \le 25$ $3 \le \alpha \le 40$	$s = 0.0869$ $\varepsilon_1 = 2.0$ $\varepsilon_2 = 2.5$

Table 6.7. *Dependence of the wave impedance $Z\ \Omega$ and of the transition attenuation constant $\alpha\,(dB)$ on ε_3 and b_1 for $w/b = 0.95$, $s/b = 1$, $a/b = 2$, $\varepsilon_1 = 10$, $\varepsilon_2 = 1$ and $t/b = 0.001$*

b_1 (cm)	$\varepsilon_3 = 4$		$\varepsilon_3 = 6$		$\varepsilon_3 = 10$		$\varepsilon_3 = 20$	
	Z	α	Z	α	Z	α	Z	α
0.002	49.69	17.03	49.65	16.76	49.62	16.49	49.56	16.24
0.005	49.41	17.84	49.27	17.78	49.02	17.73	48.63	17.67
0.010	49.43	18.07	49.20	18.04	48.88	18.00	48.26	17.92
0.050	50.00	18.22	49.59	18.17	48.90	18.04	47.52	17.70
0.100	49.73	18.07	49.21	17.95	48.20	17.69	46.25	17.07

Here we have calculated two microstrip directors, which were created on the basis of the MSLs shown in Fig.6.2. For the calculation of these microstrip directors we used the formulae that are given in Table 6.6. The schematic diagram of the MSL director functions is shown in Fig.6.3. The sizes of the MSL (Fig.6.2a) were chosen by Table 6.6. So the transition attenuation constant was $\alpha = 3\,dB$ and the measured results are given in Table 6.8. In Table 6.8 α_p is the decoupling in decibels.

Table 6.8. *The experimental data for the MSL shown in Fig.6.2a.*

MSL parameters	Frequency (*MHz*)					
	700	800	900	1000	1100	1200
α (*dB*)	3.4	3.1	3.0	3.0	3.2	3.4
α_p (*dB*)	34	34	33	31	32	32
CSWR$_1$	1.06	1.07	1.1	1.1	1.12	1.14
CSWR$_4$	1.08	1.09	1.05	1.05	1.08	1.12

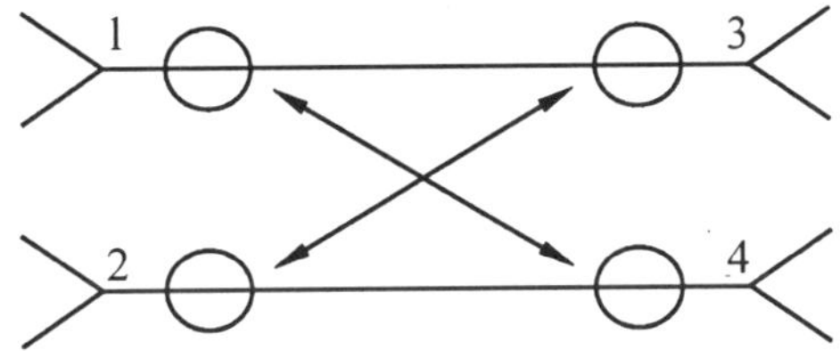

Fig. 6.3. *The schematic of the director.*

The microstrip director with the attenuation constant $\alpha = 10$ dB was calculated by the formulae (6.1) and (6.2). The experimental results for this case are presented in Table 6.9.

Table 6.9. *The experimental data for the MSL shown in Fig.6.2b.*

MSL parameters	Frequency (MHz)		
	650	850	1050
α (Ω)	10.8	10.2	10.8
α_p (Ω)	31.0	28.1	26.2
CSWR	1.24	1.1	1.2

The MSL sizes are calculated by the formulae:

$$s = -1.523 + 0.124\alpha + 5.306/\alpha , \tag{6.1}$$

$$w = 0.615 + 0.0216\alpha - 0.0609/\alpha , \qquad 9.5 \le \alpha \le 18 . \tag{6.2}$$

As we see in Tables 6.8 and 6.9, the calculated and measured results coincide within the experimental errors. The value of the Current Standing Wave Ratio (CSWR) was greater then 1.0 because there were non–uniformities in the MSL in these experiments but the non–uniformities was ignored in our calculations.

6.2. Microstrip Lines with *n–Si* Substrates and a Comparison of Our Calculated Results with the Experimental Data

MSLs with *n–Si* (electronic Silicon) substrates are used for signal processes and transmissions. To optimize the MSL electrodynamical characteristics, studies were performed on the influence of MSL sizes and its characteristics. Here we give some calculations of the MSL with an *n–Si* substrate (Fig.6.4). Our computation algorithm allowed us to take into account the metal strip thickness t and the substrate width p.

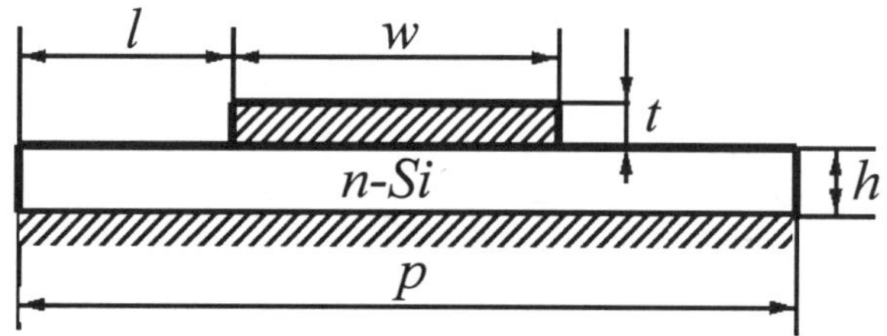

***Fig. 6.4.** Cross–section of a MSL with the n–Si substrate and designations.*

Fig.6.5 presents the dependence of the complex wave impedance $\dot{Z} = Z' - iZ''$ (the real Z' and imaginary Z'' parts) on the relative distance l/h between the metal strip and the substrate lateral edge at and the relative permittivity $\dot{\varepsilon}_{n-Si} = 11.7 - i5.202$ of the substrate $n–Si$ material when a frequency f = 2 GHz . The MSL (with a metal strip placed at the center of the substrate surface) is described by the following two parameters: the width $w/h = 0.82$ and the thickness of the metal strip $t/h = 0.02$. The sizes of the MSL were normalized to the substrate thickness h . Here the solid and dash lines are our calculated results and the circles are the experimental results [6.9] and [6.10].

A comparison between the numerical and the experimental results is presented in Fig.6.5. Here the circles denote the experimental data [6.9] and we see that the wider the MSL substrate is the larger the real Z' and imaginary Z'' parts of the complex wave impedance are. Fig.6.5 shows that the saturation of the complex wave impedance appears only at the approximate value of $l/h = 6.5$ for the MSLs that we took under consideration. We also see that if the substrate is narrower then the value $l/h = 6.5$ then the electrodynamical characteristics is strongly dependent on the substrate size. This means that we can change the MSL characteristics by making the width of the substrate smaller. For example, by shaving the lateral edges of the substrate we make the substrate narrower.

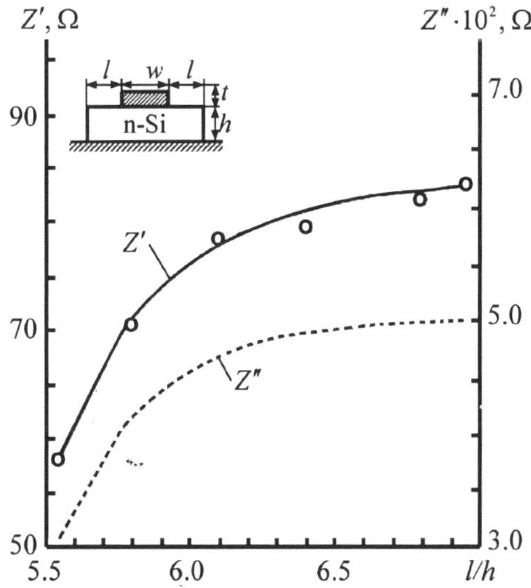

Fig. 6.5. Dependence of the MSL complex impedance $\dot{Z}$ on the relative distance l/h between the metal strip and the substrate lateral edge.

Our calculations are presented here by solid and dash lines. And the experimental results are shown by circles.

In Figs.6.6 and 6.7 we see the dependences of the complex wave impedance $\dot{Z} = Z' - iZ''$ modulus and the attenuation constant h'' of the MSL (with the semiconductor *n–Si* substrate) are greater when the resistivity ρ of the substrate material is less. The MSL is described by the following two parameters: the width $w/h = 0.82$ and the thickness of the metal strip $t/h = 0.02$. The less the value of the specific resistivity ρ of the substrate material the more the electrodynamical characteristics dependent on the frequency f. The dependences (Figs.6.6 and 6.7) are important in creating microwave microstrip devices with the substrate of a low ohmic material (for example, microwave sensors). The characteristics of these devices are sensitive to the influences of the electric field of the propagating microwave. We see that our calculations and the experiments (Figs.6.5–6.7) agree with one another.

Figs.6.8 and 6.9 show the dependences of the MSL complex wave impedance $\dot{Z} = Z' - iZ''$ on the specific resistivity ρ of the substrate material.

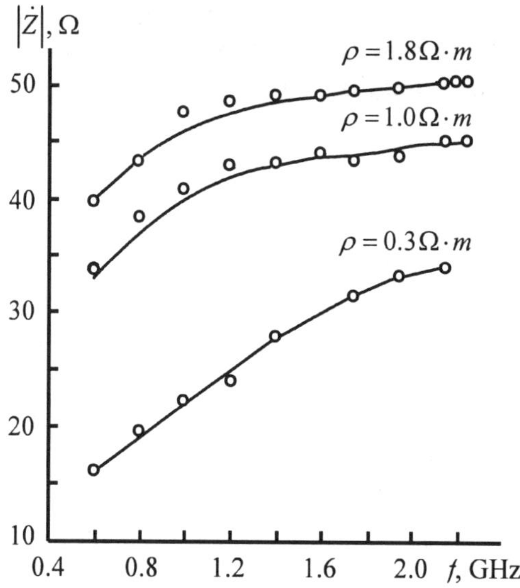

Fig. 6.6. Dependence of the complex wave impedance module of the MSL on the frequency when the specific resistivity $\rho = 0.3\ \Omega \cdot m$, $\rho = 1.0\ \Omega \cdot m$ and $\rho = 1.8\ \Omega \cdot m$ at the value $l/h = 5$. Solid lines show our calculations and the experimental data [6.9] is shown by circles.

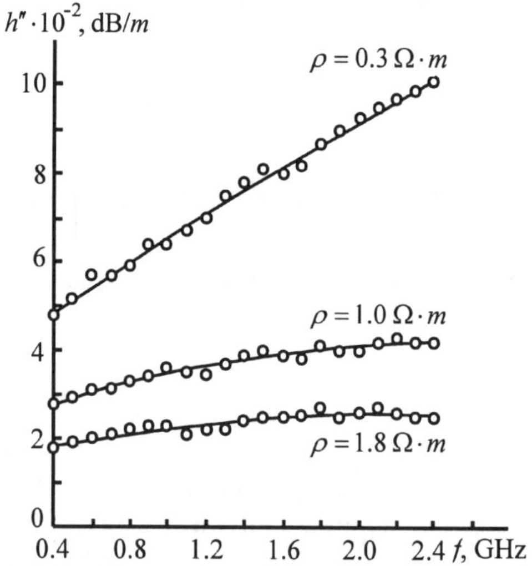

Fig. 6.7. Dependence of the attenuation constant h'' of the MSL on the frequency f for three specific resistivities: $\rho = 0.3\ \Omega \cdot m$, $\rho = 1.0\ \Omega \cdot m$ and $\rho = 1.8\ \Omega \cdot m$ at the value $l/h = 5$. Our calculations are shown by solid lines and the experimental data [6.9] is shown by circles.

These dependences (Figs.6.8 and 6.9) are particularly strong when the specific resistivity of material is in the range of $\rho = 0.05\ \Omega\cdot m$ to $\rho = 0.5\ \Omega\cdot m$. The MSL sizes are: the metal strip widths $w/h = 0.4$ and $w/h = 0.82$, their thicknesses are $t/h = 0.002$ and the distance between the metal strip and the substrate lateral edge $l/h = 1$. The real part of the relative permittivity of the substrate material is $\varepsilon'_{n-Si} = 11.7$. Our calculations were completed at the signal frequency of $f = 10\ GHz$. We see that when the MSL metal strip was wider the real part of the complex wave impedance Z' was less. We know that the larger the real part of the relative permittivity ε'_{n-Si} the smaller the value Z' becomes. In order for us to understand the dependences of an open MSL it is important to calculate (or at least to anticipate) where the electromagnetic energy of the microwave is concentrated. It is either in the MSL substrate or in the air around the metal strip and the substrate. So if the electromagnetic energy of the microwave is concentrated in the substrate then the value Z' becomes smaller when compared to the concentration of the electromagnetic energy in the air around the MSL. It is known that the impedance of a vacuum (free space) is $Z_0 \approx 377\Omega$.

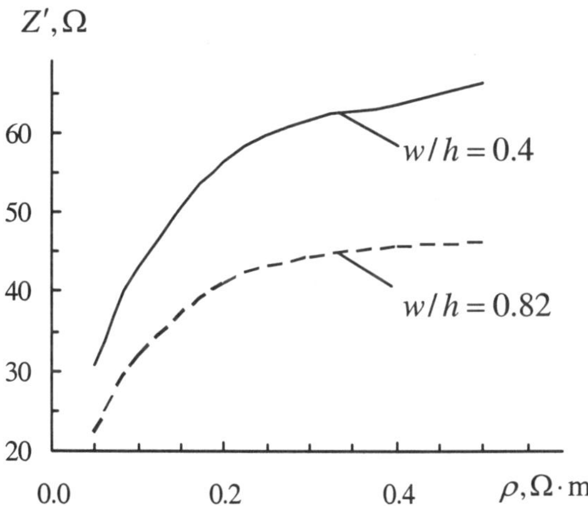

Fig. 6.8. Dependence of the real part Z' of the complex wave impedance of the MSL with the low ohmic n–Si substrate on the specific resistivity ρ of the substrate material when the metal strip widths are $w/h = 0.4$ (solid line) and $w/h = 0.82$ (dash line).

Fig.6.8 shows that the real part of the MSL complex wave impedance increased twice when the resistivity of the substrate material was from $\rho = 0.05\ \Omega\cdot m$ to $\rho = 0.5\ \Omega\cdot m$.

Fig.6.9 shows that the imaginary part of the complex wave impedance of the MSL with the n–Si substrate was changed more then two times in the range of the specific resistivity of the substrate from $\rho = 0.05\ \Omega\cdot m$ to $\rho = 0.5\ \Omega\cdot m$.

Fig.6.9 also shows that this dependence was changed nonmonotonously at the low values ρ of the specific resistivity of the n–Si substrate material.

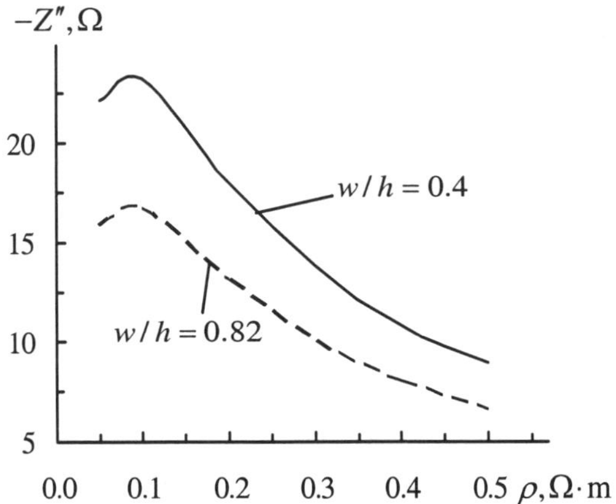

Fig. 6.9. *Dependence of the imaginary part Z'' of the complex wave impedance of the MSL with the low ohmic n–Si substrate on the specific resistivity ρ of the substrate material when the MSL sizes are: $t/h = 0.002$, $l/h = 1$, $w/h = 0.4$ (the solid line) and $w/h = 0.82$ (the dash line).*

Figs.6.10 and 6.11 show the dependence of the complex propagation constant $\dot{h} = h' - ih''$ of the MSL with the low ohmic n–Si substrate on the specific resistivity ρ of the substrate material. Here the value h' is the real part of the propagation constant (the phase constant) and h'' is the MSL attenuation constant.

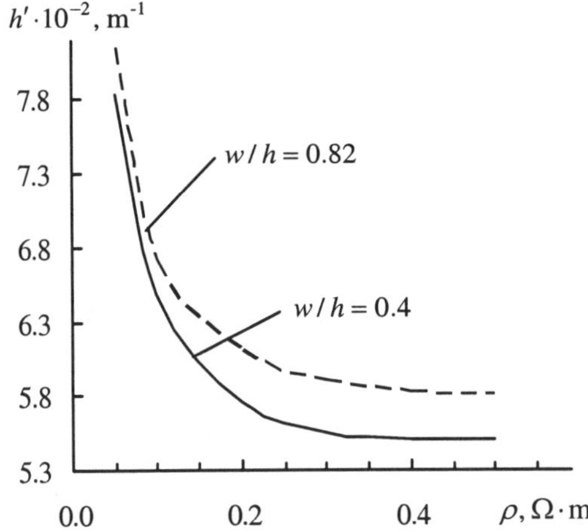

Fig. 6.10. *Dependence of the real part h' of the MSL complex propagation constant on the specific resistivity ρ of the n–Si substrate material when the MSL sizes are: $t/h = 0.002$, $l/h = 1$, $w/h = 0.4$ (the solid line) and $w/h = 0.82$ (the dash line).*

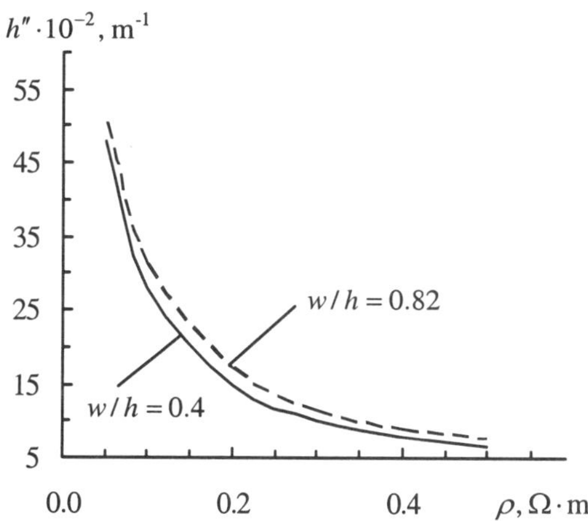

Fig. 6.11. Dependence of the imaginary part h'' of the MSL complex propagation constant on the specific resistivity ρ of the n–Si substrate material when the MSL sizes are: $t/h = 0.002$, $l/h = 1$, $w/h = 0.4$ (the solid line) and $w/h = 0.82$ (the dash line).

In this calculation the width of the metal strip is $w/h = 0.4$ or $w/h = 0.82$, its thickness is $t/h = 0.002$ and the distance between the metal strip and the substrate lateral edge is $l/h = 1$. We see that the MSL losses depend greatly on the specific resistivity ρ of the substrate material. The value h'' is changed approximately from $6.5 \cdot 10^2\,\mathrm{dB/m}$ to $50 \cdot 10^2\,\mathrm{dB/m}$. This dependence can be used for the creation of microwave attenuators, sensors and filters.

6.2.1 Calculated results for MSLs with two–layer n–Si – dielectric substrates. Here we are studying open MSLs having two–layer semiconductor–dielectric substrates. Investigations of these MSL types are important for the creation of microwave devices. In Ref. [6.11] MSLs with two–layer n–Si–dielectric substrates when the ohmic n–Si material had the specific resistivity $\rho = 0.01\,\Omega \cdot \mathrm{m}$ was studied. We analyzed dependences of the real and imaginary parts of the complex wave impedance of the TEM–wave on the relative permittivity $\dot{\varepsilon}_d = \varepsilon'_d - i\varepsilon''_d$ of the upper dielectric layer of the substrate as well as on the upper layer thickness when this thickness was changed on three orders. In our investigation the relative permittivities of the upper dielectric layer material were taken within these limits: the real part ε'_d from 1 to 6 and the imaginary part ε''_d from 0 to 10. Our numerical investigations show that it is possible to change the complex wave impedance $\dot{Z}$ more than two times by changing the dielectric permittivity and sizes of the upper layer.

Here we investigated the MSL containing a metal strip that is situated at the center of a thin dielectic layer, which is placed on top of the n–Si lower layer of the substrate (Fig.6.12). The MSL sizes are determined by these magnitudes: the metal strip width w, its thickness t, the distance between the metal strip and substrate lateral edges l, the n–Si lower substrate layer height h, and the upper dielectric layer thickness d. The width of the MSL substrate is $p = w + 2l$. The real part of the relative permittivity of n–Si material is $\varepsilon'_{n-Si} = 11.7$. The sizes of the MSL is normalized to the thickness h of the lower substrate layer.

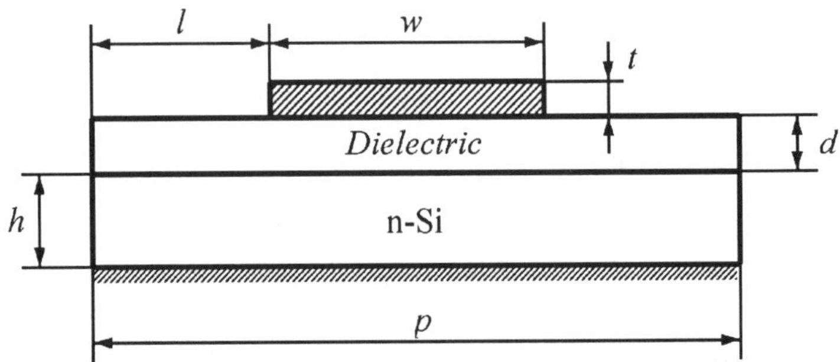

Fig. 6.12. *Cross–section of a MSL with a two–layer substrate.*

The curves in Figs.6.13–6.15 were calculated for the MSLs with normalized sizes: w/h=0.82 and t/h= 0.02. Also Figs.6.13–6.15 present the results of numerical investigations of MSLs when the specific resistivity of the substrate n–Si material is $\rho = 1.8\ \Omega\cdot m$ and the normalized distances between the metal strip and substrate lateral edges is $l/h = 0.7$.

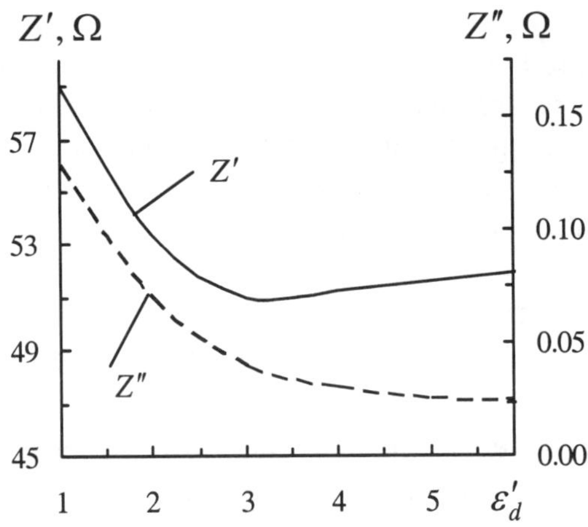

Fig. 6.13. *Dependence of the MSL complex wave impedance $\dot{Z}$ on the real part ε'_d of the relative permittivity of the upper dielectric layer at $\rho = 1.8\ \Omega\cdot m$.*

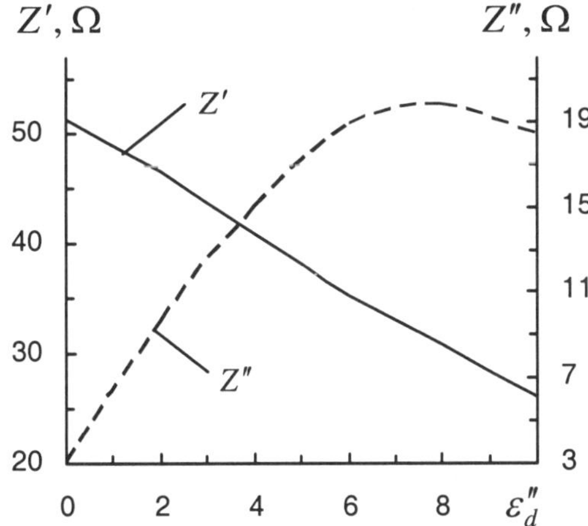

Fig. 6.14. *Dependence of the MSL complex wave impedance $\dot{Z}$ on the imaginary part ε_d'' of the relative permittivity of the upper dielectric layer at $\rho = 1.8\ \Omega{\cdot}m$.*

Figs.6.13 and 6.14 show dependences of the real and imaginary parts of the complex wave impedance $\dot{Z} = Z' - iZ''$ on the relative permittivity of the upper layer of the substrate $\dot{\varepsilon}_d = \varepsilon_d' - i\varepsilon_d''$ at a normalized thickness $d/h = 0.02$. The curves of Fig.6.13 present the complex wave impedance on the real part ε_d' of the complex relative permittivity of the upper substrate layer when the imaginary part of this layer is $\varepsilon_d'' = 5.625 \cdot 10^{-3}$. These curves show that the real part Z' has the least value when $\varepsilon_d' \approx 3$ and there is little change to its value when $\varepsilon_d' \succ 3$. The complex wave impedance is noticeably dependent on the real part ε_d' when it is within the limits $0 \prec \varepsilon_d' \prec 3$.

Fig.6.14 shows the dependence of the MSL wave impedance on the imaginary part ε_d'' of the complex relative permittivity of the upper substrate layer. We calculated the MSL wave impedance when the real part of the complex relative permittivity of the upper substrate layer was $\varepsilon_d' = 3.75$. We see that the real part of the wave impedance decreases considerably in all ranges $0 \le \varepsilon_d'' \le 10$. At the same time the value of the imaginary part Z'' has a maximum value at $\varepsilon_d'' \approx 7$.

As shown in Fig.6.15 when we increase the upper substrate layer thickness d the MSL complex wave impedance monotonically increases. When the upper layer of the MSL substrate is thicker a greater amount of electromagnetic energy is propagated in the layer. The relative permittivity of the upper substrate layer is less compared to the relative permittivity of the lower *n-Si* layer, because the MSL wave impedance increases when the upper substrate layer thickness *d* increases. As is known the lesser the relative permittivity of material around the metal strip the greater the MSL wave impedance.

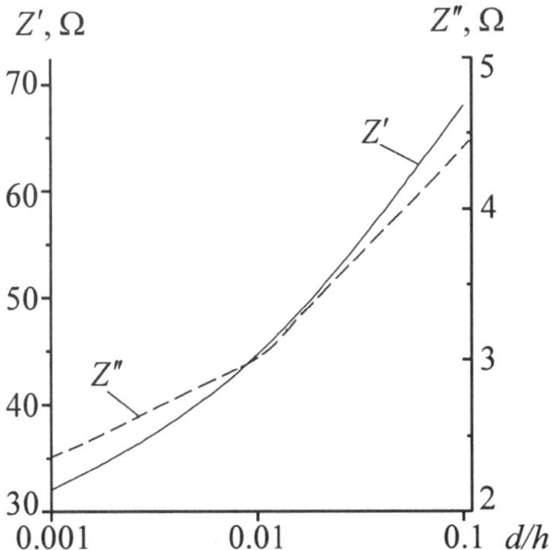

Fig. *6.15. Dependence of the MSL complex wave impedance* $\dot{Z}$ *on the upper dielectric layer thickness d.*

Our calculations show that it is possible to change the complex wave impedance of a MSL with a two–layer semiconductor–dielectric substrate more than two times by changing the dielectric permittivity and thickness of the upper substrate layer.

6.3 Microstrip Lines with *GaAs* and *GaAs–SiO₂* Substrates

The MSL with the *GaAs* (Gallium Arsenate) and *GaAs–SiO$_2$* (Silicon Dioxide) substrates are used in optical integrated circuits for signal transmissions. And investigations of MSL characteristics and their optimisation are important. Semiconductor *GaAs* material is used widely in producing MSLs because of its technological flexibility. This is especially important for hybrid optical integral circuits, where the separate microstrip elements are connected in such away as to obtain the electrodynamical characteristics needed. Different modifications of the MSL modulators, couplers and integral laser devices having the *GaAs* and *GaAs–SiO$_2$* substrates are described in [6.12]–[6.14].

In (Fig.6.16) the MSL substrate consists of two layers. The lower substrate layer is composed of the semiconductor *GaAs* and the upper substrate layer of the dielectric *SiO$_2$*. In our calculations we assumed that the metal strip is from perfect metal (that is there is no loss). The MSL attenuation constant was due only to the losses in the substrate material described by the imaginary part of permittivity at the frequency of $f = 2\ GHz$. Here we normalized all the MSL sizes to the lower layer thickness h of the substrate. Therefore, all the sizes w/h, l/h, d/h and t/h are dimensionless.

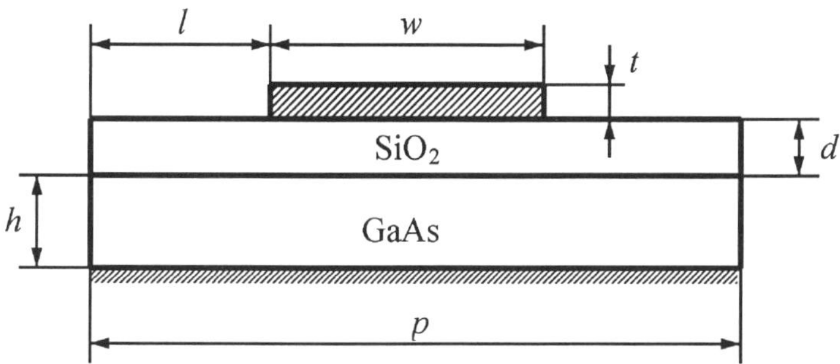

Fig. 6.16. *The MSL with the GaAs–SiO$_2$ substrate and designations.*

We studied dependencies of the several electrodynamical parameters on the metal strip width w and the distance l between the metal strip and the substitute lateral edge. These electrodynamical parameters are: the real Z' and the imaginary Z'' parts of the complex wave impedance, the MSL wave length λ, the effective permittivity ε_{MSL}^{ef}, the attenuation constant h'' and capacity C per unit length. The metal strip was placed at the substrate center [6.15]–[6.16]. The complex MSL wave impedance is expressed by $\dot{Z} = Z' - iZ''$ and the complex propagation constant is expressed by $\dot{h} = h' - ih''$, where $h' = 2\pi / \lambda$ is a phase constant.

6.3.1. The calculated results of MSLs with *GaAs* substrates. Here we calculated the electrodynamical characteristics of the MSL shown in Fig.6.16 when the upper substrate layer is absent and $d = 0$, the metal strip thickness is $t/h = 0.004$ and the *GaAs* substrate width $p/h = 2 \cdot l/h + w/h$. We used in our calculations the relative permittivity of the substrate *GaAs* material $\dot{\varepsilon}_{GaAs} = 11.2 - i0.00011$.

Fig.6.17 shows the dependence of the complex wave impedance (the real and imaginary parts) on the normalized metal strip width w/h. Changing the metal strip width w/h in the range of 0.2 to 2.2 the real wave impedance part Z' varied by approximately 3 times and the imaginary part by about 4 times. These results are important for adjusting the MSL to other waveguide tract elements.

The dependencies represented in Figs.6.18 and 6.19 show that when the metal strip width w/h increases the MSL wavelength λ decreases and three other electrodynamical parameters (effective permittivity ε_{MSL}^{ef}, the attenuation constant h'' and capacitance unit length C) also increase. This demonstrates that when the metal strip width increases a larger part of the electromagnetic energy of a microwave is propagated in the *GaAs* substrate. We also see that the metal strip width w/h is the most important size, which allows the MSL electrodynamical characteristics to change in the greatest way possible compared to other MSL sizes.

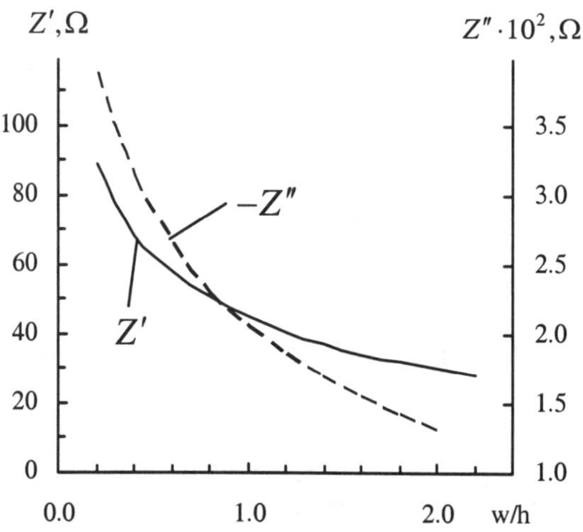

Fig. 6.17. *Dependence of the MSL complex wave impedance $\dot{Z}$ on the metal strip width w/h when its thickness $t/h = 0.004$ and GaAs substrate width is constant $p/h = 2.6$.*

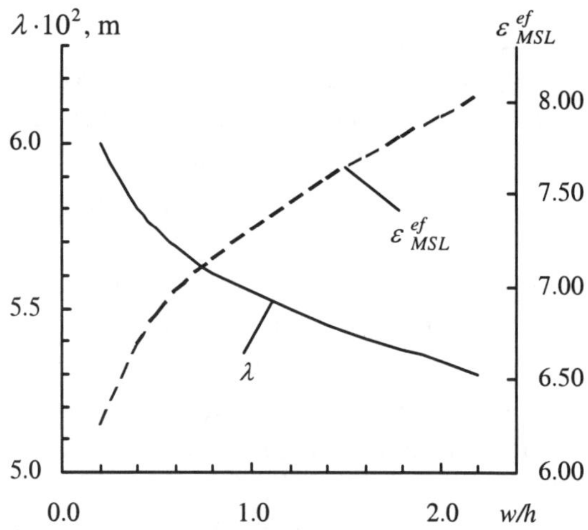

Fig. 6.18. *Dependence of the MSL wavelength λ and the effective permittivity ε_{MSL}^{ef} on the metal strip width w/h when its thickness is $t/h = 0.004$ and the GaAs substrate width is constant $p/h = 2.6$.*

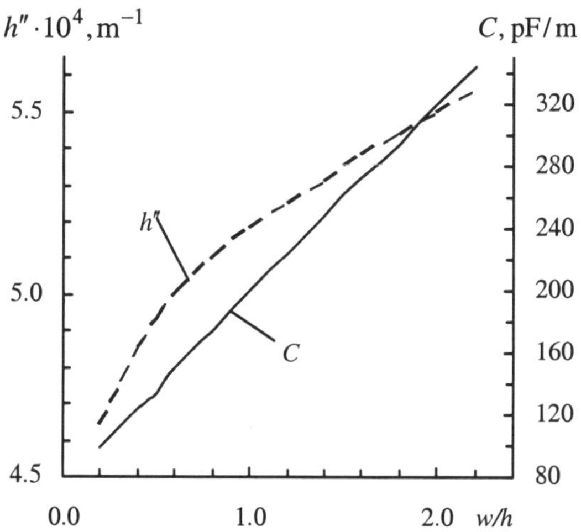

Fig. 6.19. *Dependence of the MSL attenuation constant h'' and capacitance per unit length C on the metal strip width w/h when its thickness is $t/h = 0.004$ and the GaAs substrate width is constant $p/h = 2.6$.*

One can see nonmonotonic characteristics of the MSL in Figs.6.20–6.22. The investigated electrodynamical characteristics $\dot{Z}$, λ, ε_{MSL}^{ef}, h'' and C are changed in the range of 1–2%.

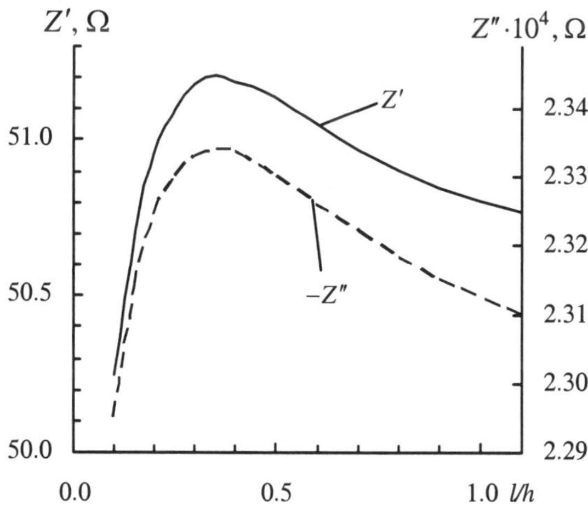

Fig. 6.20. *Dependence of the MSL complex wave impedance $\dot{Z}$ on the relative distance l/h between the metal strip and the substrate lateral edge when sizes are: $w/h = 0.8$, $t/h = 0.004$ and the GaAs substrate width p/h is not constant.*

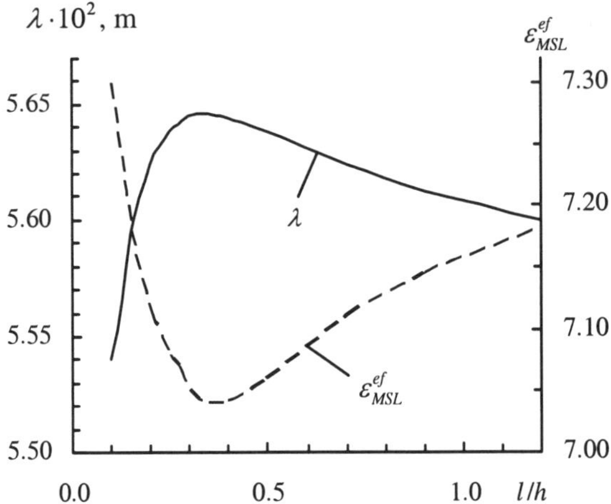

Fig. 6.21. *Dependence of the MSL wavelength λ and effective permittivity ε_{MSL}^{ef} on the relative distance l/h between the metal strip and the substrate lateral edge when sizes are: $w/h = 0.8$, $t/h = 0.004$ and the GaAs substrate width p/h is not constant.*

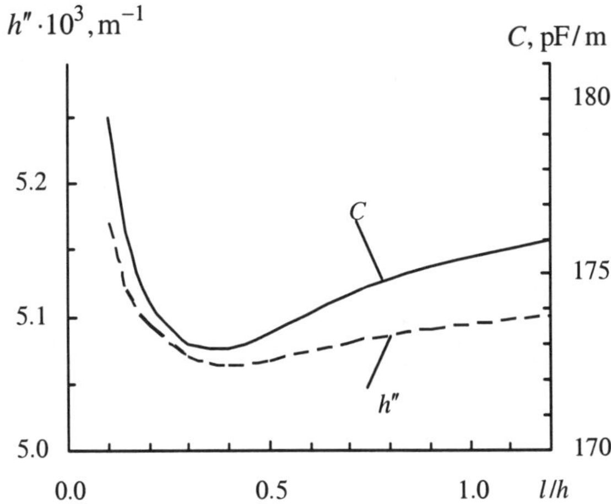

Fig. 6.22. *Dependence of the MSL attenuation constant h' and capacitance per unit length C on the relative distance l/h between the metal strip and the substrate lateral edge when sizes are: $w/h = 0.8$, $t/h = 0.004$ and the GaAs substrate width p/h is not constant.*

Quite often papers published in scientific journals (such as [6.17]) are devoted to the investigation of MSLs with substrates of infinite sizes. As they change the finite size to an infinite substrate size it is easy to calculate the characteristics of MSLs. But in the case where the substrate is of an infinite size the boundary conditions on the substrate lateral edges are not satisfied. Abstract theoretical models like these that change the real MSL geometry will occasionally lead to incorrect results.

6.3.2 The calculated results of MSLs with *GaAs–SiO₂* substrates. One of the ways to improve the MSL characteristics is to include a dielectric layer between the metal strip and the substrate [6.18]. Here we analysed the MSL with a two–layer substrate (Fig.6.16). The lower substrate layer is made of a semiconductor *GaAs* and the upper layer is made of a dielectric *SiO₂*. The relative permittivity *GaAs* is $\dot{\varepsilon}_{GaAs} = 11.2 - i0.00011$ and the *SiO₂* is $\dot{\varepsilon}_{SiO_2} = 3.75 - i0.00562$. The dependence of the MSL characteristics on the upper substrate layer thickness d, the metal strip width w and the distance l between the metal strip and the *GaAs–SiO₂* substrate lateral edges are given in Figs.6.23–6.31.

The MSL sizes are normalized to the lower *GaAs* substrate layer thickness h. The losses in the *GaAs–SiO₂* substrate materials are described through the imaginary part of the relative permittivity of these materials at the signal frequency $f = 2\ GHz$.

Fig.6.23 demonstrates the dependence of the MSL complex wave impedance on the upper substrate layer thickness d/h when the sizes are: $w/h = 1.2$, $t/h = 0.004$ and $p/h = 2.6$. The real and imaginary parts of the complex wave impedance varied substantially when we changed the MSL upper substrate layer thickness. The curves shown here are monotone, which corresponds to the data from different references. We see that the wave impedance could double when there is an additional substrate layer.

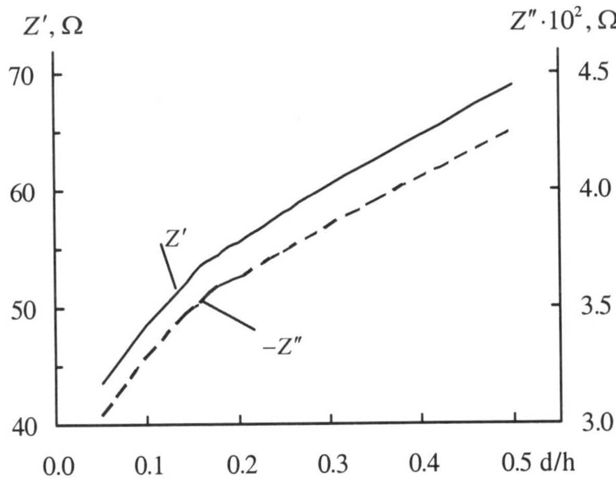

Fig. 6.23. *Dependence of the complex wave impedance $\dot{Z}$ of the MSL with the GaAs–SiO₂ substrate on its upper dielectric layer thickness d/h when sizes are: $w/h = 1.2$, $t/h = 0.004$, $l/h = 0.7$, $p/h = const$.*

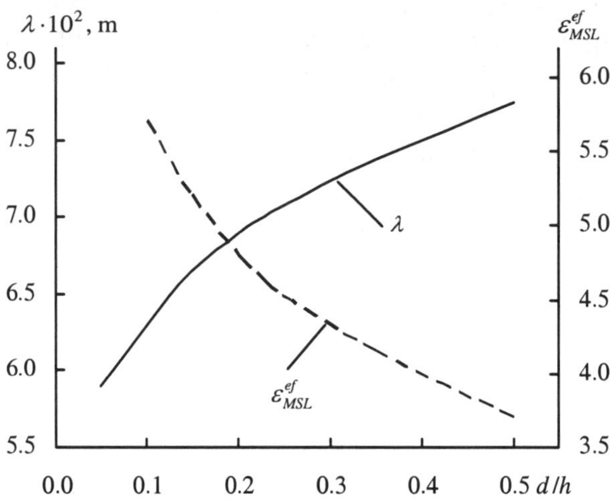

Fig. 6.24. *Dependence of the MSL wavelength and the effective permittivity ε_{MSL}^{ef} of the MSL with the GaAs–SiO$_2$ substrate on its upper dielectric layer thickness d/h when sizes are: $w/h = 1.2$, $t/h = 0.004$, $l/h = 0.7$, $p/h = const$.*

Fig.6.24 shows the MSL wavelength λ and the effective permittivity ε_{MSL}^{ef} dependence on the upper substrate layer thickness. We see that these values depend greatly on the thickness of the upper substrate layer.

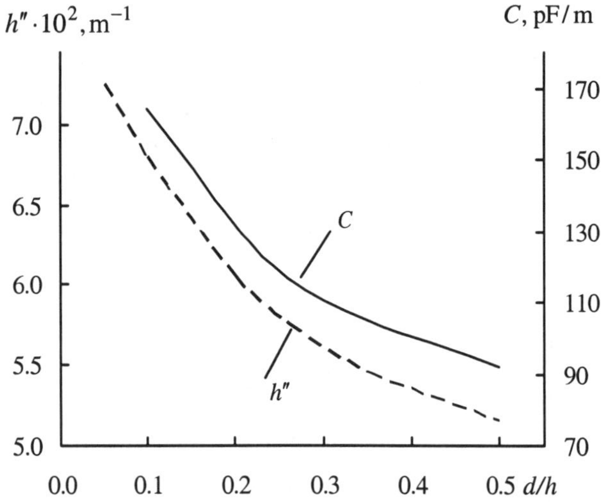

Fig. 6.25. *Dependence of the MSL attenuation constant h'' and capacitance per unit length C on its upper dielectric layer thickness d/h when sizes are: $w/h = 1.2$, $t/h = 0.004$, $l/h = 0.7$, $p/h = const$.*

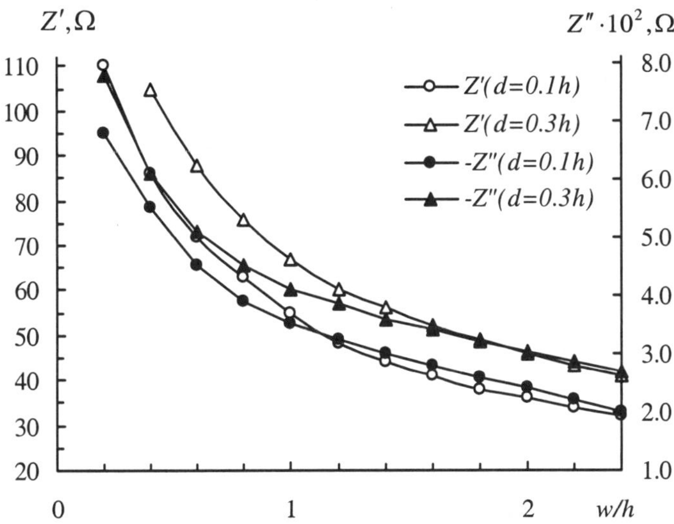

Fig. 6.26. *Dependence of the complex wave impedance $\dot{Z}$ of the MSL with the GaAs–SiO₂ substrate on the metal strip width w/h when sizes are: $t/h = 0.004$ and the substrate width is constant $p/h = 2.6$.*

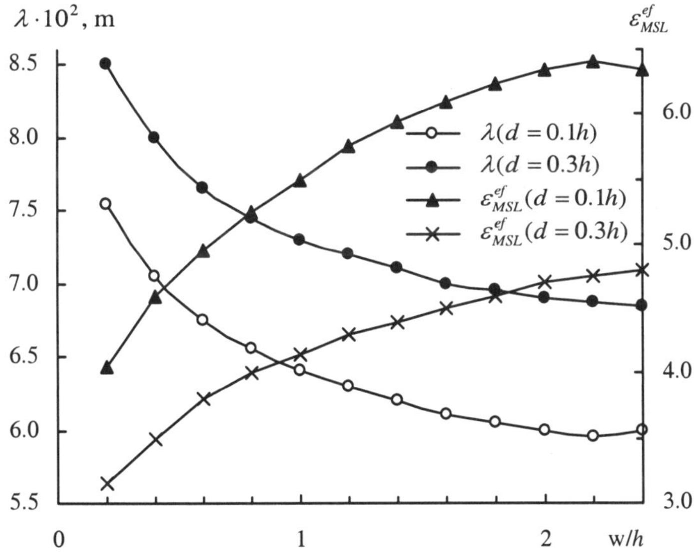

Fig. 6.27. *Dependence of the MSL wavelength λ and effective permittivity ε_{MSL}^{ef} of the MSL with the GaAs–SiO₂ substrate on the metal strip width w/h when sizes are: $t/h = 0.004$ and the substrate width is constant $p/h = 2.6$.*

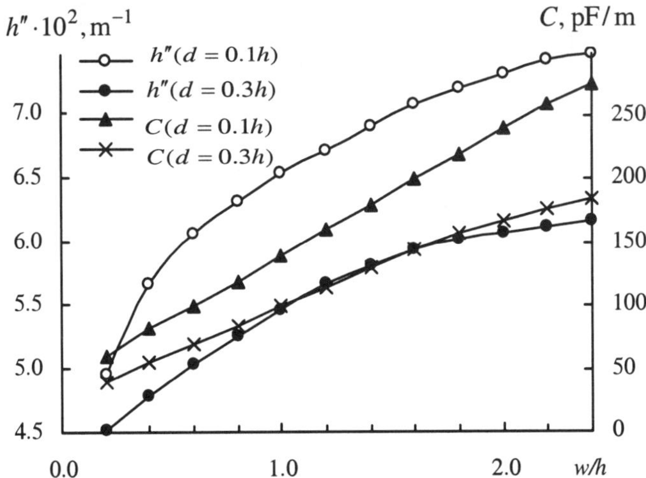

Fig. 6.28. Dependence of the attenuation constant h'' and capacitance per unit length C of the MSL with the GaAs–SiO$_2$ substrate on the metal strip width w/h when sizes are: $t/h = 0.004$ and the substrate width is constant $p/h = 2.6$.

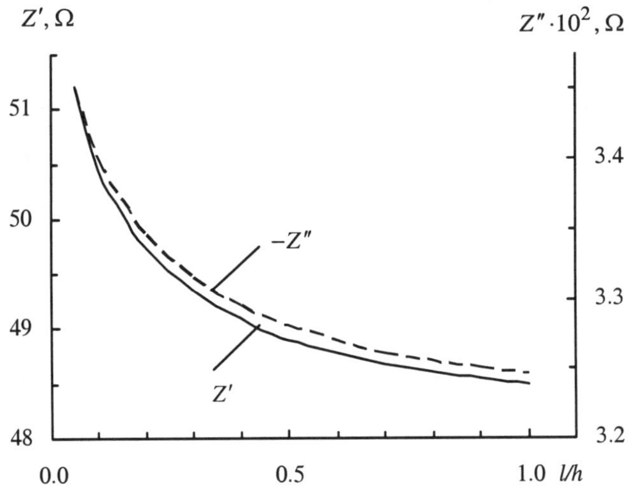

Fig. 6.29. Dependence of the complex wave impedance $\dot Z$ of the MSL with the GaAs–SiO$_2$ substrate on the relative distance l/h between the metal strip and the substrate lateral edge when sizes are: $d/h = 0.1$, $w/h = 1.2$, $t/h = 0.004$, $p/h \neq const$.

The MSL characteristics depend on the normalized metal strip width w/h as seen in Figs.6.26–6.28. Here we calculated for two substrate thicknesses: $d/h = 0.1$ and $d/h = 0.3$ when the substrate width was constant $p/h = 2.6$.

In Figs.6.29–6.31 we describe the dependence of the MSL characteristics on the magnitude l, which is the distance between the metal strip and the substrate lateral edges.

The curves in Fig.6.29 show that the complex wave impedance $\dot{Z}$ depends on the relative distance l/h when the substrate width is $p/h = w/h + 2l/h \neq const$. But we see that the influence of this parameter l on $\dot{Z}$ is less than that of the metal strip width w and of the upper layer thickness d. Also it is worthy to note that the dependence of $\dot{Z}$ on l in the MSL with a SiO_2 layer is different as compare to without the upper layer. The curves of Fig.6.29 are monotone. The corresponding dependence in the MSL without an additional upper substrate layer has a maximum wave impedance $\dot{Z}$ when the magnitude of l/h has an approximate intermediate value (Fig.6.20).

In Fig.6.30 for the MSL with the $GaAs–SiO_2$ substrate shows the dependence of the MSL wavelength λ and its effective permittivity ε_{MSL}^{ef} on the magnitude l/h. We see a greater dependence of the values λ and ε_{MSL}^{ef} on the substrate width $p/h = w/h + 2l/h$. The wider the substrate is the less λ is and the greater ε_{MSL}^{ef} becomes, because a greater part of the electromagnetic energy propagates in the MSL substrate.

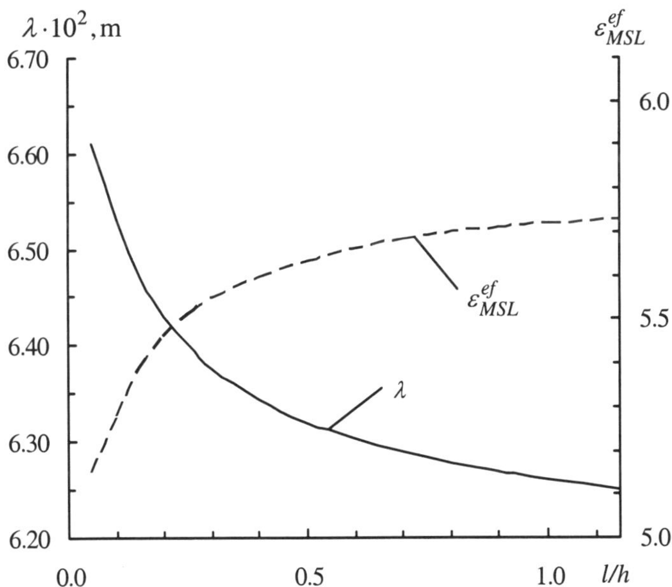

Fig. 6.30. Dependence of the MSL wavelength λ and the effective permittivity ε_{MSL}^{ef} of the MSL with the GaAs–SiO$_2$ substrate on the relative distance l/h between the metal strip and the substrate lateral edge when sizes are: $d/h = 0.1$, $w/h = 1.2$, $t/h = 0.004$.

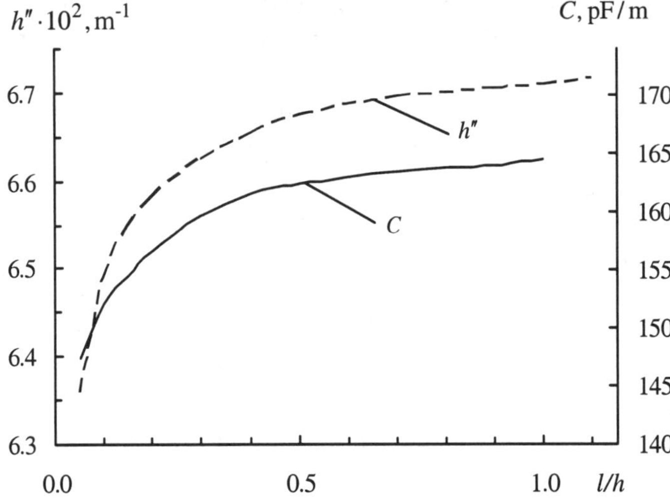

Fig. 6.31. *Dependence of the attenuation constant h'' and the capacitance per unit length C of the MSL with the GaAs–SiO$_2$ substrate on the relative distance l/h between the metal strip and the substrate lateral edge when sizes are: $d/h = 0.1$, $w/h = 1.2$, $t/h = 0.004$.*

We can see that the attenuation constant h'' for the MSL with a two–layer substrate (Fig.6.31) is greater than for the MSL without the upper SiO_2 layer (Fig.6.19). This happened because the imaginary part of the relative permittivity for the SiO_2 material is considerably larger than the one for the $GaAs$ material in our calculations.

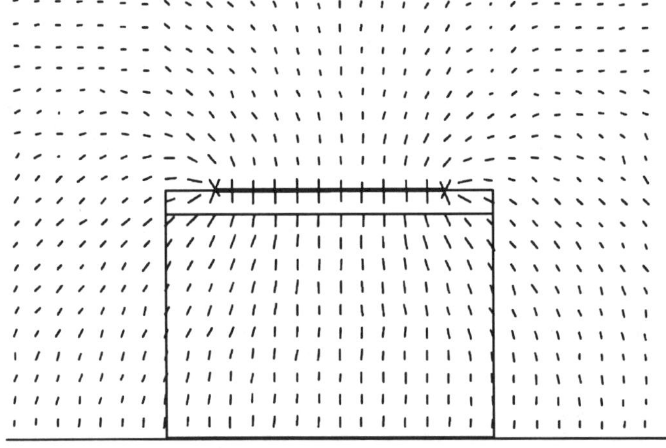

Fig. 6.32. *The quasi–TEM wave electric field distribution in the MSL with the GaAs–SiO$_2$ substrate when sizes are: $w/h = 1.2$, $l/h = 0.2$, $d/h = 0.1$, $t/h = 0.004$.*

Fig.6.32 presents the distribution of the quasi–TEM wave electric field in the MSL. One can see that the electric field has some singularities at the metal strip edges. Electric field vectors are normal to the metal surface of the metal strip and ground plate because in our computations the metal strip was made from perfect metal. Also in Fig.6.32 we see that the electric field values are largest near the metal strip edges and that they are considerably smaller further from the edges. On this graph we show the field for six hundred equally distanced points, which are in and around the MSL cross–section.

6.4 Microstrip Lines with Al_2O_3–SiO_2 Substrates

In the last few years MSLs have moved to a higher–frequency range of millimeter waves [6.19]. This was made possible only because technological progress has advanced to such a point that MSLs can now be manufactured precisely. And so we now see these MSLs are used successfully to create microelectronic devices. An open MSL with a two–layer substrate of SiO_2 (Silicon Dioxide, Silica) and Al_2O_3 (Aluminum Oxide, Alumina) employed in millimeter wave range devices were numerically studied in [6.20]. In our investigation we took into account the substrate width. And this width became a parameter allowing the desired variation of the MSL electrodynamical characteristics.

Both layers of the substrate were of isotropic materials: the upper layer of SiO_2 and the lower layer of Al_2O_3. A metal strip was placed at the center of the substrate and was described by the following parameters: the width of the metal strip w, its thickness t, its distances from the substrate edges l (right and left sides were equal), thickness of the lower substrate layer h, thickness of the upper substrate layer d and the substrate width p.

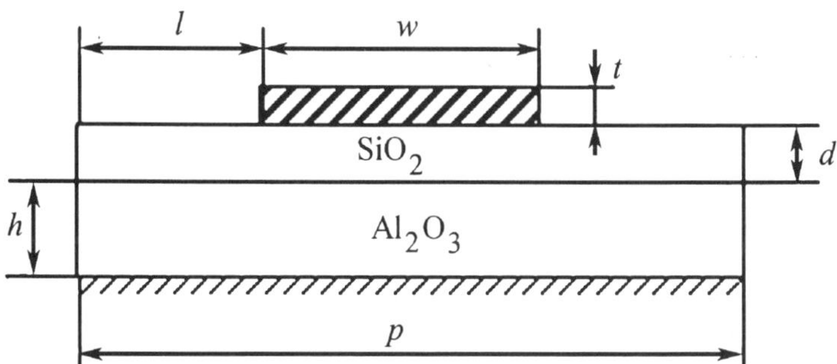

Fig. 6.33. *Cross–section of a MSL with a two–layer Al₂O₃–SiO₂ substrate and designations.*

We calculated the main characteristics of the MSL (Fig.6.33): wavelength λ, complex wave impedance $\dot{Z} = Z' - iZ''$, effective permittivity ε_{MSL}^{ef}, capacitance per unit length C and the MSL attenuation constant h''. In our calculations certain complex relative permittivity values were used: $\dot{\varepsilon}_{Al_2O_3} = 9.6 - i0.00096$ for Al_2O_3 and $\dot{\varepsilon}_{SiO_2} = 3.75 - i0.00562$ for SiO_2 at the signal frequency $f = 2$ GHz. The normalized thickness of the metal strip was $t/h = 0.004$. In quasi–TEM approximation, all sizes of the MSL are normalized on the substrate thickness h.

In our calculations h'' was only determined by the losses in the substrate. These losses were introduced by the imaginary part of the relative permittivities.

Figs.6.34–6.36 show the calculated dependences of electrodynamical characteristics of a MSL on the thickness of an upper substrate layer. These figures show that when the thickness of the upper substrate layer becomes larger, then the values Z', Z'', λ increase and the values h'', ε_{MSL}^{ef}, C decrease. The width of the substrate $p/h = 2.6$ remained constant.

When the upper layer thickness d increases the overall thickness $(h + d)$ of both substrate layers also increased. Keeping constant other normalized sizes of the MSL, the increase in the substrate thickness is equivalent to the decrease in the dimensionless parameter w/h. The narrower the metal strip w, the larger the part of the electromagnetic field energy is concentrated in the air around the metal strip. Here the effective permittivity ε_{MSL}^{ef} and losses (the attenuation constant) h'' in the MSL becomes less.

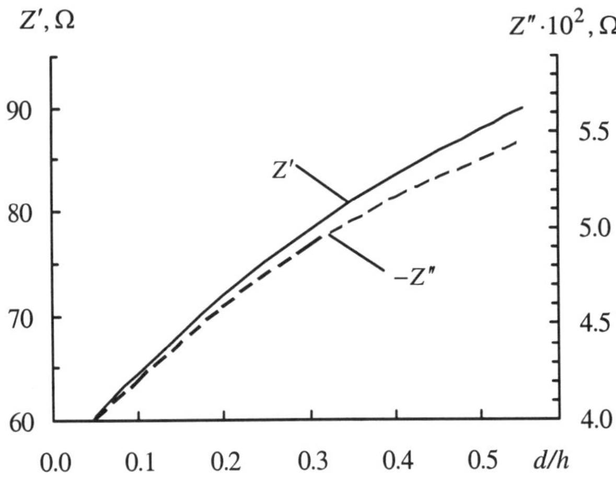

Fig. 6.34. *Dependence of a complex wave impedance $\dot{Z}$ of a MSL with the Al_2O_3–SiO_2 substrate on the upper substrate layer thickness d/h at the fixed MSL sizes: $w/h = 0.8$, $t/h = 0.004$, $l/h = 0.9$, $p/h = const$.*

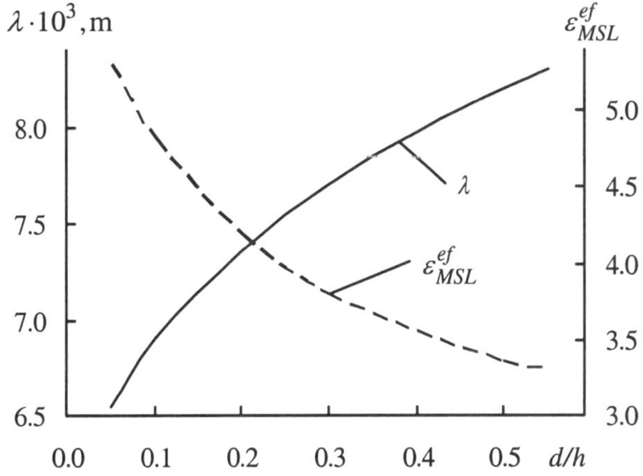

Fig. 6.35. *Dependence of the MSL wavelength λ and the effective permittivity ε_{MSL}^{ef} of the MSL with the Al₂O₃–SiO₂ substrate on the upper substrate layer thickness d/h at the fixed MSL sizes: $w/h = 0.8$, $t/h = 0.004$, $l/h = 0.9$.*

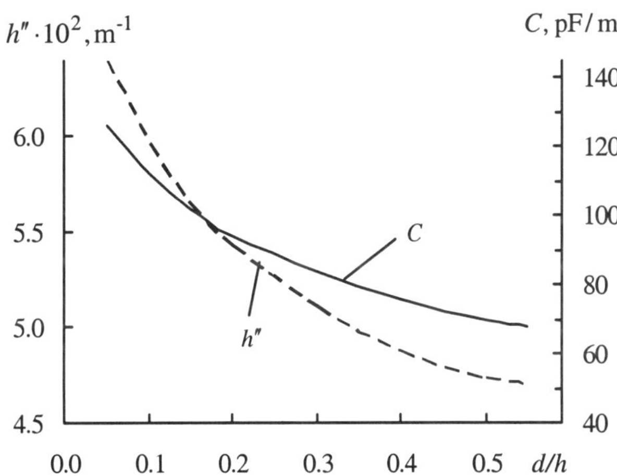

Fig. 6.36. *Dependence of the attenuation constant h'' and capacitance per unit length C of the MSL with the Al₂O₃–SiO₂ substrate on the upper substrate layer thickness d/h at the fixed MSL sizes: $w/h = 0.8$, $t/h = 0.004$, $l/h = 0.9$, $p/h = const$.*

We also studied the dependences of the above–mentioned characteristics of the MSL with the metal strip width w. These dependences are depicted in Figs.6.37–6.39.

Fig.6.37 shows the curves of the MSL complex wave impedance $\dot{Z} = Z' - iZ''$ is dependent on the metal strip width when the width of the Al₂O₃–SiO₂ substrate p is constant.

In Fig.6.38 we observed that the wider the metal strip, the larger will be the effective permittivity ε_{MSL}^{ef} and concluded that the larger part of the electromagnetic wave energy propagates underneath the metal strip inside of the substrate.

Curves in Fig.6.39 show that the attenuation constant h'' and the capacitance per unit length C increased depending on the metal strip width increase. This shows that when the metal strip width increases, a greater part of electromagnetic energy propagates inside of the Al_2O_3–SiO_2 substrate.

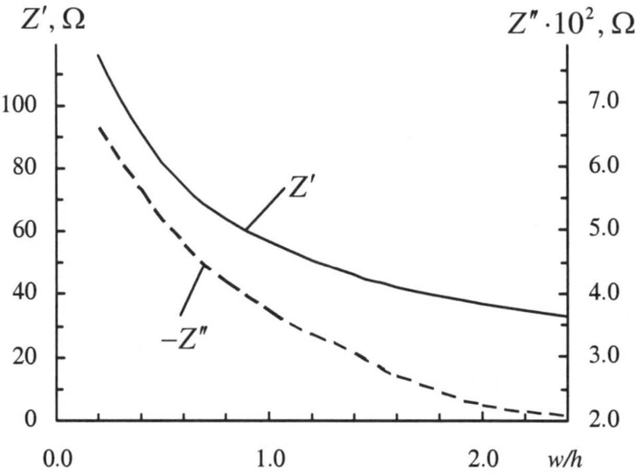

Fig. 6.37. Dependence of the complex wave impedance $\dot{Z}$ of the MSL with the Al_2O_3–SiO_2 substrate on the metal strip width w at the fixed MSL sizes: $d/h = 0.1$, $t/h = 0.004$, when the substrate width is constant $p/h = 2.6$.

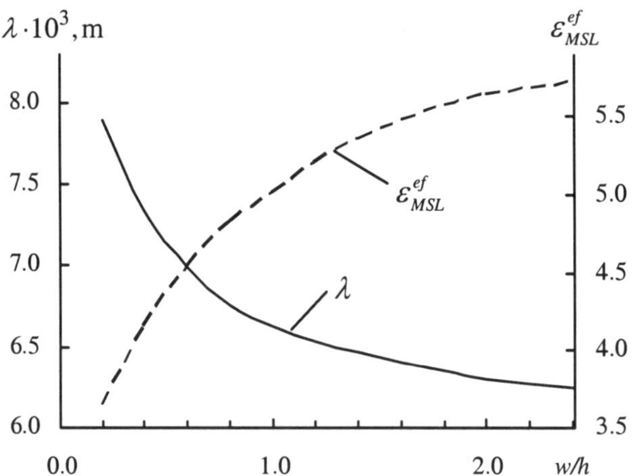

Fig. 6.38. Dependence of the MSL wavelength λ and effective permittivity ε_{MSL}^{ef} of the MSL with the Al_2O_3–SiO_2 substrate on the metal strip width w/h at the fixed MSL sizes $d/h = 0.1$, $t/h = 0.004$, when the substrate width is constant $p/h = 2.6$.

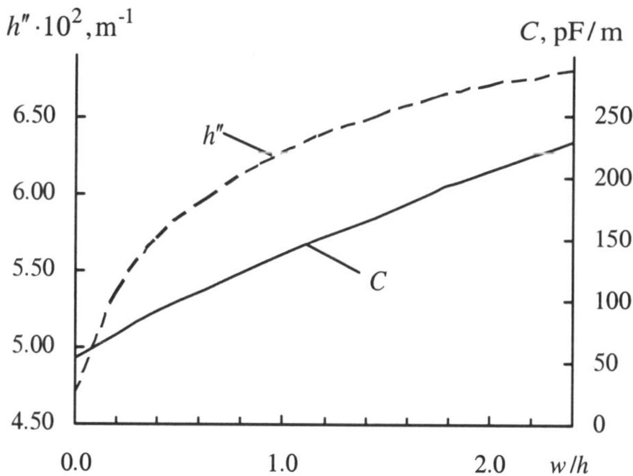

Fig. 6.39. *Dependence of the MSL attenuation constant* h'' *and capacitance per unit length* C *of the MSL with the* Al_2O_3–SiO_2 *substrate on the metal strip width* w/h *when sizes are:* $d/h = 0.1$, $t/h = 0.004$, *when the substrate width is constant* $p/h = 2.6$.

Fig.6.40 shows the dependences of the complex wave impedance $\dot{Z} = Z' - iZ''$ on the relative distance l/h between the metal strip and the substrate lateral edges of the MSL at the frequency $f = 2\ GHz$. The MSL with the Al_2O_3–SiO_2 substrate is described by the following three normalized sizes: the width $w/h = 1.2$, the thickness of the metal strip $t/h = 0.004$ and the upper substrate layer thickness $d/h = 0.1$. The sizes of the MSL were normalized to the substrates thickness h. The width of the substrate p/h was not constant in this case. If the MSL substrate increases then the complex wave impedance will decrease. It is interesting to note that the dependences presented in Fig.6.40 are different when compared to the same dependences for others MSLs. For example, if we compare the same dependences on the relative distance l/h for MSLs with different substrates: Al_2O_3–SiO_2 (Fig.6.40), $GaAs$–SiO_2 (Fig.6.29), $GaAs$ (Fig.6.20) and n–Si (Fig.6.5), we see the dependences are quite different. We also see that the relative distance l/h between the metal strip and the substrate lateral edges are an important parameter that allows the MSL characteristics to change in unpredictable way. To make the assumption that the MSL substrate is infinite could lead to incorrect calculations. This shows that in some MSL calculations it is necessary to take into account the real sizes of the substrate.

Figs.6.41 and 6.42 present the dependences of the MSL wavelength λ, effective permittivity ε_{MSL}^{ef} attenuation constant h'', and capacitance per unit length C on the ratio of the distance l between the metal strip edge and substrate lateral edge and also the lower layer of the substrate thickness h.

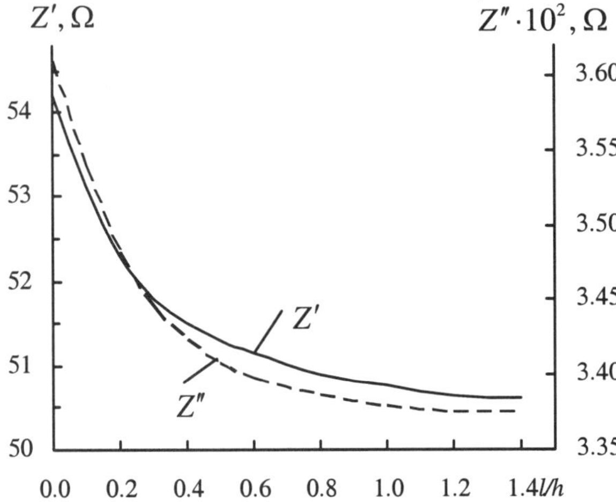

Fig. 6.40. *Dependence of the complex wave impedance $\dot{Z}$ of the MSL with the Al_2O_3–SiO_2 substrate on the relative distance l/h between the metal strip and the lateral substrate edges when sizes are: $d/h = 0.1$, $w/h = 1.2$, $t/h = 0.004$ and the substrate width p/h is not constant.*

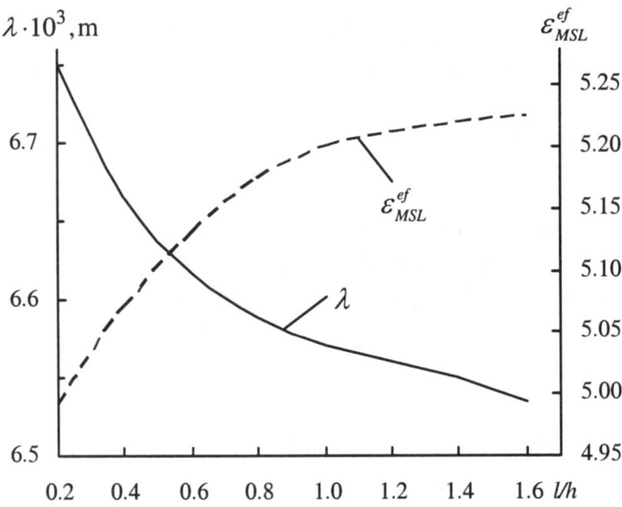

Fig. 6.41. *Dependence of the MSL wavelength λ and effective permittivity ε_{MSL}^{ef} of the MSL with the Al_2O_3–SiO_2 substrate on the relative distance l/h between the metal strip and the substrate lateral edge when sizes are: $d/h = 0.1$, $w/h = 1.2$, $t/h = 0.004$ and the substrate width p/h is not constant.*

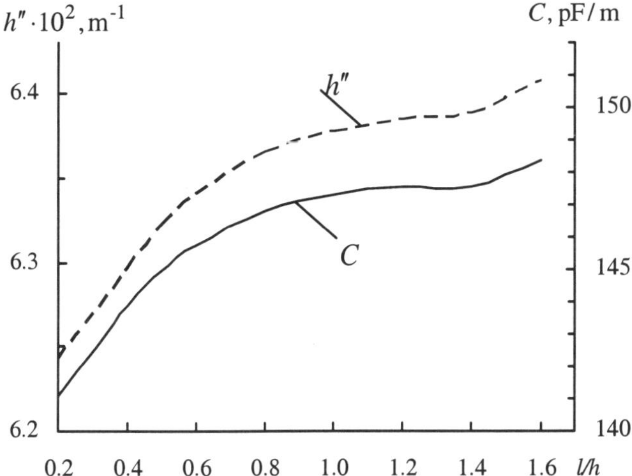

Fig. 6.42 *Dependence of the attenuation constant h" and capacitance per unit length C of the MSL with the Al₂O₃–SiO₂ substrate on the relative distance l/h between the metal strip and the substrate lateral edge when sizes are: d/h = 0.1, w/h = 1.2, t/h = 0.004 and the substrate width p/h is not constant.*

We can see that the electrodynamical characteristics of the MSL with the Al_2O –SiO_2 substrate have these ordinarily features.

6.5 Microstrip Lines with *LiNbO₃* Substrates

Devices of integral optoelectronics are manufactured on the basis of open MSLs with thin *LiNbO₃* substrates [6.21] and [6.22]. The real parts of the permittivity tensor of the *LiNbO₃* substrate material are $\varepsilon_{xx} = \varepsilon_{yy} = 43$ and $\varepsilon_{zz} = 28$.

Here we study the characteristics of open MSLs with *LiNbO₃* substrates (Fig.6.43). Our algorithms enable one to view the real geometry of metal strip thickness, substrate width and the anisotropy of substrate material [6.23]–[6.25].

Fig.6.44 presents the dependence of the real part Z' of the MSL complex wave impedance on the metal strip width w at two metal strip thicknesses t. The sizes of the MSLs (Fig.6.43) are: the height of the substrate $2h = 0.5 \cdot 10^{-3}$ m, the distances between the metal strip and the substrate lateral edges are $l_1/h = l_2/h = l/h = 0.5$ and the substrate width $p/h = (l_1 + l_2 + w)/h = 2.6$ is constant. Our calculations were compared with experimental data given in Ref. [6.24]. The dependence of electrodynamical characteristics on the metal strip width w is important because it allows us to choose the real part Z' of the MSL complex wave impedance within a greater range.

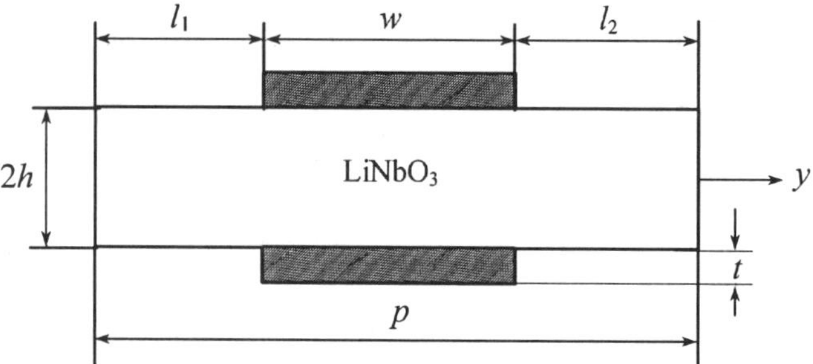

Fig. 6.43. *Cross–section view of a MSL with the LiNbO₃ substrate and designations.*

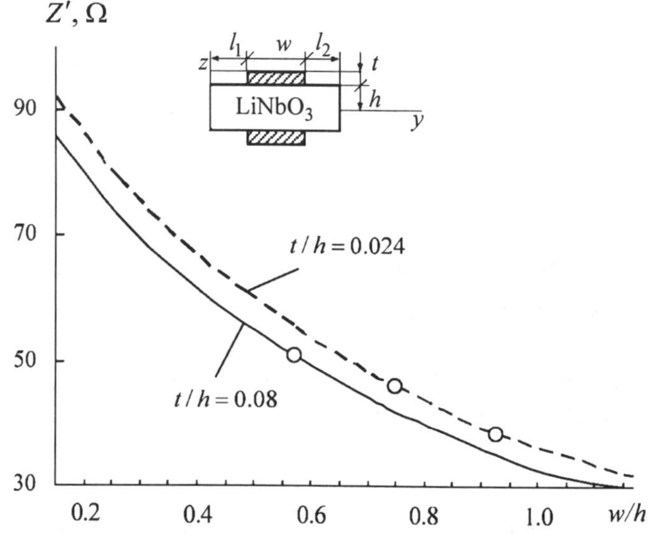

Fig.6.44. *Dependence of the real part* Z' *of the MSL complex wave impedance on the metal strip width* w *for two MSLs with different metal strip thicknesses* $t/h = 0.024$ *(the dash line) and* $t/h = 0.008$ *(the solid line). Circles are the experimental values. The LiNbO₃ substrate width is constant* $p/h = (2 \cdot l + w)/h = 2.6$.

Here we investigated how different locations of the metal strip on the surface of the substrate influenced the real part Z' of the MSL complex wave impedance (Fig.6.45). The width of the MSL substrate was constant $p/h = (2 \cdot l + w)/h = 2.6$. Our calculations show that by moving the metal strip from the center (where $l_1 = l_2$) to the edge (where $l_2 = 0$) of the substrate it is possible to change the value of the real part Z' of the MSL complex wave impedance by about 7%. We can use different locations of the metal strip on the substrate when we want to change the value of the MSL complex wave impedance without changing the sizes of the metal strip.

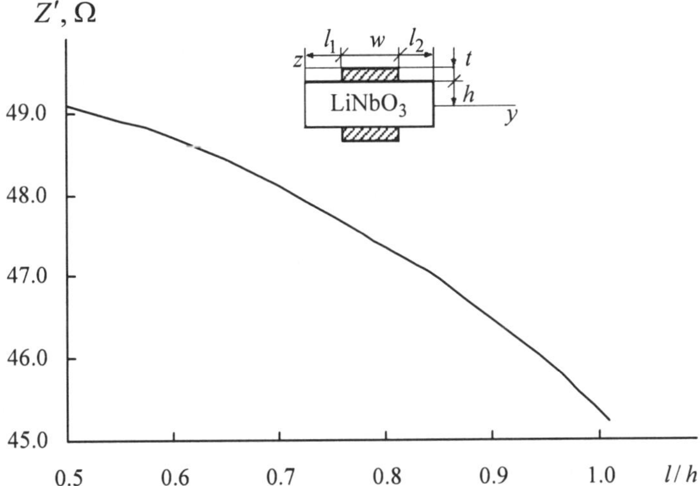

Fig. 6.45. *Dependence of the real part* Z' *of the MSL complex wave impedance on the displacement of the metal strip on the substrate when the LiNbO₃ substrate width is* $p/h = const$, $w/h = 1.6$ *and* $t/h = 0.04$.

Here the dependence was calculated (Fig.6.46) when the distances between the metal strip and the lateral substrate edges were equal ($l_1 = l_2 = l$). The width of the *LiNbO₃* substrate changed and $p = 2 \cdot l + w$ is not constant.

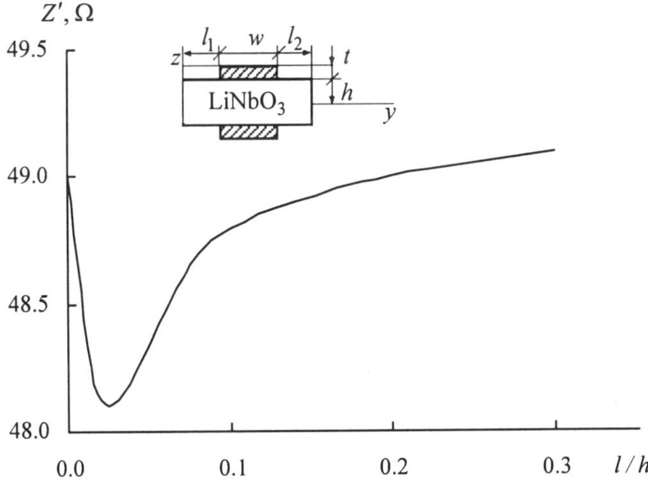

Fig. 6.46. *Dependence of the real part* Z' *of the MSL complex wave impedance on the distance between the metal strip and the lateral edges of the substrate when* $l_1 = l_2 = l$, $w/h = 1.6$, $t/h = 0.04$. *The LiNbO₃ substrate width is not constant* $p/h \neq const$.

We see that the dependence is nonmonotonic. When we change the distance between the metal strip and the substrate lateral edges it is possible to improve the real part Z' of the MSL complex wave impedance. The dependence when the *LiNbO₃* substrate is narrow ($l/h \prec 0.05$) is obvious and when the metal strip is placed near the substrates lateral edge then the electromagnetic wave energy is concentrated near that edge. In this case small chances in the substrate width will noticeably influence the electrodynamical characteristics. When we increase the distance (such as between the metal strip and the substrate lateral edge $l/h \succ 0.05$) there is less electromagnetic energy that is concentrated near the substrate lateral edge. But in this case the influence of the substrate width is insignificant. And we see that the MSL dependence on the width of the *LiNbO₃* substrate are unpredictable.

Fig.6.47 presents the dependence of the real part Z' of the MSL complex wave impedance on the metal strip thickness t. The sizes of the MSLs are: the height of the substrate $2 \cdot h = 0.5 \cdot 10^{-3}$ m, the width of the metal strip $w/h = 1.6$, the distances between the metal strip and lateral substrate edges $l_1/h = l_2/h = l/h = 0.5$ and $p/h = const$. When we increase the metal strip thickness from $t/h = 0.02$ $(t = 5\mu m)$ to $t/h = 0.14$ $(t = 35\mu m)$ the value of Z' increased about 15%.'

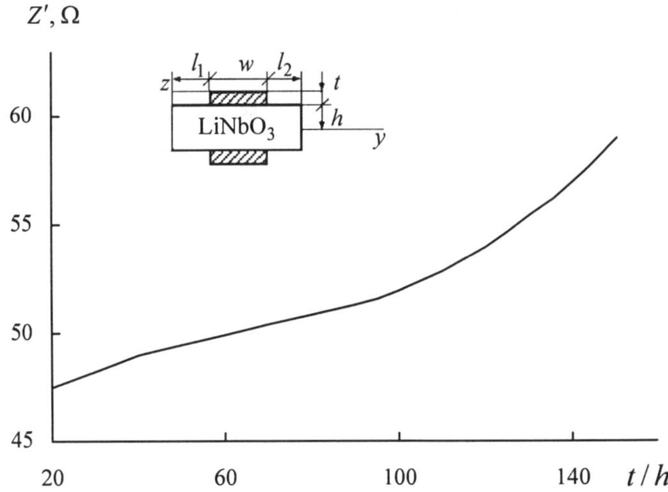

Fig. 6.47. Dependence of the real part Z' of the MSL complex wave impedance on the metal strip thickness t when the substrate width is constant $p/h = 2.6$.

The dependence of the real part Z' of the MSL complex wave impedance on the component ε_{zz} of the permittivity tensor is shown in Fig.6.48. The sizes of the MSL are: half the height of the substrate $h = 0.25 \cdot 10^{-3}$ m, the metal strip width $w/h = 1.6$, the distances between the metal strip and lateral substrate edges

$l_1/h = l_2/h = l/h = 0.5$ and the thickness of the metal strip $t/h = 0.04$. The real part Z' of the MSL complex wave impedance decreases about 60% when the component ε_{zz} increases from 10 to 60.

We can see that it is possible to change the real part Z' of the MSL complex wave impedance not only by changing the w but also by changing t, l_1 or l_2 (when $l_1 \neq l_2$), $p = l_1 + l_2 + w$ and the electrophysical parameter of the substrate material. Our algorithm allows us to take into consideration the real geometry of the MSL and the anisotropy of the substrate material.

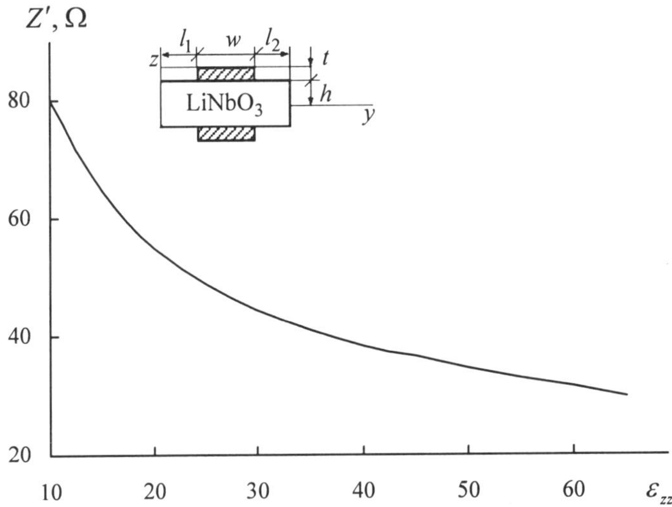

Fig. 6.48. Dependence of the real part Z' of the MSL complex wave impedance on the component ε_{zz} of the permittivity tensor when $\varepsilon_{xx} = \varepsilon_{yy} = 43$.

The dependence of the value Z' on the MSL sizes can be explained by the distribution of the electric field inside and outside of the MSL. A large part of the electromagnetic wave energy is concentrated in the substrate under the metal strip and a small part is distributed in the air around the MSL [6.25]. The metal strip width is an important parameter because the wider the metal strip the greater the electromagnetic wave energy is propagated inside of the substrate. When the metal strip is nearer the lateral substrate edge, a large part of the electromagnetic wave energy will propagate near the edge and the size of the substrate influence on the MSL dependences is obvious. But only a small part of the electromagnetic wave energy will propagate near the lateral substrate edge when the metal strip is far from the substrate edge. The influence on the substrate lateral edge is insignificant when this happens.

Now we will analyse MSLs with an anisotropic *LiNbO₃* substrate as in Fig.6.49. We investigated the dependences of the complex wave impedance $\dot{Z} = Z' - iZ''$, MSL wavelength λ (the signal frequency was $f = 2\ GHz$), effective permittivity ε_{MSL}^{ef}, attenuation constant h'' and capacity per unit length C on the metal strip width w and the distance l between the metal strip and the lateral substrate edges. The complex propagation constant is $\dot{h} = h' - ih''$. When computing the electrodynamical characteristics of the MSL shown in Figs.6.50–6.52, we used the complex diagonal components of the permittivity tensor of the substrate anisotropic material *LiNbO₃* which are: $\dot{\varepsilon}_{xx} = \dot{\varepsilon}_{zz} = 43 - i0.0043$ and $\dot{\varepsilon}_{yy} = 28 - i0.0028$.

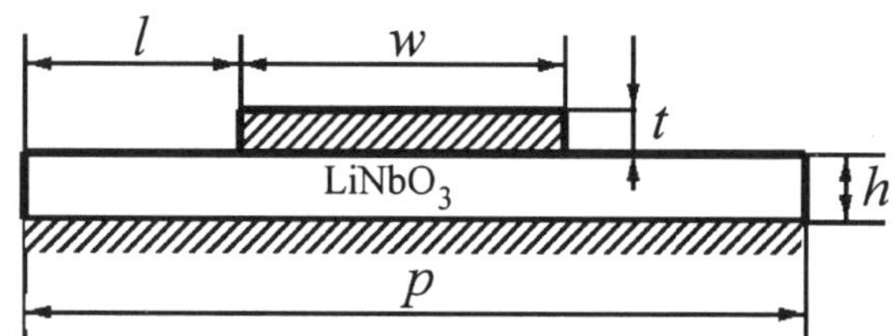

Fig. 6.49. *The MSL with an anisotropic LiNbO₃ substrate and designations.*

Fig.6.50 shows the dependence of the wave impedance $\dot{Z} = Z' - iZ''$ of the MSL with the *LiNbO₃* substrate on the ratio w/h. The width of the substrate $p/h = 2 \cdot l + w = 2.6$ is constant in this case, which means that $w/h \neq const$ and $l/h \neq const$. We see the MSL wave impedance will change significantly as a function of the metal strip width.

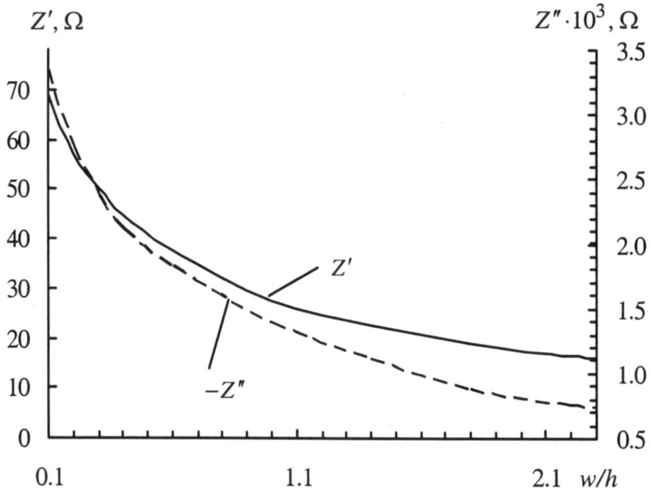

Fig. 6.50. *Dependence of the MSL complex wave impedance $\dot{Z}$ on the metal strip width w/h when sizes are: $t/h = 0.04$ and the LiNbO₃ substrate width is constant $p/h = 2.6$.*

The dependencies presented in Figs.6.51 and 6.52, show that when the metal strip width w/h increases, the MSL wavelength λ decreases and values h'', ε_{MSL}^{ef} and C increased. This demonstrates that when the metal strip width is increased then a larger part of the TEM–wave energy is propagated inside the $LiNbO_3$ substrate under the metal strip.

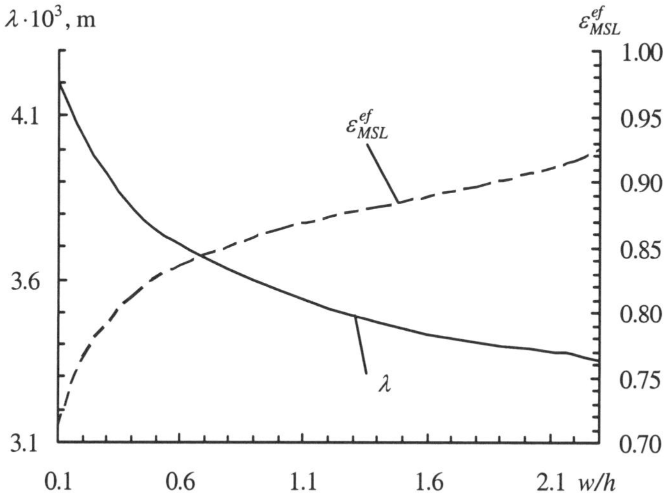

Fig. 6.51. *Dependences of the MSL wavelength λ and the attenuation constant h'' on the metal strip width w/h when sizes are: $t/h = 0.04$ and the LiNbO₃ substrate width is constant $p/h = 2.6$.*

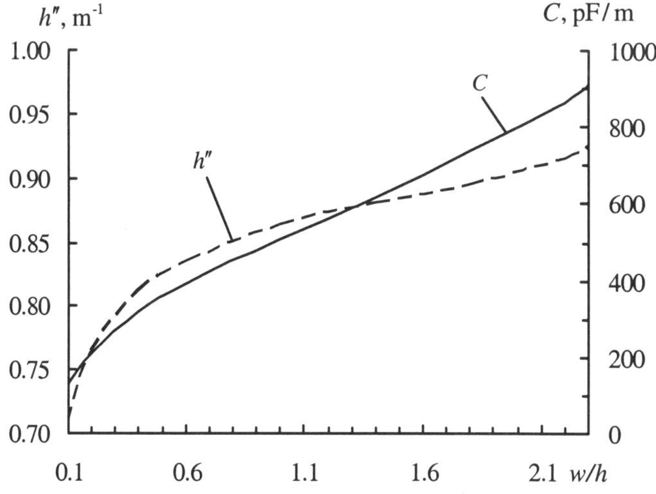

Fig. 6.52. *Dependence of the effective permittivity ε_{MSL}^{ef} and capacitance per unit length C on the metal strip width w/h when sizes are: $t/h = 0.04$ and the LiNbO₃ substrate width is constant $p/h = 2.6$.*

The wider the metal strip, the greater the part of the electromagnetic wave energy is concentrated in the substrate and a lesser part is distributed in the air around the MSL. And here we see that the ratio of the metal strip width w and the substrate thickness h has a great influence on the MSL with a one–layer $LiNbO_3$ substrate.

Here we will investigate dependences of the MSL electrodynamical characteristics on a substrate width. We see in Figs.6.53–6.55 that the substrate width p/h is the parameter, which will change the MSL characteristics. In this calculation the values are: $p/h \neq const$, $l/h \neq const$ and $w/h = const$.

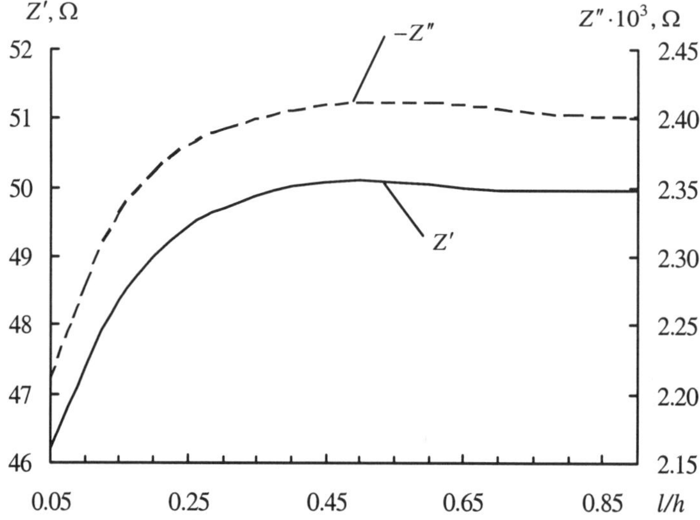

Fig. 6.53. *Dependence of the complex wave impedance $\dot{Z}$ of the MSL with the one – layer LiNbO$_3$ substrate on the relative distance l/h between the metal strip and the lateral substrate edges when sizes are: $w/h = 0.3$, $t/h = 0.04$, $p/h \neq const$.*

In Figs.6.53–6.55 the substrate width $p/h \neq const$ is the parameter that will change the MSL characteristics. To receive the needed complex wave impedance, complex propagation constant and other MSL electrodynamical characteristic one must change the MSL substrate width. If one wants to calculate MSLs we know it is important to take into account the real sizes of the substrate. But it is most critical when the substrate is narrow.

In studying the MSLs on an anisotropic two–layer $LiNbO_3$–SiO_2 substrate (Fig.6.56) we used the SIE method in TEM approximation and here all sizes of the MSL was normalized on the lower layer $LiNbO_3$ substrate thickness h.

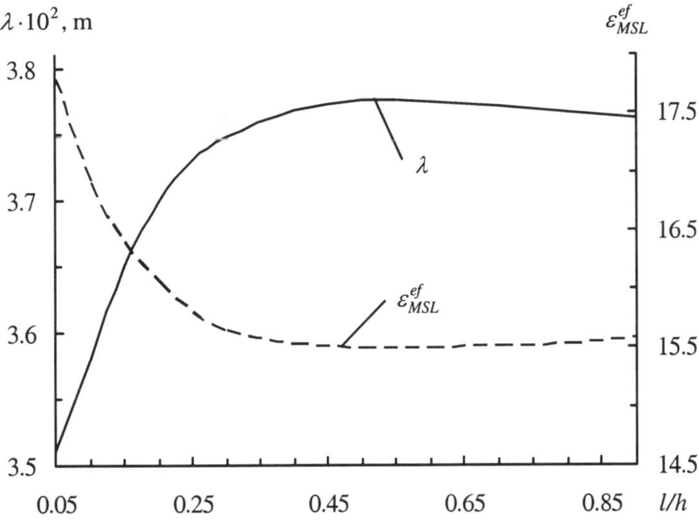

Fig. 6.54. *Dependence of the MSL wavelength λ and the effective permittivity ε_{MSL}^{ef} of the MSL with the one–layer LiNbO₃ substrate on the relative distance l/h between the metal strip and the lateral substrate when sizes are:* $w/h = 0.3$, $t/h = 0.04$, $p/h \neq const$.

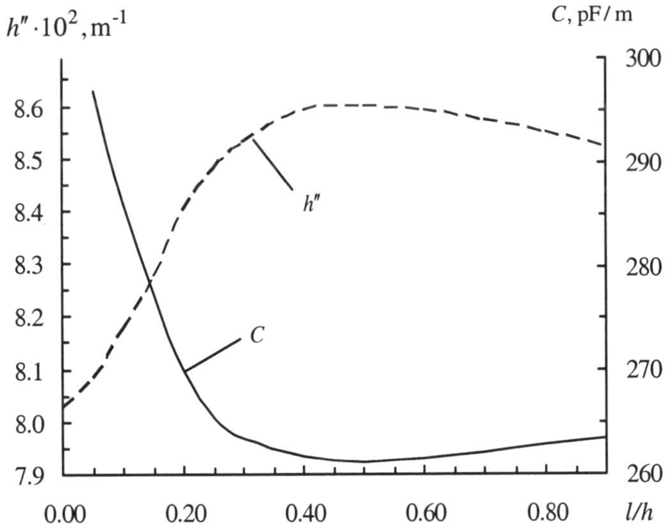

Fig. 6.55. *Dependence of the MSL attenuation constant h'' and capacitance per unit length C of the MSL with the one–layer LiNbO₃ substrate on the relative distance l/h between the metal strip and the lateral substrate when sizes are:* $w/h = 0.3$, $t/h = 0.04$, $p/h \neq const$.

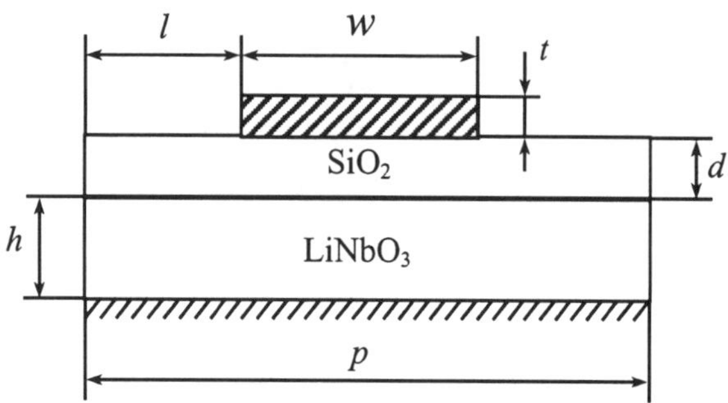

Fig. 6.56. *Cross–section of the MSL with a two–layer LiNbO$_3$–SiO$_2$ substrate and designations.*

Here we analysed the main electrodynamical characteristics of the MSL (Fig.6.56). The upper substrate layer *SiO$_2$* is isotropic material and the lower substrate layer *LiNbO$_3$* is anisotropic material. A metal strip was placed at the center of the substrate and was described by the following parameters: the metal strip width w, its thickness t, and its distances from the substrate edges l. The thickness of the lower substrate layer is h, thickness of the upper *SiO$_2$* substrate layer is d and substrate width is $p = 2 \cdot l + w$.

When computing the electrodynamical characteristics of the MSL shown in Figs.6.57–6.65, we used the complex diagonal components of the permittivity tensor of the substrate anisotropic material *LiNbO$_3$* equal $\dot{\varepsilon}_{xx} = \dot{\varepsilon}_{zz} = 43 - i0.0043$ and $\dot{\varepsilon}_{yy} = 28 - i0.0028$ as well as $\dot{\varepsilon}_{SiO_2} = 3.75 - i0.00562$ for the *SiO$_2$* at the signal frequency $f = 2$ GHz. We see here that the losses of the upper layer *SiO$_2$* ($\varepsilon'' = 0.00562$) is greater than the losses of the lower layer *LiNbO$_3$* ($\varepsilon'' = 0.0028 \div 0.0043$).

Figs.6.57–6.59 show the calculated dependences of the electrodynamical characteristics of the MSL on the thickness of the upper substrate layer. In these figures the normalized width of the metal strip was $w/h = 1$, its thickness was $t/h = 0.004$ (here t/h was ten times less then before) and the width of the *LiNbO$_3$–SiO$_2$* substrate $p/h = (2 \cdot l + w)/h = 3$ was constant.

We see the dependence of the MSL complex wave impedance (Fig.6.57) on the thickness of the upper *SiO$_2$* layer is significant. The real Z' and imaginary Z'' parts of the complex wave impedance change in a similar way. The Z' and Z'' values decrease at $d/h \prec 0.03$ and then they increase at $d/h \succ 0.03$. These nonmonotonic characteristics of the MSL show that when the upper *SiO$_2$* layer is $d/h \prec 0.03$ then a large part of the electromagnetic energy propagates around the surface border between the upper *SiO$_2$* and lower *LiNbO$_3$* layers of the MSL substrate. The thicker the upper layer becomes the greater the part of the electromagnetic energy propagates in the *LiNbO$_3$* layer. We see in Fig.6.57 that when the upper layer thickness increases the MSL wave impedance $\dot{Z}$ becomes less at $d/h \prec 0.03$. When the upper SiO_2 layer becomes $d/h \succ 0.03$ then the electromagnetic energy propagates more in the upper SiO_2 layer. We see that the value $\dot{Z}$ increases at $d/h \succ 0.03$.

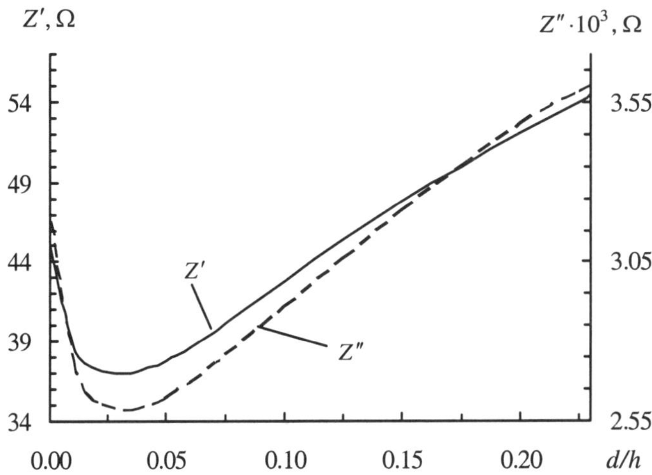

Fig. 6.57. *Dependence of the complex wave impedance $\dot{Z}$ of the MSL with the two–layer $LiNbO_3$–SiO_2 substrate on the upper substrate layer thickness d/h at the fixed MSL sizes: $w/h = 1.0$, $t/h = 0.004$, $l/h = 1.0$, $p/h = const$.*

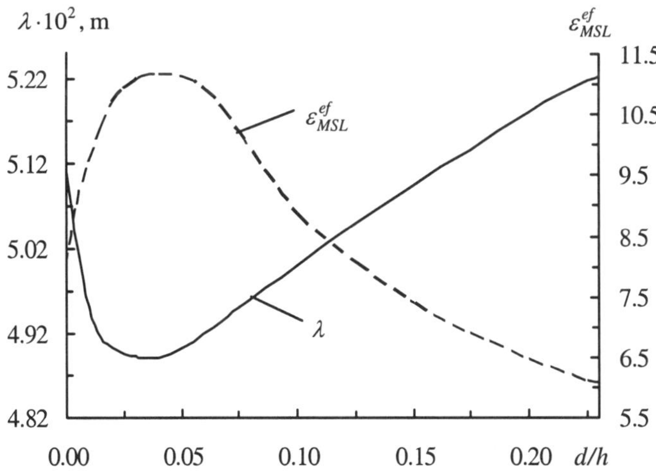

Fig. 6.58. *Dependence of the MSL wavelength λ and the effective permittivity ε_{MSL}^{ef} of the MSL with the two–layer $LiNbO_3$–SiO_2 substrate on the upper substrate layer thickness d/h at the fixed MSL size: $w/h = 1.0$, $t/h = 0.004$, $l/h = 1.0$, $p/h = const$.*

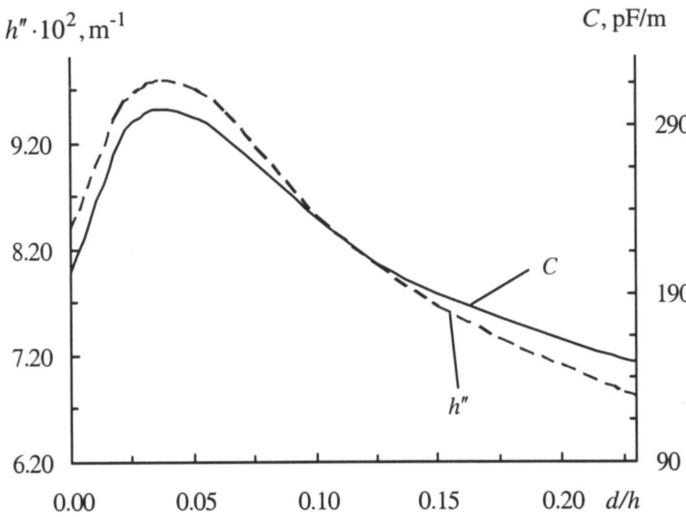

Fig. 6.59. *Dependence of the attenuation constant h'' and capacitance per unit length C of the MSL with the two–layer $LiNbO_3$–SiO_2 substrate on the upper substrate layer thickness d/h at the fixed MSL sizes $w/h = 1.0$, $t/h = 0.004$, $l/h = 1.0$, $p/h = const$.*

Figs.6.58 and 6.59 show that the dependences of the MSL wave length λ, the effective permittivity ε_{MSL}^{ef}, the attenuation constant h'' and the capacitance unit length C on the thickness d/h, which have extremes at about $d/h \approx 0.05$. We can explain these characteristics when we realize, which substrate layers the electromagnetic energy is concentrated in.

In our calculations, losses h'' were only determined by the losses in the substrate. The losses were introduced by the imaginary part of the relative permittivity of the substrate materials: $LiNbO_3$ ($\varepsilon_{xx}'' = \varepsilon_{zz}'' = 0.0043$, $\varepsilon_{yy}'' = 0.0028$) and SiO_2 ($\varepsilon_{SiO_2}' = 0.00562$).

When the upper layer thickness d increases, the overall thickness $(h+d)$ of the substrate also increases. Keeping other normalized sizes of the MSL constant, the increase in the common substrate thickness is equivalent to the decrease in the dimensionless parameter w/h, t/h and l/h. When the upper layer thickness d/h increases the dependences of the electrodynamical characteristics on it has an extreme point at about $d/h = 0.05$.

The MSL electrodynamical characteristics on the normalized metal strip width w/h are shown in Figs.6.60–6.62. Calculations were completed for the substrate thickness $d/h = 0.1$, the metal strip thickness $t/h = 0.004$ and the substrate width $p = 2 \cdot l + w = 2.6$. In this investigation the value l/h was not constant.

The narrower the metal strip w, the greater the part of the electromagnetic energy is concentrated in the air around the metal strip and the MSL effective permittivity ε_{MSL}^{ef} and losses (the attenuation constant) h'' in the MSL will become less (Figs.6.61 and 6.62).

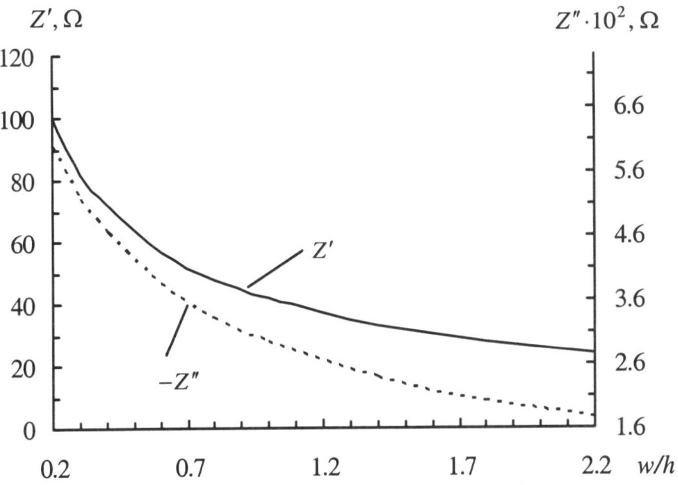

Fig.6.60. *Dependence of the MSL complex wave impedance $\dot{Z}$ on the metal strip width w/h when sizes are: $d/h = 0.1$, $t/h = 0.004$ and the two–layer LiNbO₃–SiO₂ substrate width is constant $p/h = 2.6$.*

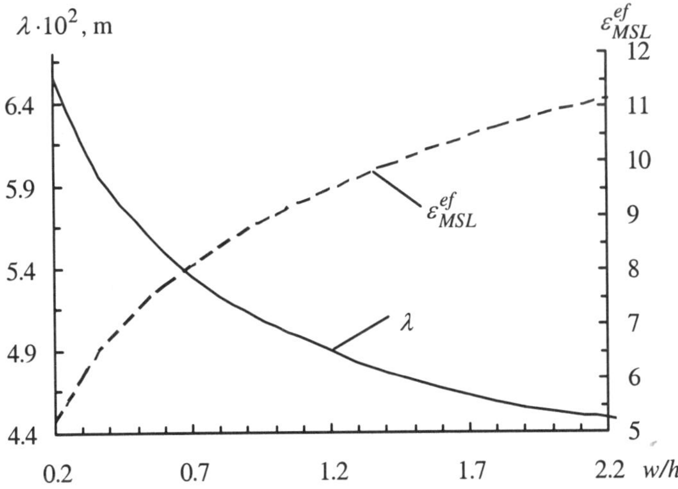

Fig. 6.61. *Dependence of the MSL wavelength λ and effective permittivity ε_{MSL}^{ef} on the metal strip width w/h when sizes are: $d/h = 0.1$, $t/h = 0.004$ and the two–layer LiNbO₃–SiO₂ substrate width is constant $p/h = 2.6$.*

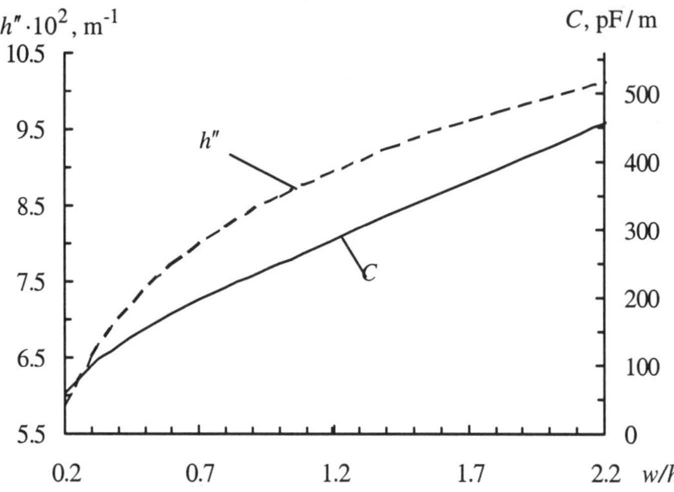

Fig. 6.62. *Dependence of the attenuation constant h'' and capacitance per unit length C of the MSL on the metal strip width w/h when sizes are: $d/h = 0.1$, $t/h = 0.004$ and the two–layer LiNbO$_3$–SiO$_2$ substrate width is constant $p/h = 2.6$.*

Changing the metal strip width w/h in the range of 0.2 to 2.2 the real and imaginary parts of the complex wave impedance $\dot{Z}$ decreased more than 3 times in Fig.6.60. The dependences of the value $\dot{Z}$ on the metal strip width w are important, because it allows us to choose the real part Z' of the impedance within a great range of Ohms. The dependence of the electrodynamic characteristics of the metal strip width w for the MSL (Fig.6.56) is the most important parameter for adjusting the other waveguide tract elements.

The substrate width of the MSL becomes an additional parameter allowing the desired variation of its electrodynamical characteristics (Figs.6.63–6.65). So to get the needed complex wave impedance and complex propagation constant one might only have to change the MSL substrate width. The wider the MSL substrate, the less the real Z' and imaginary Z'' parts of the complex wave impedance is for the investigated MSLs. It is interesting to note that the dependences of Figs.6.63–6.65 are different when we compared them to the same dependences of other MSL in this chapter. We see that the distance l between the metal strip and the lateral substrate edges is also an important parameter that allows the MSL characteristics to change in an unpredictable way.

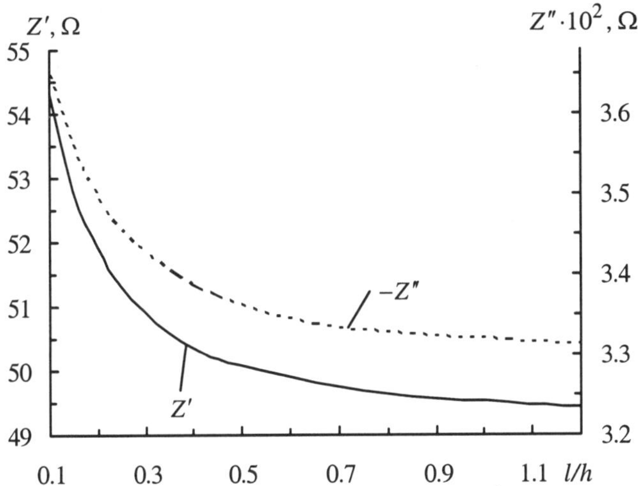

Fig. 6.63. *Dependence of the MSL complex wave impedance* $\dot{Z}$ *on the relative distance* l/h *between the metal strip and the lateral substrate when sizes are:* $w/h = 0.8$, $t/h = 0.004$ *and the two–layer* $LiNbO_3$–SiO_2 *substrate width* p/h *is not constant.*

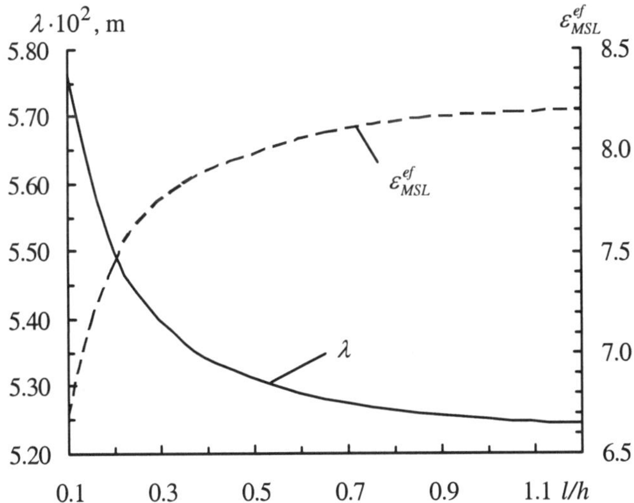

Fig. 6.64. *Dependence of the MSL wavelength* λ *and effective permittivity* ε_{MSL}^{ef} *on the relative distance* l/h *between the metal strip and the lateral substrate when sizes are:* $w/h = 0.8$, $t/h = 0.004$ *and the two–layer* $LiNbO_3$–SiO_2 *substrate width* p/h *is not constant.*

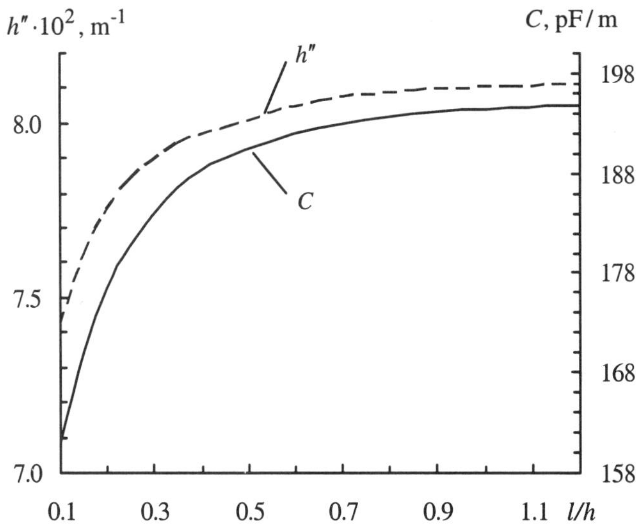

Fig. 6.65. *Dependence of the MSL attenuation constant h'' and capacitance per unit length C on the relative distance l/h between the metal strip and the lateral substrate when sizes are: $w/h = 0.8$, $t/h = 0.004$ and the two–layer LiNbO$_3$–SiO$_2$ substrate width p/h is not constant.*

The dependences on the relative distance l/h for different MSLs illustrate that if someone were to make the assumption that the MSL substrate is infinite this could possibly lead to the wrong calculation particularly if the MSL has a narrow substrate. This also shows that in some calculations it would be best to take into account the real sizes of the narrow substrate.

In this chapter we analysed MSLs with isotropic and anisotropic crystal (*LiNbO$_3$*) substrates. Gyrotropic media are a certain kind of anisotropic media. In the next chapter we will consider MSLs with gyroelectric and gyromagnetic substrates. The common name of gyroelectric and gyromagnetic media are gyrotropic media. Magenized semiconductors and magnetized ferrites are examples of gyrotropic media. So in the next chapter we will analyse electrodynamical characteristics of MSLs with magnetised semiconductors and ferrite substrates.

7. ANALYSIS OF MICROSTRIP LINES WITH GYROTROPIC SUBSTRATES IN TEM–APPROXIMATION BY THE SIE METHOD

7.1. Microstrip Lines with Longitudinally Magnetized Semiconductor Substrates

Longitudinally magnetized semiconductors (semiconductor plasma) are called gyrotropic media, which are characterize by the tensor of permittivity as was shown in chapter 4.

Designing Super High Frequency (SHF) microstrip devices and executing experiments with semiconductor plasma, one often uses electron Indium Antimonide (n–$InSb$) [7.1]. Due to the fact that among the many available materials, n–$InSb$ has a comparatively high free charge carrier mobility (can be up to $\mu_{el} = 80$ m^2/(V·s)) at a temperature of 77 K. Also in our investigations of MSLs with gyrotropic substrates we use semiconductor n–$InSb$ materials. We admitted that electrophysical parameters of n–$InSb$ were: an effective mass $m^* = 0.014 m_{el}$, relative permittivity of lattice $\varepsilon_L = 17.8$ and had the free charge carrier mobility of $\mu_{el} = 40$ m^2/(V·s)).

Since the losses in the magnetized n–$InSb$ are considerable we decided to analyze how MSL losses depend on its geometry and on the magnetic induction B of the external constant magnetic field (see [7.2]–[7.4]). When we analyzed the MSL losses we found an unusual behavior of electrodynamical characteristics of two metal strips MSL with magnetized n–$InSb$ substrate. For example, we discovered the MSL with lossy gyrotropic substrate (chapter 5 section 5.4) has no simple relations between capacitance matrix elements C_{ik} even when the MSL has a geometrical symmetry. To prove the last statement we investigated a MSL (Fig.7.1a) with the sizes: $w/h = 0.5$, $l/h = 2$, $s/h = 0.2$ and $t/h = 0.005$. Calculations were carried out for $B = 1.0$ T, $f = 40$ GHz and $n = 2.5 \cdot 10^{20}$ m^{-3}. Here permittivity tensor components n–$InSb$ substrate were $\varepsilon_{xx} = 18.36 - i0.71$ and $\varepsilon_{xy} = -22.51 - i0.035$. Here we calculated the matrix C_{ik} elements for a two metal strips geometrically symmetric MSL (Fig.7.1a):
$C_{11} = -2.85 + i1.49$, $\qquad$ $C_{22} = 15.86 - i0.77$, $\qquad$ $C_{12} = -10.33 + i5.02$ $\qquad$ and $C_{21} = -8.97 - i5.75$. We see that the elements C_{11} compared with C_{22} and C_{12}

compared with C_{21} were different [7.3]. The difference grows, when the gyrotropy coefficient $G \equiv \left| \varepsilon_{xy} / \varepsilon_{xx} \right|$ grows. In the equation (5.59):

$$LCV = \frac{h^2}{\omega^2} V \, , \qquad (7.1)$$

the matrix LC for a simple two metal strips MSL is:

$$LC = \begin{Vmatrix} a & b \\ c & d \end{Vmatrix} ,$$

where:

$$a = L_{11}C_{11} + L_{12}C_{21} , \quad b = L_{11}C_{12} + L_{12}C_{22} ,$$

$$c = L_{21}C_{11} + L_{22}C_{21} , \quad d = L_{21}C_{12} + L_{22}C_{22} . \qquad (7.2)$$

The elements C_{ik} and L_{ik} are different. We find in (7.1) the eigenvalues:

$$\frac{h^2}{\omega^2} = \frac{1}{2}(a+d) \pm \sqrt{\frac{1}{4}(a-d)^2 + bc} \, . \qquad (7.3)$$

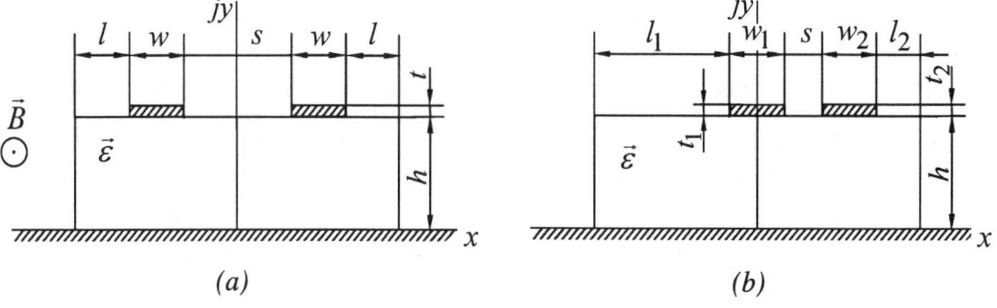

Fig. 7.1. *The open two metal strips MSL with gyrotropic substrate:*
(a) geometrically symmetric and (b) geometrically asymmetric.

In this particular case when: $C_{11} = C_{22}$, $L_{11} = L_{22}$, $C_{12} = C_{21}$, $L_{12} = L_{21}$, then equations $a = d$ and $b = c$. And we write:

$$\left(h/\omega \right)^2 = a \pm b \, . \qquad (7.4)$$

The upper sign in formula (7.4) corresponds to the even wave, when the voltage (and current) in each metal strip are equal ($V_1 = V_2$). The lower sign in formula (7.4) corresponds to the odd wave, when the voltage in each metal strip has the opposite signs:

$$V_1 = -V_2 \, .$$

From equations (7.1), we determine the voltage coupling coefficient:

$$g = \frac{V_2}{V_1} = \frac{\left(h/\omega \right)^2 - a}{b} = \frac{c}{\left(h/\omega \right)^2 - d}$$

for each eigenwaves in (7.3). Similarly, we define the current coupling coefficient $p = I_2/I_1$. The relations:

$$\frac{h}{\omega}V = LI , \qquad \frac{h}{\omega}I = CV ,$$

are often written in this form:

$$V = ZI \qquad \text{or} \qquad I = YV ,$$

defined by the matrices of impedance Z and admittance Y . To illustrate the MSL characteristic dependence on material and geometric asymmetry, Fig. 7.2 shows the dependence of matrix $\dot{C} = C'_{ik} - iC''_{ik}$ elements on the ratio lengths l_2/l_1 for the MSL (Fig. 7.1b). Here we see that for the symmetric MSL with a gyrotropic substrate, when $l_2 = l_1$, the corresponding matrix elements are different. The MSL sizes (Fig. 7.1b) are determined by the ratios $w_1/h = w_2/h = 0.1$, $t_1/h = t_2/h = 0.002$, $s/h = 0.1$ and $l_1/h = 1$. Calculations were carried out for the values $B = 0.4\,\text{T}$, $f = 70\,\text{GHz}$ and $n = 20^{20}\,\text{m}^{-3}$.

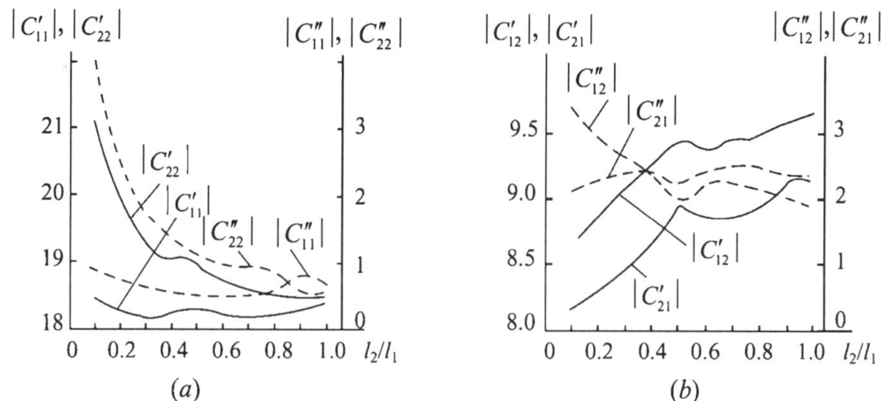

Fig. 7.2. *Dependence of absolute values of real and imaginary parts of the matrix $\dot{C} = C'_{ik} - iC''_{ik}$ elements for geometrically asymmetric MSL (Fig.7.1b).*

7.1.1. One metal strip MSL. The characteristics of microstrip devices can be improved by choosing optimal MSL sizes. The width of the metal strip is one of the parameters that the MSL characteristics depend on.

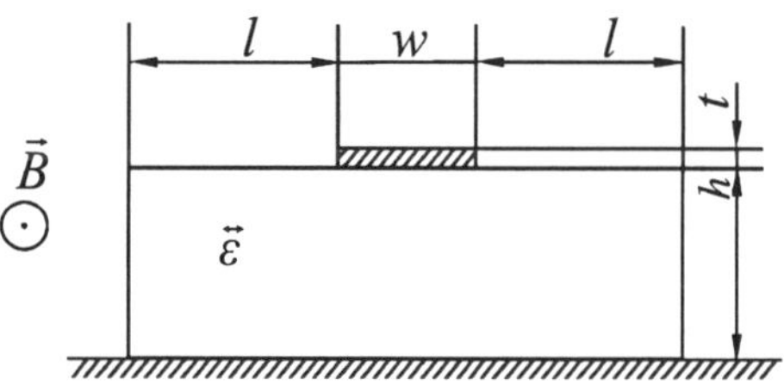

Fig. 7.3. *One metal strip MSL with a longitudinally magnetized semiconductor substrate.*

The dependences of the relative phase shift $\nabla\varphi = \left(\dot{h_1} - \dot{h_2}\right)\Delta L$ and of the losses h'' in the MSL (Fig.7.3) on the metal strip width is shown in Fig.7.4. Here the propagation constant is $\dot{h} = h' - ih''$. The calculation was executed for the magnetic inductions $B_1 = 0.4\,\text{T}$ and $B_2 = 0.8\,\text{T}$. The value h_1' corresponds to the value B_1 and the value h_2' corresponds to the value B_2. The magnitude ΔL is the MSL section length. Calculations of the characteristics were executed for the MSL sizes: $l/h = 1$, $t/h = 0.002$ at the frequency $f = 70\,\text{GHz}$ with an electron concentration $n = 5 \cdot 10^{20}\,\text{m}^{-3}$ in the semiconductor material of the substrate.

In Fig.7.4 we see that the attenuation h'' is lower and the relative phase shift $\Delta\varphi$ is larger for the more narrow metal strips. The MSL losses h'' depend on the electric loss angle δ_{angle} of the substrate material and the area of the contact between the metal strip and the substrate. As it is known the ratio $\varepsilon''/\varepsilon'$ is called a loss tangent $tan\,\delta_{angle} = \varepsilon''/\varepsilon'$. If there is an increase in the width of the metal strip w then the MSL losses h'' will increase. We explain here that the wider the metal strip the greater part of the electromagnetic wave will interact with the substrate material. Because the electric loss angle δ_{angle} of the substrate semiconductor material is greater when compared to the value δ_{angle} of the air around the open MSL.

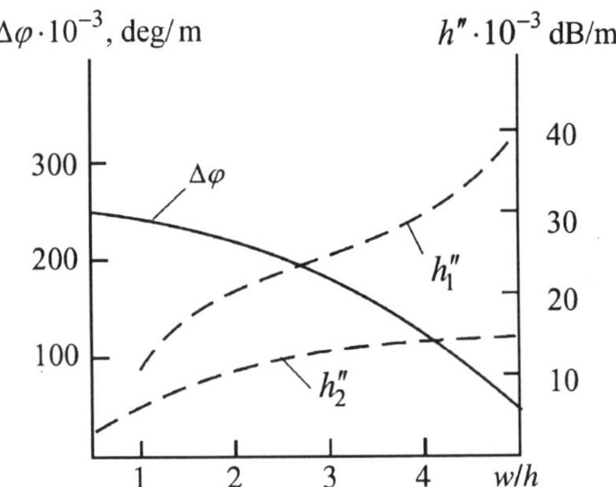

Fig. 7.4. *Dependence of the relative phase shift $\nabla\varphi$ (solid line) and attenuation h_1'' and h_2'' (dash lines) on the normalized width of the metal strip w/h.*

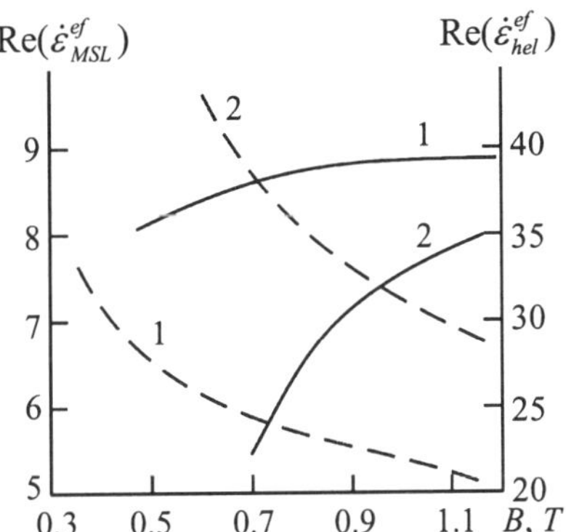

Fig. 7.5. *Dependence of the real part of the effective permittivity* Re($\dot{\varepsilon}^{ef}_{MSL}$) *(solid line) of the MSL (Fig.7.3) and the real part of effective permittivity helicon wave* Re($\dot{\varepsilon}^{ef}_{hel}$) *(dash line) on the value of the magnetic induction B .*

In Fig.7.5, we compare the characteristics of the wave in the MSL (Fig.7.3) and of the helicon wave propagating in the boundless semiconductor. We analyzed the MSLs with the sizes $w/h = 0.2$, $l/h = 1.2$ and $t/h = 0.0025$ at a frequency $f = 35\,\text{GHz}$. The MSL substrate semiconductor material had two electron concentrations $n_1 = 0.5 \cdot 10^{20}\,\text{m}^{-3}$ (line 1) and $n_2 = 1.5 \cdot 10^{20}\,\text{m}^{-3}$ (line 2). Fig.7.5 shows the dependence of the real parts of the effective permittivity Re($\dot{\varepsilon}^{ef}_{MSL}$) of a one metal strip MSL and a helicon wave Re($\dot{\varepsilon}^{ef}_{hel}$) (calculated by the formulae [4.11]) on the value of the magnetic induction B . When the magnetic induction B of the external constant magnetic field increases one observes a saturation of the dependences Fig.7.5, which is greater when the electron concentration is less. We see that the character of these curves Re($\dot{\varepsilon}^{ef}_{MSL}$) and Re($\dot{\varepsilon}^{ef}_{hel}$) are different in the fact that they run in the opposite direction from each other. When the magnetic induction increases the Re($\dot{\varepsilon}^{ef}_{hel}$) of the helicon wave decreases and the Re($\dot{\varepsilon}^{ef}_{MSL}$) of the MSL increases [7.5].

7.1.2. Comparison of our calculated results with experimental data. In Fig.7.6 we compare the calculated results and the experimental data [7.6]. Here we see that the results agreed with each other.

We investigated three MSLs each having the same sizes $w = 0.13 \cdot 10^{-2}\,\text{m}$, $l = 0.52 \cdot 10^{-2}\,\text{m}$, $t = 10^{-5}\,\text{m}$, the length of these MSLs (Fig.7.3) was $\Delta L = 2 \cdot 10^{-2}\,\text{m}$. However, the charge carrier concentration as well as the substrate thickness was different for each MSL. The first MSL (on the top of Fig.7.6) had parameters $n = 10^{20}\,\text{m}^{-3}$ and $h = 0.8 \cdot 10^{-3}\,\text{m}$. The second MSL had parameters of $n = 10^{20}\,\text{m}^{-3}$ and $h = 0.13 \cdot 10^{-2}\,\text{m}$.

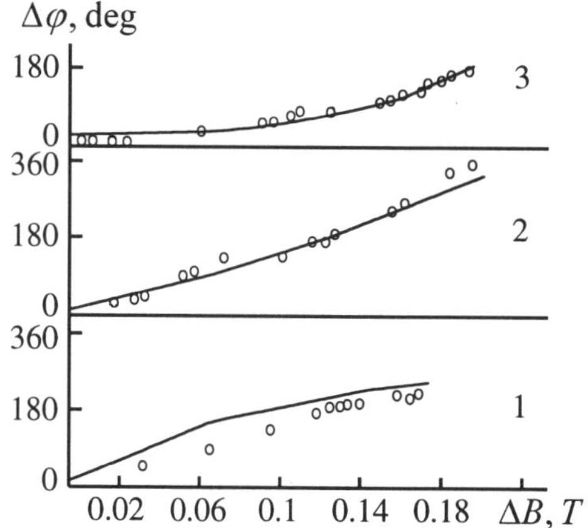

Fig. 7.6. *Dependence of the relative wave phase shift* $\Delta\varphi$ *on the magnetic induction change* ΔB *for three MSLs.*

The third MSL had parameters of $n = 0.5 \cdot 10^{20}$ m^{-3} and h $= 0.13 \cdot 10^{-2}$ m. In Fig.7.6 our calculations are shown by lines and the experimental results are designated by circles. The theoretical determination of the relative phase shift consists of the calculation of the wave propagation constants and the application of the formula $\Delta\varphi = (h_1' - h_2')\Delta L$. The values h_1' and h_2' are the real parts of the propagation constant at two magnetic induction values B_1 (starting point of our counting) and B_2, where $\Delta B = B_1 - B_2$ is the change of the magnetic induction B of the external constant magnetic field.

The experimental points of Fig.7.6 are determined by the interference picture of two harmonic signals, where the phase of the extreme point of the picture at the largest magnetic induction is the zero phase count point. Measurements were executed by use of the interference method [7.7] at the frequency 9.3 GHz and a temperature of 77 K. The interference characteristics of two signals were measured as a function of the magnetic induction B of the external constant magnetic field. One of the signals was constant and the other signal, which passed through the MSL, was not constant. The other signal was changed by the magnetic field. We chose a counting point in the experiment interference picture to begin comparing our calculations. This count point corresponded with the gyrotropy of the MSL substrate, when the gyrotropy was the least. The gyrotropy is less when the magnetic induction B of the external constant magnetic field is greater. So the count point corresponded with the greatest value of the magnetic induction in our experiments. Our calculations and experiments agreed.

7.1.3. Coupled MSLs. Here we will consider the characteristics of the coupled geometrically symmetric two metal strips MSL with the longitudinally magnetized semiconductor substrate as a function of the distances between the metal strips at

$f = 40\,\text{GHz}$, $B = 1\,\text{T}$ and $n = 10^{20}\,\text{m}^{-3}$. The MSL (Fig.7.1$a$) relative sizes are: $w/h = 0.5$, $l/h = 2$ and $t/h = 0.005$. As you know the coupled MSL when it has an even excitation the charges of both metal strips have the same signs. And the coupled MSL when it has an odd excitation the charges of both metal strips have the opposite signs.

Fig.7.7 shows the dependence of the eigenwave propagation constants $\dot{h}_{1,2}$ on the value s/h [7.3]. As one can see from Fig.7.7, when the distances between the metal strips are small (in the case of a strong coupling) the losses of the first wave are larger than that of the second wave. Here the first wave turns into the even wave after taking off the external constant magnetic field.

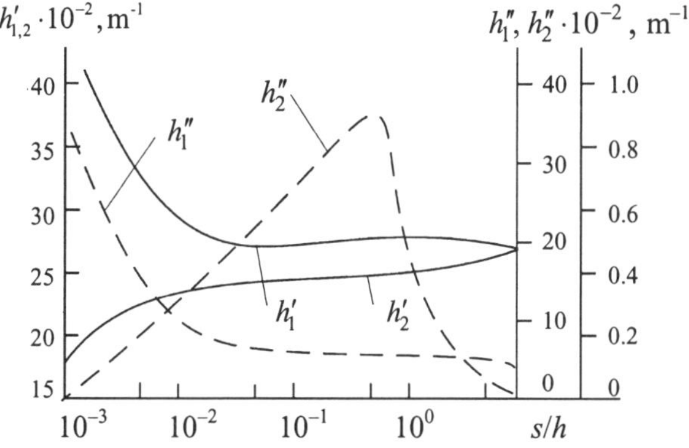

Fig. 7.7. *Dependence of eigenwave propagation constants* $\dot{h}_{1,2} = h'_{1,2} - ih''_{1,2}$ *on the normalized distance* s/h *between the metal strips.*

For an even excitation of a two metal strips MSL a greater part of the wave energy is propagating in the volume between the earth plate and the metal strips MSL. In the semiconductor (dissipative plasma) substrate the losses of the first eigenwaves are considerable. In case of an odd excitation the wave energy is concentrated mainly in the gap between the metal strips (having opposite charges), which lead to the lower losses as compared to the first wave. Here Fig.7.7 shows that the second wave losses have a maximum. The growth of the losses before the maximum is explained by the diminishing of the coupling between the metal strips, when the distance s is increased. This results in the wave being drawn into the semiconductor, where it is intensively absorbed. In the region after the maximum the coupling between the metal strips become weak and the MSL can be considered as two not interacting one metal strip MSLs. In these two MSLs the losses are smaller than in the strongly coupled MSLs. In this last circumstance we see that in a one metal strip MSL the wave is concentrated in the volume under the strip but in the coupled two metal strips MSL it is concentrated under the metal strips and above the gaps between the metal strips.

For $s/h \to \infty$ the characteristics of both propagating waves merge and become the same as that of a one metal strip MSL.

7.1.4. Losses in geometrically asymmetric MSLs. The investigation of geometrically asymmetric MSLs with dielectric substrates were investigated in [7.8]–[7.10]. Here we analyzed the losses in geometrically asymmetric MSLs with magnetized semiconductor substrates. On the bases of these MSLs it was possible to created devices controlled by magnitude of the external constant magnetic field [7.4].

When we investigated these MSLs (Fig.7.8) we saw it was possible to optimize the electrodynamical characteristics of certain devices, which were created on the basis of these MSLs. In optimizing these MSLs we had to choose certain widths or thicknesses of metal strips at a fixed magnitude of the external constant magnetic field.

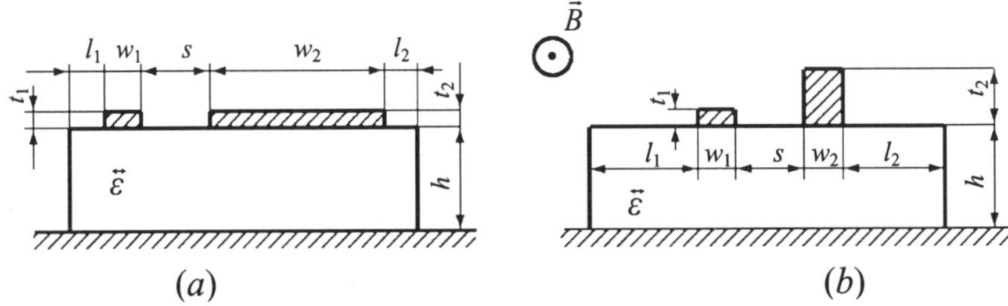

Fig. 7.8. Cross–section of the geometrically asymmetric MSL with the longitudinally magnetized semiconductor substrate (a) with metal strips of different widths and (b) with metal strips of different heights.

Fig.7.9 demonstrates the dependences of eigenwaves losses on the ratio of the metal strip widths w_2/w_1 for the MSL with the normalized sizes $w_1/h = 0.1$, $l_1/h = l_2/h = 1$, $t_1/h = t_2/h = 0.002$, $s/h = 0.1$ (Fig.7.8a) at $B = 0.4T$, $f = 70\,\text{GHz}$, $n = 10^{20}\,\text{m}^{-3}$. One can see from Fig.7.9 that choosing the ratio w_2/w_1 allows us to regulate the eigenwave losses in the MSL.

The dependence of eigenwave losses in the asymmetric MSL (Fig.7.8b) on the ratio t_2/t_1 is shown in Fig.7.10. The MSL is characterized by the ratios $w_1/h = w_2/h = 0.1$, $l_1/h = l_2/h = 1$, $t_1/h = 0.002$ and $s/h = 0.1$. The values B, f and n are the same as in the previous case ($B = 0.4T$, $f = 70\,\text{GHz}$, $n = 10^{20}\,\text{m}^{-3}$). It follows from Fig.7.10 that at $t_2/t_1 \approx 30$ the losses of eigenwaves differ by three orders. When the ratio t_2/t_1 is larger then 30 one of these eigenwaves will propagate with very small losses, while the other eigenwaves increases rapidly.

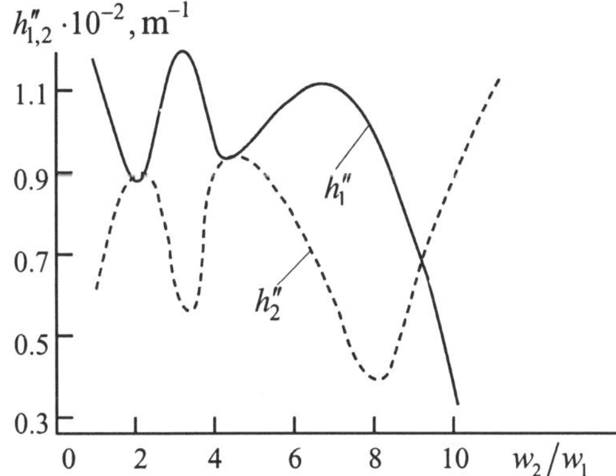

Fig. 7.9. *Dependence of eigenwave losses h_1'' and h_2'' in the asymmetric MSL (Fig.7.1a) on ratio of the metal strip widths w_2/w_1.*

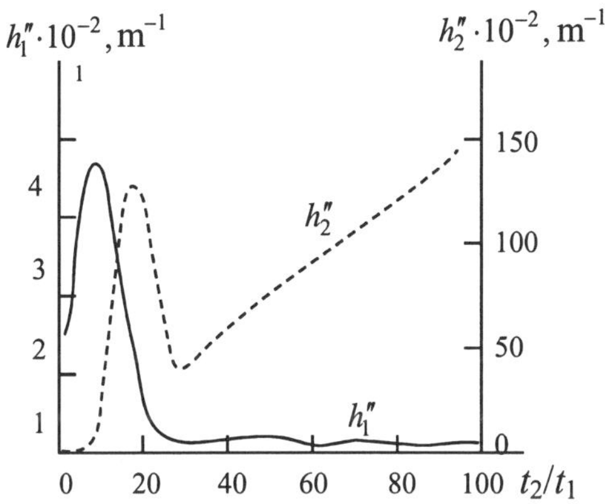

Fig. 7.10. *Dependence of eigenwave losses h_1'' and h_2'' in the asymmetric MSL (Fig. 7.1b) on the ratio of the metal strip thickness t_2/t_1.*

7.2. Meander Line on a Ferrite Substrate

The meander line (Fig.7.11) consisting of N periods on the ferrite substrate is a structure controlled by the magnitude of the external constant magnetic field H_0. We will now consider the case of the longitudinal magnetization when the external constant magnetic field is directed along the length l. To calculate the signal phase shift of the meander line we studied a regular MSL having three interacting metal strips on a longitudinally magnetized ferrite substrate (Fig.7.12) in TEM–approximation. The inductances per unit length L_{ik} (i and $k = 1,2,3$) are depend on the MSL geometry, the signal frequency and the external constant magnetic field. The permeability tensor elements are functions of the frequency f and the external constant magnetic field strength H_0.

When we have the full matrix of the inductances per unit length, it is easy to determine propagation constants of eigenwaves and the speed of the signal propagation along the meander line. Thus the signal propagation speed in the meander line along the metal strip is calculated when we taken into account the two neighboring metal strips on the left side and the two neighboring metal strips on the right side. Naturally the speed depends on the position of the cross–section $z = \mathrm{const}$ in the meander line (Fig.7.11).

The regular MSL with the ferrite finite size substrate and with three strips of finite thicknesses as shown in Fig.7.12 were calculated by the SIE method. The comparison of calculated results and experimental data prove the correctness of the TEM–approximation [7.11].

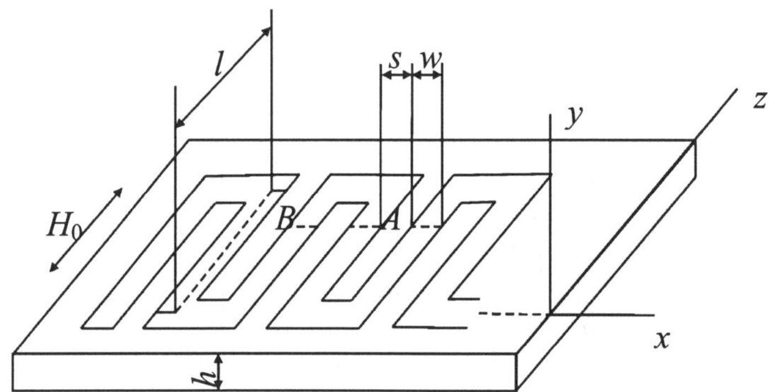

Fig. 7.11. *The microstrip meander line on the ferrite substrate.*

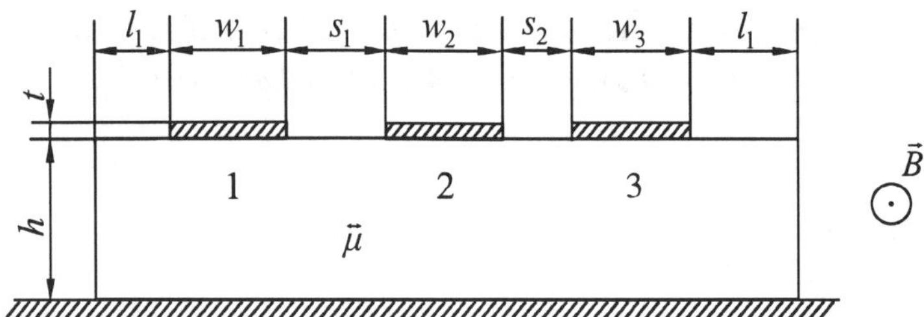

Fig. 7.12. *Cross–section of a regular MSL having three interacting metal strips (where $w_1 = w_2 = w_3 = w$ and $s_1 = s_2 = s$).*

The ferrite gyrotropy $\ddot{\mu}$ of MSL substrate causes the inductance matrix elements to become complex values $\dot{L}_{ik} = L'_{ik} - iL''_{ik}$. When the external constant magnetic field is absent the tensor (4.8) changes into the diagonal form and all the elements $L''_{ik} = 0$. The external magnetic field H_0 causes a separate direction for the electromagnetic wave to propagate in the ferrite substrate. Therefore, the middle strip "feels" its left and right metal strip adjacent to it in different ways, because $L_{ik}'' = -L_{ki}''$ for $H_0 \neq 0$.

Fig.7.13 shows the dependences of certain elements of inductance matrix on the value of the external constant magnetic field strength H_0 for a MSL with a ferrite substrate. The ferrite material of the substrate was 40SC-2. The elements of the relative permeability tensor (4.8) were calculated by use of the formulae [7.12]. We did not take into account the losses of the substrate material. When we calculated the phase shift of the signal that passed though the meander line (with n-periods) we admitted that all meander periods were in the same identical condition. The largest error in our calculations is in the beginning and ending periods of the meander line because they only had adjacent metal strips on one side. The propagation constant and phase shift of the meander sections that were normal to the external constant magnetic field strength H_0 were calculated as a one metal strip MSL with a demagnetized ferrite substrate. We use this approach for two reasons: 1) the length of these sections $\left(2N\left(w+s\right)\right)$ was less by one order than the length $\left(2N\cdot l\right)$ of the main meander elements; 2) according to [7.13], the phase shift in the case of the transversal magnetization of a MSL only depends a little on the external constant magnetic field strength H_0. The full phase shift of the signal after passing through the meander line was determined from the phase of its complex amplitude output (when the signal phase in input is zero). We calculated the amplitudes of all the eigenwaves that were described in chapter 5 (section 5.7).

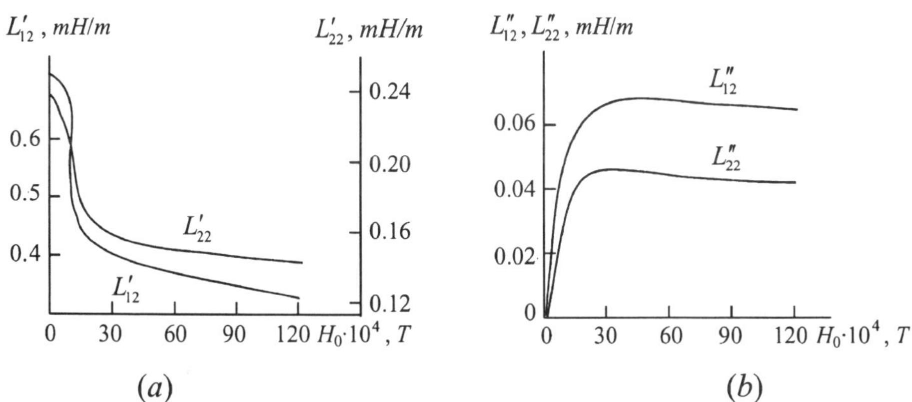

Fig. 7.13. *Dependence of the inductance matrix elements on the external constant magnetic field H_0 for the MSL (Fig.7.12) when the sizes are: $w/h = 0.7$, $t/h = 0.005$ and $s/h = 0.3$.*

In this way, we calculated the phase shift φ_0 when the substrate ferrite was in the demagnetized state. Then we calculated the phase shift φ_1 for the direction of the external constant magnetic field H_0. If this field increased then the phase shift would also increase. In addition, we calculated the phase shift φ_2 when the external constant magnetic field H_0 went in the opposite direction. The full phase shifts were calculated by the formulae: $\varphi^+ = \varphi_0 - \varphi_1$ and $\varphi^- = \varphi_0 - \varphi_2$.

Fig.7.14 shows the calculated signal propagation constant (obtained from the energy speed) dependence $h' = h'(z)$ on the position of the cross–section for the meander line on the substrate of the ferrite 60 SCh in three magnetization states. In Fig.7.14 the difference of the propagation constants for two opposite external constant magnetic field directions, (which defines the value of the controlled phase shift), is changed along the distance l and there is a maximum of the phase shift in the middle of the long section of the meander line. The result is explained by the change of the SHF–field polarization ellipticity along the direction of the external constant magnetic field.

Fig.7.15 presents the experimental and calculated phase shifts φ^+ and φ^- for the meander line on the substrate of the ferrite 40 SCh–2 with $N = 15$ as a function of the external constant magnetic field strength. In Fig.7.16 we show a similar dependence for the meander line on the substrate of the ferrite 60 SCh with $N = 14$. In Fig.7.17 we show the phase shifts $\varphi_\Sigma = \varphi^+ - \varphi^-$ as a function of the SHF–signal wavelength.

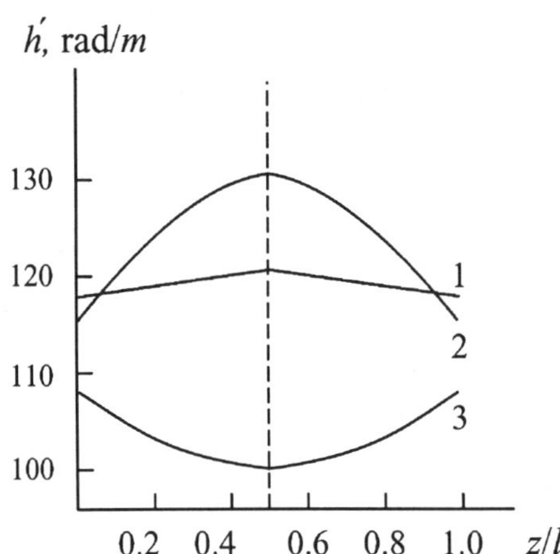

Fig. 7.14. Dependence of the propagation constant in the meander line (Fig.7.11) on the coordinate z: curve 1 was calculated for $H_0 = 0$, curve 2 was calculated $H_0 = +30 \cdot 10^{-4} T$ and curve 3 was calculated $H_0 = -30 \cdot 10^{-4} T$. The sizes of the meander line are: $w/h = 0.6$, $t/h = 0.005$, $s/h = 0.3$ and $l = 0.108\lambda_0$.

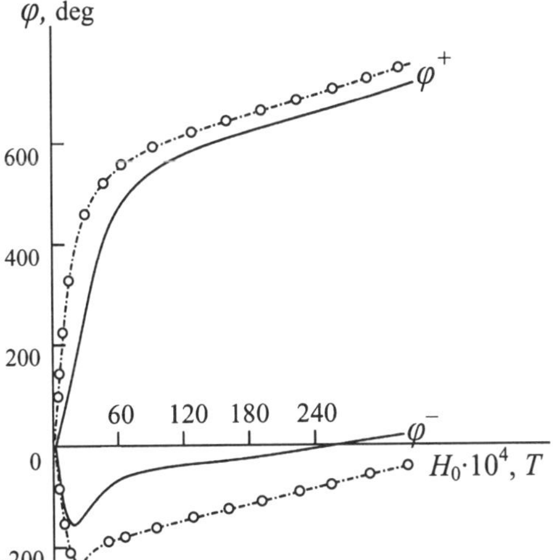

Fig. 7.15. *The meander line phase shift dependence on the external constant magnetic field H_0. Our calculations are solid lines and experimental results are the dash lines with circles. The sizes of the meander line are: $w/h = 0.7$, $s/h = 0.3$, $t/h = 0.005$, $l = 0.1065\lambda_0$ and $h = 0.015\lambda_0$.*

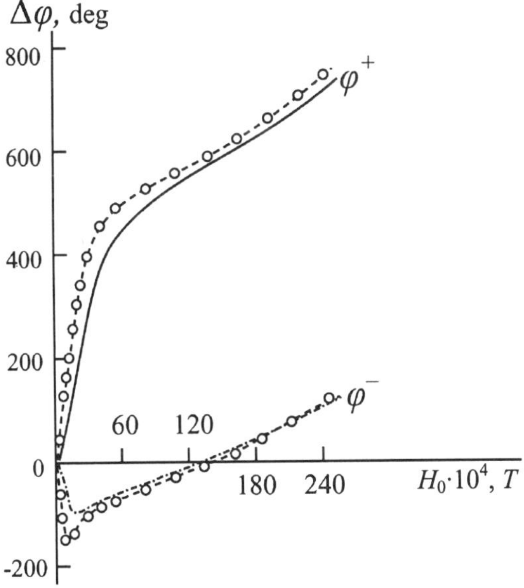

Fig.7.16. *Dependence of the phase shift in the meander line on the external constant magnetic field H_0. Our calculations are solid lines and the experimental results are the dash lines with circles. The sizes of meander line are: $w/h = 0.6$, $s/h = 0.3$, $t/h = 0.005$, $h = 0.0087\lambda_0$, $l = 0.1075\lambda_0$.*

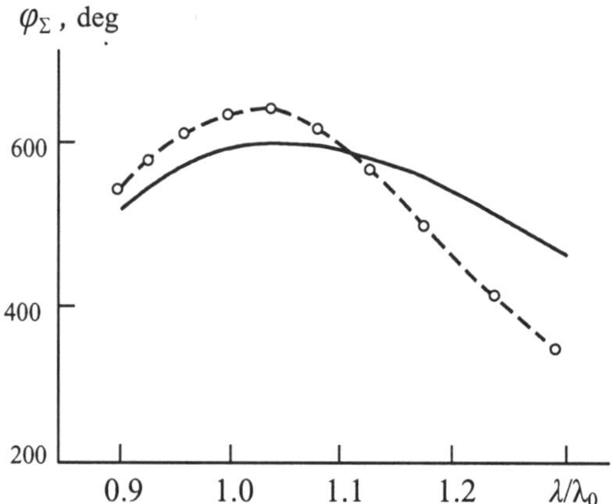

Fig. 7.17. *Dependence of the phase shift φ_Σ on the wavelength ratio λ/λ_0 at the external constant magnetic field $H_0 = 60 \cdot 10^{-4} T$. Our calculations are a solid line and the experimental result is the dash line with circles. The sizes of meander line are: $w/h = 0.6$, $s/h = 0.3$, $t/h = 0.005$, $h = 0.0087\lambda_0$ and $l = 0.1075\lambda_0$.*

The input impedance Z_0 in the meander line depends on its geometry, the load impedance Z_1 and the magnetization extent and frequency. At a certain frequency, which is called the cutoff frequency, the input impedance Z_0 becomes an imaginary number (a reactance) and the current standing–wave ratio (CSWR) increases sharply. And here Z_1 is a real number (a resistance). Fig.7.18 shows the theoretical and experimental dependence of the cutoff frequency on the length l. Table 7.1 compares the calculated and measured cutoff frequencies at the different magnitudes of the external constant magnetic field for the meander line on the substrate of the ferrite 60 SCh when sizes are: $h = 10^{-3} m$, $w/h = 0.7$, $s/h = 0.5$, $l/h = 11.4$ and $t/h = 0.007$. It is worthy to note that $\lambda_{cut}/4 \prec l$.

The SIE method enables one to determine the value and orientation of the magnetic induction vector at any point of the meander line cross–section. The density of the equivalent current (which is concentrated on the metal surface) is proportional to the tangent component of the magnetic field (as in [7.14]).

Fig.7.19 shows the current density distribution (in relative units) on the lower and upper strip sides and on the metal plate in the cross–sections $z = l/2$ (a) and $z = 3l/4$ (b) for three values of the external constant magnetic field. The meander line sizes are: $w/h = 0.6$, $s/h = 0.3$, $t/h = 0.005$ and $l/h = 12$. The substrate was made of ferrite material 60 SCh.

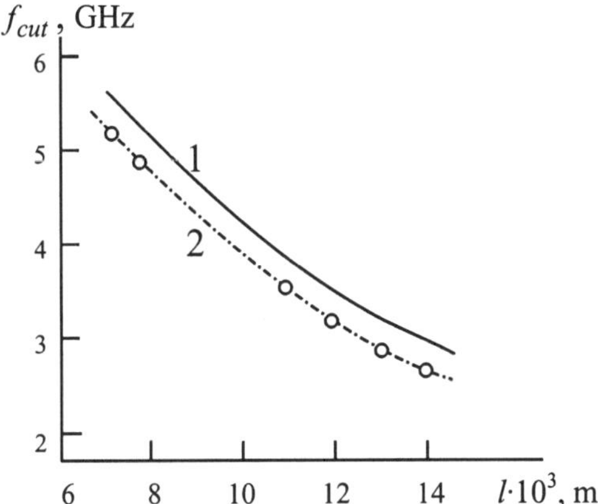

Fig. 7.18. *The results of calculations and experiments for the meander line with parameters* $w/h = 0.7$, $t/h = 0.007$, $s/h = 0.5$ *and* $h = 10^{-3}$ m. *Calculations are shown by the solid line and the experimental result is dash line with circles.*

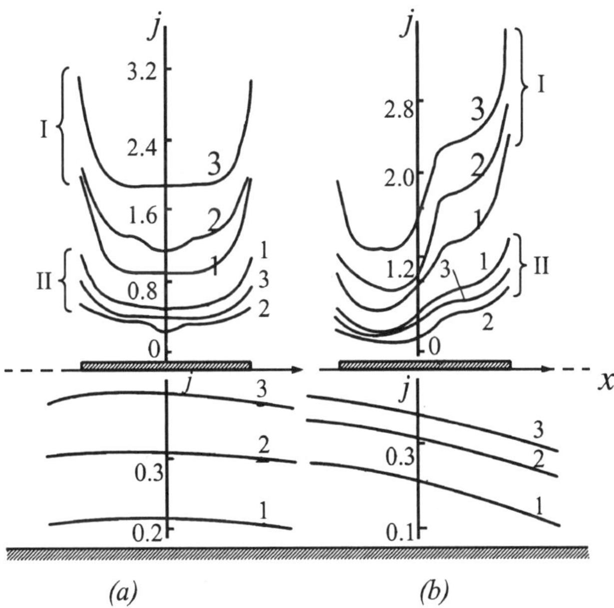

Fig. 7.19. *The current density distribution on the lower strip side is designated as Roman No I and the upper strip sides are designated as Roman No II. The current density distribution on the metal plate correspond as: curve 1,2,3 when* $H_0 = 0$; $H_0 = -120 \cdot 10^{-4}$ T *and* $H_0 = +120 \cdot 10^{-4}$ T.

Table 7.1. The cutoff frequencies at different magnitudes of the external constant magnetic field strength H_0 for the meander line on the ferrite substrate.

The external constant magnetic field strength H_0 (T)	f_{cut} (GHz)	
	calculated	experiment
0	3.6	3.4
$+30 \cdot 10^{-4}$	3.8	3.55
$-30 \cdot 10^{-4}$	4.0	3.7

Here the sharp breach of the symmetry in the current density distribution for the cross–section is displaced from the symmetry plane $z = l/2$. Approaching the metal strip edge the current density must increase by the formula $r^{-\psi}$, $0 \prec \psi \prec 1$, where r is the distance from the metal strip edge. These calculations are only approximate and we see that the dependences near the metal strip edge are smooth.

If the current distribution over the conductor surface is known the attenuation coefficient is determined by the formula [7.15]:

$$h'' = \frac{8.68R}{2Z} \int_L \frac{|j_d|^2 \, ds}{|I|^2} \cdot 10^2 \quad (dB/m) \, . \tag{7.5}$$

Here Z is the characteristic line wave impedance and $R = \sqrt{\pi f \mu_{met} \rho}$ is the surface impedance of the metal skin–layer (Ohms/square). In formula (7.5) the value j_d is the current density and I is the full current. Here μ_{met} is the metal permeability, ρ is the metal specific resistivity and f is the frequency. It was assumed that the conductor thickness was several times larger than the skin–layer thickness. The condition is satisfied in this case, because the strips are made of copper and they thicknesses are $t/h = 0.007$, where $h = 10^{-3}$ m. The integration in the formula (7.5) is carried out over the strip contour and under the strip contour.

Since the current density distribution and wave impedance depend on the cross–section position in the meander line, then the attenuation coefficient depends on the coordinate z (Fig.7.11). This dependence appeared to be weak and it was enough to calculate the coefficient α at three points on the long section of the meander line ($z = l/4$, $z = l/2$, $z = 3l/4$) and to take the average value h_l''. In the normal section of the meander line, the attenuation coefficient was calculated as for the one metal strip MSL with a magnetodielectric substrate. The attenuation coefficient h_c'' of the total meander line was determined as the average:

$$h_c'' = \frac{h_l'' l + h_s'' s}{l + s} \cdot 10^2 \quad (dB/m) \, .$$

The comparison of calculated losses in metal conductors with the measured total losses of the meander line at two frequencies is shown in Fig.7.20. One can see that the losses in metal make up a considerable part of the total losses but the cause of the intersection at the experimental curve is not clear. And that the total losses (at the small value distance s) at a lower frequency is higher compared to the losses at a higher frequency. Note that here the losses in the ferrite SCh–B and the radiation losses were not taken into consideration.

An experimental determination of the controlled phase shift was calculated on the measurement stand (Fig.7.21). The external constant magnetic field oriented along the section of the length l in the substrate plane Fig.7.11. The value of the controlled phase shift $\varphi_{control}$ was determined by the displacement of a minimum standing wave in the measuring line,

$$\varphi_{control} = \frac{720^\circ}{\lambda}\Delta x,$$

where λ is a wavelength in the measuring line and Δx is the displacement of the standing wave minimum.

The measurements were calculated for two positions of the expanding coaxial line which lengths were different from each other by $\lambda/2$, then the arithmetic average of the two measurements were taken. The mean–square error of the controlled phase shift measurement did not exceed $\pm 3^\circ$. The qualitative and quantitative concordance of the calculation and the experiment were within the norm.

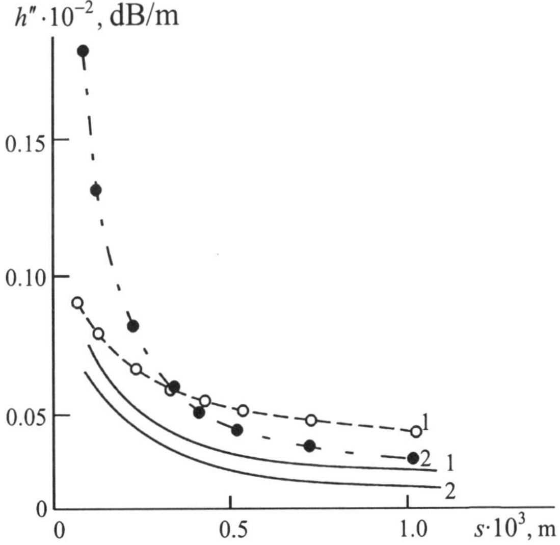

Fig. 7.20. Dependences of losses in metal conductors as a function of distances s at two corresponding wavelengths: $\lambda = \lambda_0$ (curve 1) and $\lambda = 1.2\lambda_0$ (curve 2). Our calculations are shown as solid lines and the experimental results are seen as dash lines with circles and points. The MSL sizes are: $w/h = 0.6$, $t/h = 0.005$, $l/h = 12$ and $h = 10^{-3}$ m.

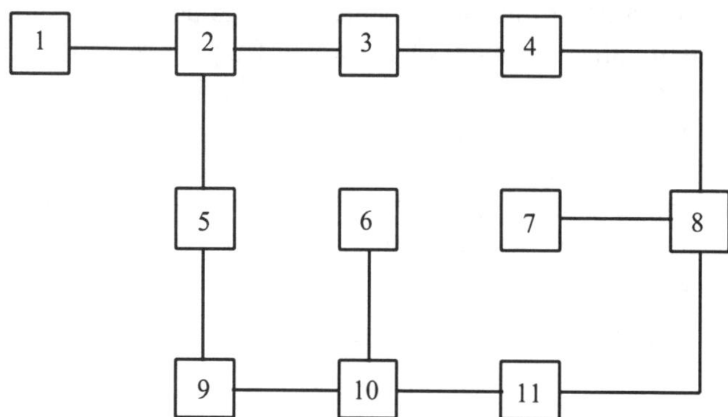

Fig. 7.21. *Schematic structure to measure the phase characteristics of a meander line: 1 is a generator of the SHF signal; 2 is the coaxial T–junction; 3, 9, 11 are the coaxial valves; 4 is the expanding coaxial line; 5 is the attenuator; 6 is the indicator; 7 is the voltage source; 8 is the meander structure and 10 is the measuring line.*

The measurements of SHF–energy losses and of CSWR on the meander line were calculated with standard equipment like the panorama CSWR measurer of the type P2–38 and P2–42. In this last case, the waveguide–microstrip transitions were used.

The main sources of divergence of the calculated and experimental data are: 1) the approximate character of the formulae by which the permeability tensor elements $\ddot{\mu}_{ik}$ (on the signal frequency and the value of the magnetic field strength H_0) were calculated; 2) the imaginary parts $\mathrm{Im}\,\dot{C}_{ik}$ of the capacitance matrix were equalized to zero; 3) the dependence of the phase shift on the magnetic field strength H_0 was not taken into account for meander sectors that were normal to H_0; 4) the meander line corners were not taken into account.

7.3. Microstrip Lines with Transversally Magnetized Ferrite Substrates in the Approximation of Harmonic Functions

There are several papers on the theoretical study of MSLs in the case of magnetization, which is normal to the metal plate (Fig.7.22a). In [7.16] and [7.17] the presence of a screen is necessary. It is assumed in [7.16] that the metal conductor is $w \gg h$ and that the field structure does not depend on the coordinate y. The problem is simplified after these approaches and the solution to this problem allows one to determine the propagation constants exactly. However, one can only make a general conclusion about the field structure. In article [7.17] to determine the propagation constant they used the "mode matching" method [7.18]. Also in [7.17] besides the screen they required the strip to be infinitely thin and the ferrite to have infinite sizes in the direction of the x axis. In [7.19] they applied two methods (the partial region method with Fourier transformation, the Galerkin method) and the same restrictions are imposed here.

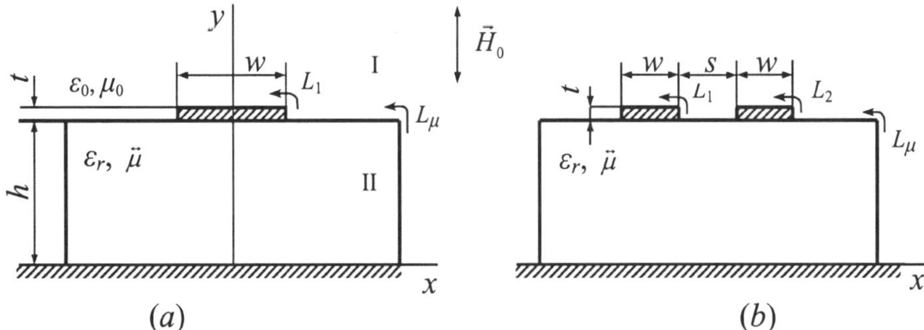

Fig. 7.22. *Cross–section of (a) one metal strip MSL and (b) two metal strips MSL with the transversally magnetized ferrite substrate.*

We will now use the SIE method to solve this problem. In the case of the transversal magnetization, it is necessary to take into account the longitudinal components of the electric and magnetic fields. The transversal components from Maxwell's equations are expressed through the longitudinal components. The longitudinal components in ferrite medium satisfy the system of coupled differential equations of the second order. The TEM–approximation describes the electromagnetic process well for some (low enough) frequencies. One neglects in this system of coupled differential equations all the terms apart from those having the highest derivatives and only the Laplace's operator in the cross–section plane remains. The harmonic functions are used as the solutions in this case. The qualitative valuation of the terms in the differential equations for the longitudinal components shows that the harmonic functions can be used as the solutions. Thus the longitudinal components are written in the form of the contour integrals, which satisfies Laplace's equation and have the unknown function μ of the distribution density. The boundary conditions on the metal surfaces and the surfaces dividing the ferrite and the air lead to the SIE method, which was solved by the Krylov–Bogoliubov method [3.5]. The solvability condition is written down equalizing the determinant of the corresponding system of the linear homogeneous equations to zero. The root of the obtained equation determines the propagation constant depending on the MSL geometry, the value of the external constant magnetic field strength H_0 and the frequency f (Fig.7.22).

This method enables one to calculate the field configuration in the cross–section of the MSL. Note once more that our solution algorithm does not depend on the quantity of the metal strips and the form of the MSL cross–section. This means that we can calculate MSLs with very complicated cross–sections.

7.3.1. The differential and integral equations. The longitudinal field components obey in the region I (Fig.7.22a) the differential equations (1.4), where $\varepsilon = \mu = 1$ are the relative permittivity and permeability of a vacuum. In the region II, the ferrite permeability (4.8) is described by the tensor in this form:

$$\ddot{\mu} = \mu_0 \begin{pmatrix} \mu_1 & 0 & i\mu_a \\ 0 & \mu_y & 0 \\ -i\mu_a & 0 & \mu_1 \end{pmatrix}. \tag{7.6}$$

For the region II, we find from Maxwell's equations:

$$\frac{\partial^2 E_z}{\partial y^2} + \frac{k^2 \varepsilon_r \mu_1 - h^2}{k^2 \varepsilon_r \mu_y - h^2} \cdot \frac{\partial^2 E_z}{\partial x^2} - \omega\mu_0\mu_a \frac{\partial H_z}{\partial y} +$$

$$+ \frac{\omega h \mu_0 (\mu_1 - \mu_y)}{k^2 \varepsilon_r \mu_y - h^2} \cdot \frac{\partial^2 H_z}{\partial x \partial y} + (k^2 \varepsilon_r \mu_1 - h^2) E_z = 0, \tag{7.7}$$

$$\frac{\partial^2 H_z}{\partial x^2} + \frac{k^2 \varepsilon_r \mu_1 - h^2}{k^2 \varepsilon_r \mu_y - h^2} \cdot \frac{\partial^2 H_z}{\partial y^2} + \frac{\omega\varepsilon_0 \varepsilon_r \mu_a}{\mu_1} \frac{\partial E_z}{\partial y} +$$

$$+ \frac{\omega h \varepsilon_0 \varepsilon_r (\mu_1 - \mu_y)}{\mu_1 (k^2 \varepsilon_r \mu_y - h^2)} \cdot \frac{\partial^2 E_z}{\partial x \partial y} + (k^2 \varepsilon_r \frac{\mu_1^2 - \mu_a^2}{\mu_1} - h^2) H_z = 0, \tag{7.8}$$

where ε_r is the relative ferrite permittivity. Note that for the saturated ferrite magnetization (when $M = M_s$) the tensor components $\mu_1 = \mu_y$. For $M \prec M_s$ one can see (from the formulae for the nonsaturated ferrite tensor elements [7.12]) that the ratio $(\mu_1 - \mu_y)/\mu_1$ is always less then 0.06. Therefore it is reasonable to assume in the equations (7.7) and (7.8) that $(\mu_1 = \mu_y)$. Then we obtained:

$$\frac{\partial^2 E_z}{\partial x^2} + \frac{\partial^2 E_z}{\partial y^2} - \omega\mu_0\mu_a \frac{\partial H_z}{\partial y} + (k^2 \varepsilon_r \mu_1 - h^2) E_z = 0,$$

$$\frac{\partial^2 H_z}{\partial x^2} + \frac{\partial^2 H_z}{\partial y^2} - \omega\varepsilon_0 \varepsilon_r \frac{\mu_a}{\mu_1} \frac{\partial E_z}{\partial y} + (k^2 \varepsilon_r \frac{\mu_1^2 - \mu_a^2}{\mu_1} - h^2) H_z = 0. \tag{7.9}$$

The transversal components of the electromagnetic field in ferrite are expressed through the longitudinal components by the formulae:

$$E_x = \frac{1}{k^2 \varepsilon_r \mu_1 - h^2} \left(-i\omega\mu_0\mu_1 \frac{\partial H_z}{\partial y} - ih \frac{\partial E_z}{\partial x} \right),$$

$$E_y = \frac{1}{k^2 \varepsilon_r \mu_1 - h^2} \left(i\omega\mu_0\mu_1 \frac{\partial H_z}{\partial x} + ih\omega\mu_0\mu_a H_z - ih \frac{\partial E_z}{\partial y} \right),$$

$$H_x = \frac{1}{k^2 \varepsilon_r \mu_1 - h^2} \left(-ih \frac{\partial H_z}{\partial x} + ik^2 \varepsilon_r \mu_a H_z + i\omega\varepsilon_0 \varepsilon_r \frac{\partial E_z}{\partial y} \right),$$

$$H_y = \frac{1}{k^2 \varepsilon_r \mu_1 - h^2}\left(-ih\frac{\partial H_z}{\partial y} - i\omega\varepsilon_0\varepsilon_r\frac{\partial E_z}{\partial x}\right). \tag{7.10}$$

The transversal components in region I are determined by the formulae (1.3), which coincided with the formulae (7.10), if we use $\mu_a = 0$. We used the harmonic functions (satisfying Laplace's equation) as the solutions of the equations (7.9). When verifying the accuracy of our approximation all the terms in the equations (7.9) were calculated and compared with each other. The assumption taken here proved correct for the band of the frequencies we used. We presented the components E_z and H_z as the harmonic functions:

$$E_z = \mathrm{Re}\int_L h^E(t)\ln(z-t)ds ,$$

$$H_z = \mathrm{Im}\int_L h^H(t)\ln(z-t)ds . \tag{7.11}$$

Here $t = x + iy$ is the complex coordinate of the contour point. The value $L = \sum_j L_j + \sum_\mu L_\mu$, where L_j are the contours of the metal conductors and L_μ are the contours dividing the ferrite and air. The value $z = x_0 + iy_0$ is the complex coordinate of the point in which the field is determined and $ds = |dt|$. The values $h^E(t)$ and $h^H(t)$ are the unknown real functions that must be determined.

It is evident that the presentation of the electric and magnetic fields (7.11) are harmonic functions and therefore satisfies Laplace's equation [3.4]. The equations for determining the functions $h^E(t)$, $h^H(t)$ are obtained from the boundary conditions on the perfect metal contours L_j :

$$E_z = 0, \qquad E_s = 0$$

and on contours L_μ dividing the ferrite and air:

$$E_{z1} = E_{z2}, \qquad H_{z1} = H_{z2},$$

$$E_{s1} = E_{s2}, \qquad H_{s1} = H_{s2}.$$

Here E_{si}, H_{si} and $i = 1,2$ are the tangent components of the electric and magnetic fields in the regions I and II.

Since:

$$E_S = \frac{1}{k^2\varepsilon_r\mu_1 - h^2}\left(-ih\frac{\partial E_z}{\partial s} - i\omega\mu_0\mu_1\frac{\partial H_z}{\partial n} + ih\omega\mu_0\mu_a H_z\sin\theta_0\right), \tag{7.12}$$

then on the perfect metal $E_z = 0$ and so $\partial E_z/\partial s = 0$. The condition of the vanishing tangent components of the electric field at the points of the metal surfaces gives two equations:

$$E_z = 0, \qquad \frac{\partial H_z}{\partial n} = h \frac{\mu_a}{\mu_1} H_z \sin \theta_0 \ .$$

(7.13)

Here θ_0 is the angle between the contour tangent and the x axis. If the metal strip is thin then we did not take points on the vertical surface of this strip and $\sin \theta_0 = 0$. From the boundary condition $E_{s1} = E_{s2}$ we have:

$$\frac{1}{k^2 \varepsilon_r \mu_1 - h^2} \left\{ h \frac{\partial E_{z1}}{\partial s} + \omega \mu_0 \mu_1 \frac{\partial H_{z1}}{\partial n} - h \omega \mu_0 \mu_a H_z \sin \theta_0 \ \right\} =$$

$$= \frac{1}{k^2 - h^2} \left\{ h \frac{\partial E_{z2}}{\partial s} + \omega \mu_0 \frac{\partial H_{z2}}{\partial n} \right\} .$$

(7.14)

In a similar way from the boundary condition $H_{s1} = H_{s2}$ we obtained:

$$\frac{1}{k^2 \varepsilon_r \mu_1 - h^2} \left\{ \omega \varepsilon_0 \varepsilon_r \frac{\partial E_{z1}}{\partial n} - h \frac{\partial H_{z1}}{\partial s} - k^2 \varepsilon_r \mu_a H_{z1} \sin \theta_0 \right\} =$$

$$= \frac{1}{k^2 - h^2} \left\{ \omega \varepsilon_0 \frac{\partial E_{z2}}{\partial n} - h \frac{\partial H_{z2}}{\partial s} \right\}$$

(7.15)

at the points of the contours L_μ. Note that the form and the number of the contours L_j and L_μ can be arbitrary.

7.3.2. The numerical solution of the integral equations. The Krylov–Bogoliubov method reduces the integral equations to the homogeneous system of the linear equations. The condition of solvability is obtained equalizing the determinant of the system to zero:

$Det(h') = 0$.

The root of the equation obtained determines the propagation constant h'. Thus one can obtain the dependence of h' on the line geometry and on the value of the external constant magnetic field. After obtaining the value h' the determination of the electric field of the MSL structure is not difficult. Since the fields we determined with the exactness of a free factor, the values $h^H(t_1)$ and $h^E(t_1)$ were chosen arbitrarily at one of the points on the contour L, which was subdivided into segments. The others values $h^H(t_j)$ and $h^E(t_j)$, when $j \neq 1$ are expressed through $h^H(t_1)$ and $h^E(t_1)$. Afterwards one is able to calculate the field components.

First, we used our SIE method to calculate electrodynamical characteristics of the MSL (Fig.7.22a) with the dielectric substrate. This MSL dependence of the propagation constant h' on the value of the substrate material relative permittivity ε_r is shown in Fig.7.23. The dependence obtained in the TEM–approximation is presented here and the difference is within 5%. In Fig.7.24 the transversal electric field force lines of the MSL are shown. The calculated magnetic field components over the one metal strip MSL on the straight line $y = h + 0.19w$ were compared to the experimental data [7.20] in Fig.7.25. The field distribution is shown here only for $x \succ 0$ because the fields are symmetric with respect to the plane $x = 0$.

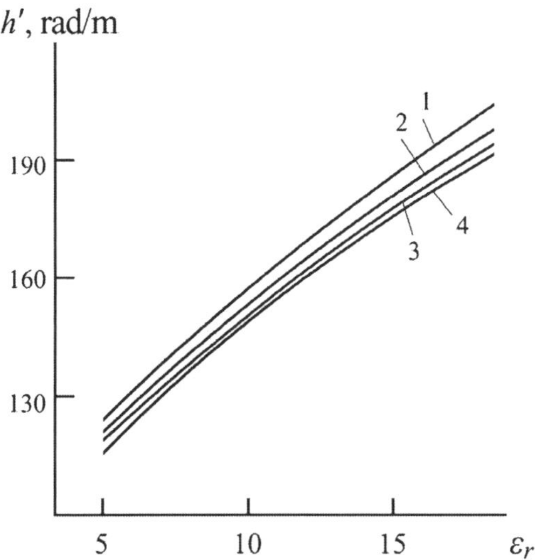

Fig. 7.23. *Dependence of the propagation constant h' of the MSL (Fig.7.22a) on the relative permittivity ε_r of the substrate material. Curves 1 and 2 were calculated according to the known formulae for MSL; Curves 3 and 4 we calculated by the SIE method. The normalized width of the metal strip: curves 1 and 3 correspond to $w/h = 0.5$ and curves 2 and 4 correspond to $w/h = 1$.*

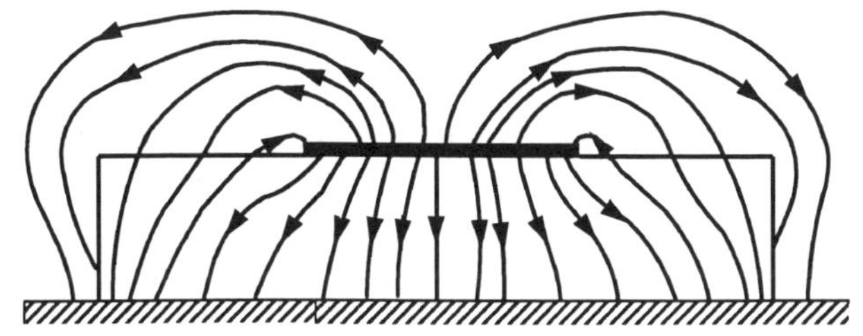

Fig. 7.24. *Distribution of the electric field force lines in the MSL shown in Fig.7.22a at the constitutive parameters $\mu_r = 1$ and $\varepsilon_r = 5$.*

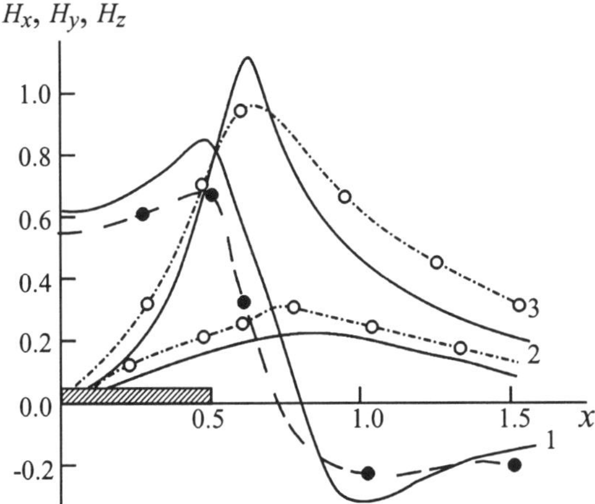

Fig. 7.25. *Dependence of the magnetic field components on the transversal coordinate* x *for the MSL (Fig.7.22a) with the parameters:* $\varepsilon_r = 5$ *and* $w/h = 0.5$. *Curve 1 shows* H_x *distribution; curve 2 shows* H_z *distribution and curve 3 shows* H_y *distribution. Our calculations are the solid lines and the experimental results of [7.20] are the dash lines with circles and points.*

Secondly, we used our SIE method to calculate electrodynamical characteristics of the MSL (Fig.7.22a,b) on the ferrite substrate in the case of the transversal magnetization (Figs.7.26–Fig.7.28). In Fig.7.26, we see the calculated phase shift:

$$\varphi = \frac{\left(h' - h_0'\right) \cdot L \cdot 180°}{\pi}$$

in a one metal strip MSL (Fig.7.22a) dependence on the value of the external constant magnetic field strength H_0 at the frequency $f = 3\,\text{GHz}$. Here the length of the MSL is $L = 0.14\,\text{m}$. The value h_0' is the propagation constant when the external constant magnetic field strength is $H_0 = 0$. Also Fig.7.26 shows the experimental dependence $\varphi(H_0)$. When the external constant magnetic field goes in the opposite direction the dependence $\varphi(H_0)$ is symmetric.

Fig.7.27 shows the vectors of the transversal magnetic field strength (the numbers mean the relative value of the strength H_0 in the corresponding point) for the MSL with the ferrite 60 SCh substrate at the frequency $f = 2.5\,\text{GHz}$ and $w/h = 1$. The ferrite is magnetized up to the saturation. The magnetic and electric fields of the MSL loose its symmetry with respect to the y axis in the presence of the external constant magnetic field. The distribution of the electric field components above the metal strip on the straight line $y = h + 0.3w$ is shown in Fig.7.27.

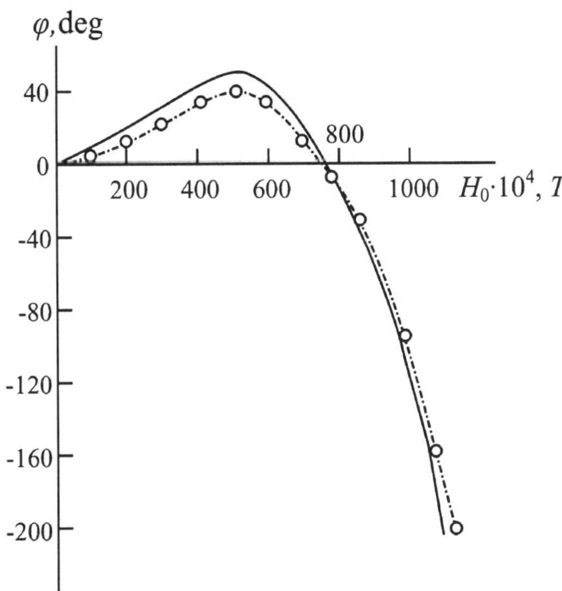

Fig. 7.26. Dependence of the phase shift on the external constant magnetic field for the MSL (Fig.7.22a) with $w/h = 1.2$. Our calculations are the solid lines and the experimental results of [7.20] are the dash line.

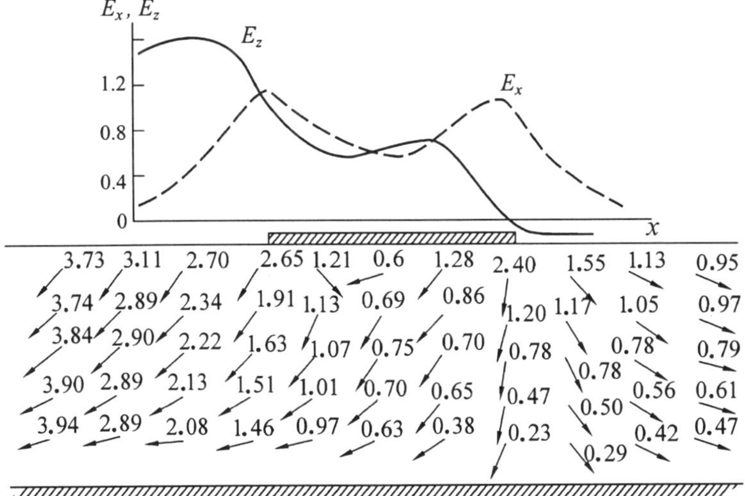

Fig. 7.27. Distribution of the electric field in a one metal strip MSL.

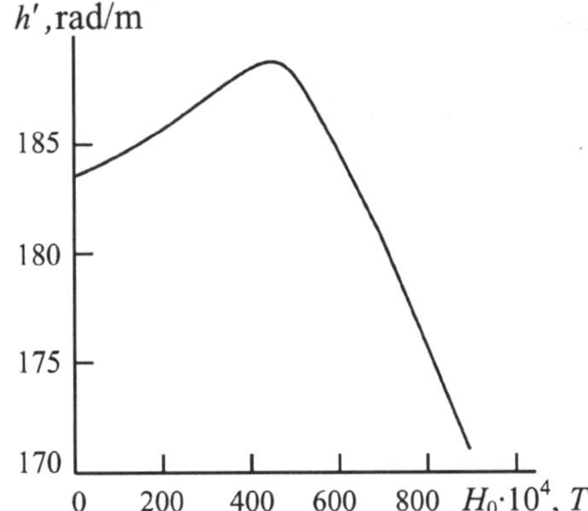

Fig. 7.28. *Dependence of the propagation constant* h' *on the value of the external constant magnetic field strength* H_0.

Fig.7.28 shows the dependence of the propagation constant h' on the value of the external constant magnetic field strength H_0 at the frequency $f = 3\,\text{GHz}$ for the MSL with the ferrite $60\,\text{SCh}$ substrate and the MSL (Fig.7.22b) sizes: $w/h = 1$, $s/h = 05$ and $t/h = 0.003$ [7.21].

7.4. Microstrip Lines with the Logitudinally Magnetized Layer $\tilde{\varepsilon}$ – Gyrotropic Substrates

The study of MSLs with a layer substrate having different electrophysical parameters allowed us to simulate "artificial" materials with required properties. When the MSL substrate consisted of semiconductor material, we introduce a dielectric layer to diminish the losses.

7.4.1. A one metal strip MSL with a layer semiconductor–dielectric substrate. We can improve the MSL characteristics and match it to outside feeders by introducing into the substrate a dielectric layer separating the metal strip from the semiconductor substrate [7.2]. Here we will consider the influence of a dielectric layer on the MSL characteristics. The MSL (Fig.7.29) under consideration has the sizes $w/h = 1.75$, $t/h = 0.0053$, $l/h = 2$ and the length of MSL section is $\Delta L = 2.28 \cdot 10^{-2}\,\text{m}$. The calculated results, which we show in Figs.7.30 and 7.31, were obtained at $B = 1\,\text{T}$, $f = 36\,\text{GHz}$ and two values of free charge carrier concentration in the semiconductor layer.

Fig.7.30 presents the dependence of the complex wave impedance $\dot{Z} = Z' - iZ''$ and the MSL propagation constant $\dot{h} = h' - ih''$ on the thickness d of the dielectric layer with $\varepsilon_r = 10$. The real part of the wave impedance increases when the dielectric layer thickens. The imaginary part of the wave impedance

(Fig.7.30a) and the propagation constant (Fig.7.30b) are raising to the extremes at the normalized thickness of the dielectric layer $d/h = 0.09$.

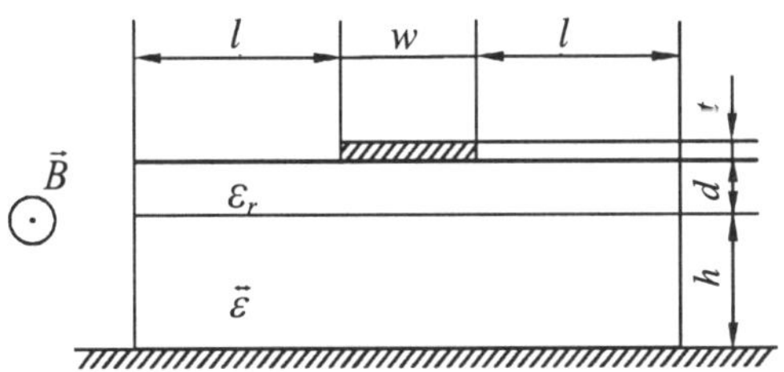

Fig. 7.29. *Cross–section of a one metal strip MSL with the longitudinally magnetized layer semiconductor–dielectric substrate.*

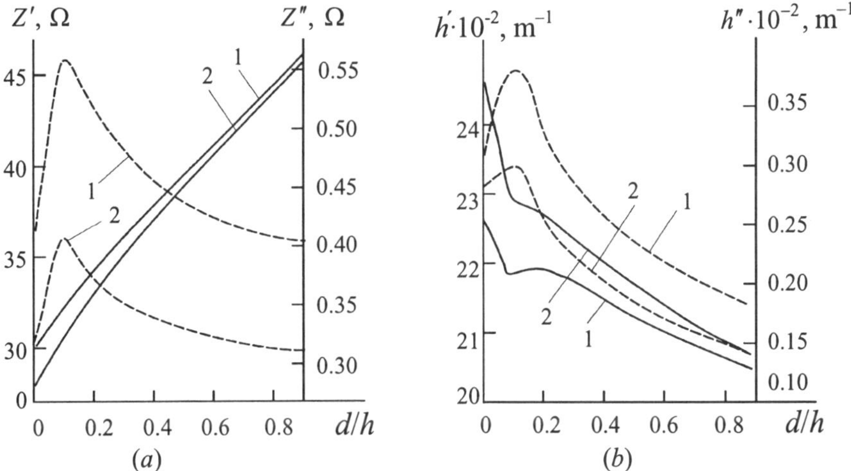

Fig. 7.30. *Dependence of (a) the complex wave impedance $\dot{Z}$; (b) the propagation constant $\dot{h}$ on the dielectric layer thickness d for the MSL shown in Fig.7.29. Our calculations of the real parts of the values $\dot{Z}$ and $\dot{h}$ are the solid lines and the imaginary parts are the dash lines. Curve 1 corresponds to the value of $n = 2.44 \cdot 10^{20}$ m^{-3} and curve 2 corresponds to the value of $n = 2.0 \cdot 10^{20}$ m^{-3}.*

Fig.7.31 presents the dependence of values $\dot{Z} = Z' - iZ''$ and $\dot{h} = h' - ih''$ on the relative permittivity ε_r of the dielectric layer (Fig.7.29) with the same MSL parameters and except the MSL parameter is different at $d/h = 0.35$. As one can see from Fig.7.31b the MSL losses increase as the permittivity ε_r increases. Note also that while permittivity ε_r is increasing the absolute phase shift $\varphi = h'\Delta L$ is also increasing.

The dependence of the complex wave impedance $\dot{Z}$ and of the propagation constant $\dot{h}$ on the frequency is presented in Fig.7.32. The MSL parameters are the same as in the previous case except the MSL parameter is different at $\varepsilon_r = 10$.

The MSL under consideration has the sizes: $w/h = 1.75$, $t/h = 0.0053$, $l/h = 2$, $d/h = 0.35$ at $B = 1\,\mathrm{T}$, $f = 36\,\mathrm{GHz}$ and two values of free charge carrier concentration n in the semiconductor layer.

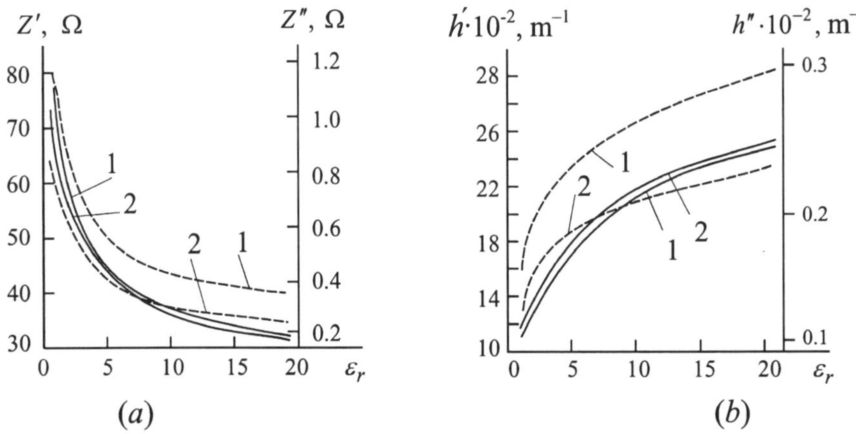

Fig. 7.31. *Dependence of (a) the complex wave impedance $\dot{Z}$ and (b) the propagation constant $\dot{h}$ on the relative permittivity ε_r of the upper layer for the MSL (Fig.7.29). The solid lines are the real parts and the dash lines are the imaginary parts of the values $\dot{Z}$ and $\dot{h}$. Curve 1 corresponds to the value of $n = 2.44 \cdot 10^{20}\,\mathrm{m}^{-3}$ and curve 2 corresponds to the value of $n = 2.0 \cdot 10^{20}\,\mathrm{m}^{-3}$.*

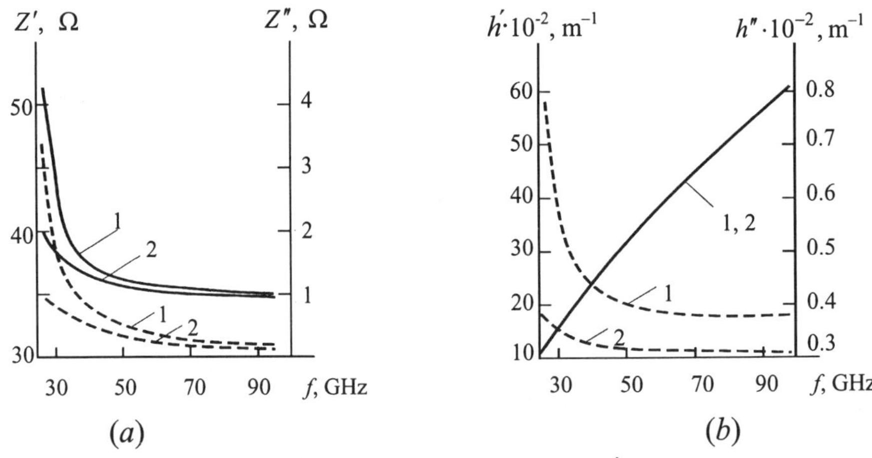

Fig. 7.32. *Dependence of (a) the complex wave impedance $\dot{Z}$ and (b) the propagation constant $\dot{h}$ on the frequency for the MSL shown in Fig.7.29. Solid lines are the real part and dash lines are the imaginary parts. Curve 1 corresponds to the value of $n = 2.44 \cdot 10^{20}\,\mathrm{m}^{-3}$ and curve 2 corresponds to the value of $n = 2.0 \cdot 10^{20}\,\mathrm{m}^{-3.}$*

The MSL complex wave impedance $\dot{Z}$ decreases when the frequency increases (Fig.7.32*a*). The values h' are the same for both concentrations of the semiconductor material (Fig.7.32*b*). Here we notice the features of dependences of the MSL losses h'' on the frequency where it has the opposite character when compared to similar dependence for the helicon wave propagating in the boundless semiconductor under the same conditions [4.11]. When the frequency increases (in the frequency range under consideration) the MSL losses diminish.

7.4.2. Comparison of our calculated results with the experimental data. Fig.7.33*a,b* presents the calculated dependence of the complex wave impedance $\dot{Z} = Z' - iZ''$ as well as the wave propagation constant $\dot{h} = h' - ih''$. Fig.7.33*c* presents the calculated and experimental dependence of the MSL phase shift on the value of the magnetic inductance B of the external constant magnetic field.

The MSL (Fig.7.29) main parameters corresponds to the ones in the preceding section and an additional size is the height of substrate, which is now equal to $h = 0.57 \cdot 10^{-3} \text{m}$. Here the MSL under consideration has the parameters: $w/h = 1.75$, $t/h = 0.0053$, $l/h = 2$, $d/h = 0.44$, $\varepsilon_r = 10$, $\Delta L = 2.28 \cdot 10^{-2} \text{m}$ and the concentration $n = 2.44 \cdot 10^{20} \text{ m}^{-3}$ in the semiconductor layer.

Fig.7.33*a,b* shows the magnetic induction B increasing while the MSL losses are diminishing similar to the losses of the helicon wave in the boundless semiconductor. The wavelength $\lambda = 2\pi/h'$ in the MSL increases if the magnetic induction increases. It is contrary to the increasing wavelength of the helicon wave if the magnetic induction increases [4.11].

Fig.7.33*c* presents the dependence of the relative phase shift after passing through the MSL on the value of the magnetic induction at the frequencies $f = 36\,\text{GHz}$ (curve 1) and $f = 29.65\,\text{GHz}$ (curve 2). The phase shift measurement was carried out at a temperature of 77 K by the method [7.7]. The MSL was placed between two channel–bar waveguides. The interference of two harmonic signals (one signal passed though the MSL and the other was a controlled signal) were analyzed as a function of the magnetic inductance value [7.2].

Experimental points were determined from the interference picture of the two harmonic signals. The phase counting zero point was taken as the phase of the extreme picture's point at the largest value of the magnetic inductance. For the theoretical determination of the phase shift $\Delta\varphi$ we calculated the real part of the wave propagation constant for the MSL (see section 7.1.2). Curves 1 and 2 (Fig.7.33*c*) show that the theoretical and experimental results coincide with each other. The mean deviation of the calculated data from this experimental does not exceed 2%.

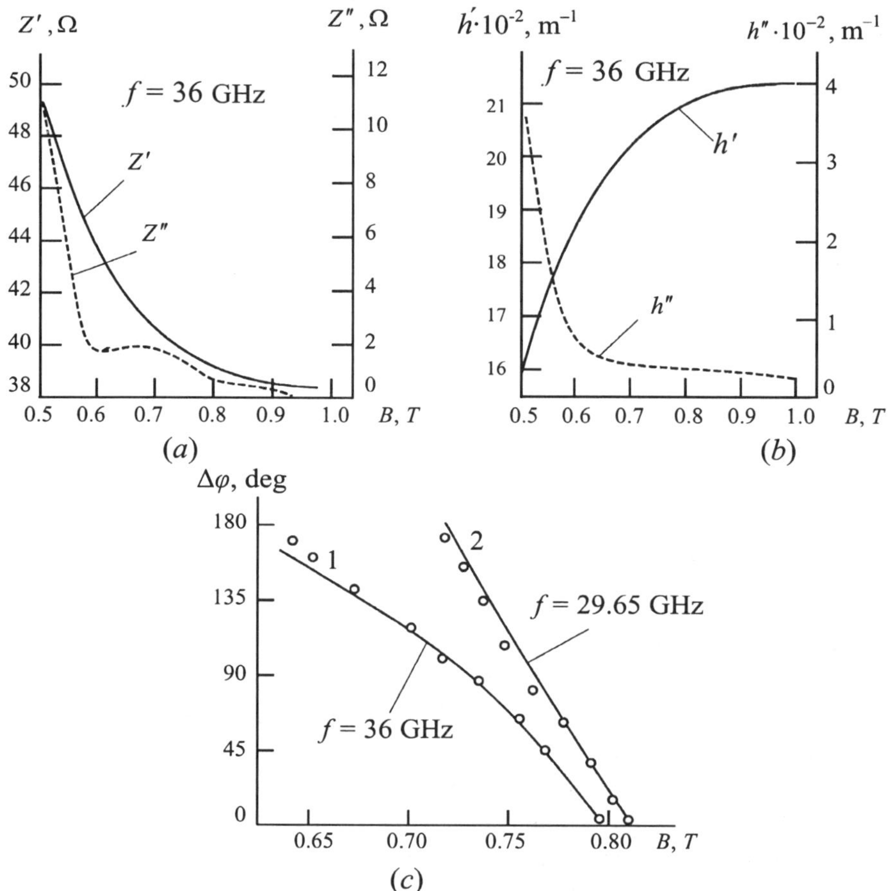

Fig. 7.33. *Dependence of (a) the complex wave impedance* $\dot{Z}$ *; (b) the propagation constant* $\dot{h}$ *and (c) the values of the relative phase shift* $\Delta\varphi$ *on the magnetic induction* B *for the MSL (Fig.7.29). Our calculations are the solid line and the experimental results are circles.*

7.4.3. The resonance wave attenuation in MSLs on the metal strip thickness and comparison of our calculated results with experimental data. The study of the influence of the metal strip thickness on the complex wave impedance and the propagation constant in the screened MSL with the dielectric substrate was presented in a number of papers (see [7.22] – [7.26]). The paper [6.7] presents the dependence of the losses on the strip width at several values of its thickness for similar MSLs. These papers draw attention to the monotonic character of these dependences.

Here we determined the dependence of losses on the metal strip thickness for the MSLs with gyrotropic substrates (Figs. 7.34 and 7.35). The analysis of such dependence is especially necessary when the substrate has semiconductor material. Because the semiconductor usually has large losses. The SIE method enables one to study the dependence of the MSL electrodynamical characteristics on the metal strips thickness.

When analyzing the metal strip thicknesses influence on the characteristics of the MSL with the substrate having the longitudinally magnetized semiconductor it reveals the resonance phenomena in the MSL. The presence of dielectric layer does not change the character of MSL dependences on the metal strip thickness. The placement of this layer is necessary for the matching of the MSL complex wave impedance with a standard impedance of a feeding waveguide.

Fig.7.34 presents the dependences of the effective permittivity $\dot{\varepsilon}_{MSL}^{ef}$, the complex wave impedance $\dot{Z} = Z' - iZ''$ and the propagation constant $\dot{h} = h' - ih''$ on the normalized metal strip thickness. The calculation of the MSL characteristics (depending on the strip thickness) was obtained for the MSL (Fig.7.29) on the semiconductor–dielectric substrate with these sizes: $w/h = 1.75$, $d/h = 0.35$, $l/h = 2$ at $B = 1\,\mathrm{T}$, $f = 36\,\mathrm{GHz}$ and $n = 2.44 \cdot 10^{20}\,\mathrm{m}^{-3}$. The relative permittivity of the substrate dielectric layer was $\varepsilon_r = 10$.

The graphs (Fig.7.34) show the resonance character of the dependence of the complex wave impedance $\dot{Z}$ and the propagation constant $\dot{h}$ on the metal strip thickness. We note that when approaching the value $t/h = 0.59$ the active component of the resistant increases sharply reaching the maximum at $450\ \Omega$ and at the same time, its reactive component turns to zero at this point. At this resonance, the MSL losses h'' increase sharply. Note that at $t/h = 0.59$ the real and imaginary parts of the MSL effective permittivity $\dot{\varepsilon}_{MSL}^{ef}$ passes through zero and the value $\dot{\varepsilon}_{MSL}^{ef}$ becomes minus (Fig.7.34a). If the magnetic induction B of the external constant magnetic field and the frequency f decreases, and the charge carrier concentration n increases, then the crossing point of the curves $\mathrm{Re}(\dot{\varepsilon}_{MSL}^{ef})$ and $\mathrm{Im}(\dot{\varepsilon}_{MSL}^{ef})$ moves to the left (to a thinner metal strip).

Consideration of the EM waves propagating in the boundless semiconductor along the magnetization direction shows that at $\dot{\varepsilon}^{ef} = 0$ appears the magneto–plasmatic resonance of the ordinary wave [4.11]. This resonance depends on the semiconductor material electrophysical parameters, the external constant magnetic field strength H_0 and the signal frequency f. The resonance, which we observed in the MSL at $\dot{\varepsilon}_{MSL}^{ef} = 0$ depends on its sizes including the metal strip thickness. The resonance in the MSL with a semiconductor substrate appears when the magnetic field value is less as compared to the boundless semiconductor material. In our comparisons the electrophysical parameters of semiconductors of a substrate and boundless material were the same and the frequency f was fixed. The value H_0 for the resonance in the boundless semiconductor can be about twice as much as the resonance in the MSLs. The thinner the metals strip then the less the external constant magnetic field strength H_0 has to be in order to the resonance to appear in this MSL.

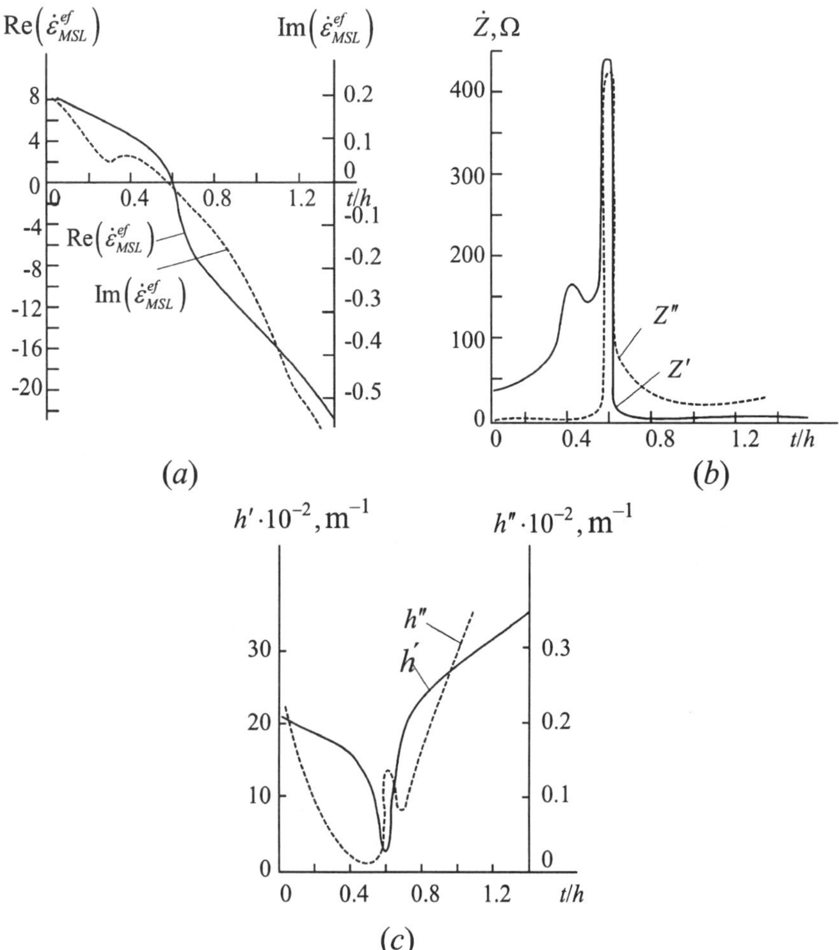

Fig. 7.34. *Dependence of (a) the effective permittivity $\dot{\varepsilon}_{MSL}^{ef}$ (b) the complex wave impedance $\dot{Z}$ and (c) the propagation constant $\dot{h}$ for the MSL (Fig.7.29) on the normalized thickness t/h of the metal strip.*

The resonance attenuation of the wave in the MSL with the longitudinally magnetized semiconductor–dielectric substrate was observed in the MSL (Fig.7.29). MSL sizes were: $h = 0.5 \cdot 10^{-3}$ m, $d = 0.25 \cdot 10^{-3}$ m, $w = 10^{-2}$ m at the relative permittivity $\varepsilon_r = 9.8$ and two concentrations : $n = 1 \cdot 10^{20}$ m^{-3} and $n = 2.0 \cdot 10^{20}$ m^{-3}. The results of our calculations (curves) and experiments (circles) for the frequencies $f = 30.3\,\text{GHz}$ (Fig.7.35a) and $f = 34.02\,\text{GHz}$ (Fig.7.35b) coincide qualitatively. As it was already mentioned, the losses in the MSL strongly depend on the metal strip sizes and the substrate physical properties. One can see in Fig.7.35 that for the thicker strip (curve 2 and 5) and for the smaller charge carrier concentration n (curve 3), the character of the dependence appears to be monotonic and not resonant. In our calculations we have not taken into account the reflection of the TEM mode at the end of the MSL where it joins with the waveguide tract elements. This explains why there is some difference between the experimental results and our calculations.

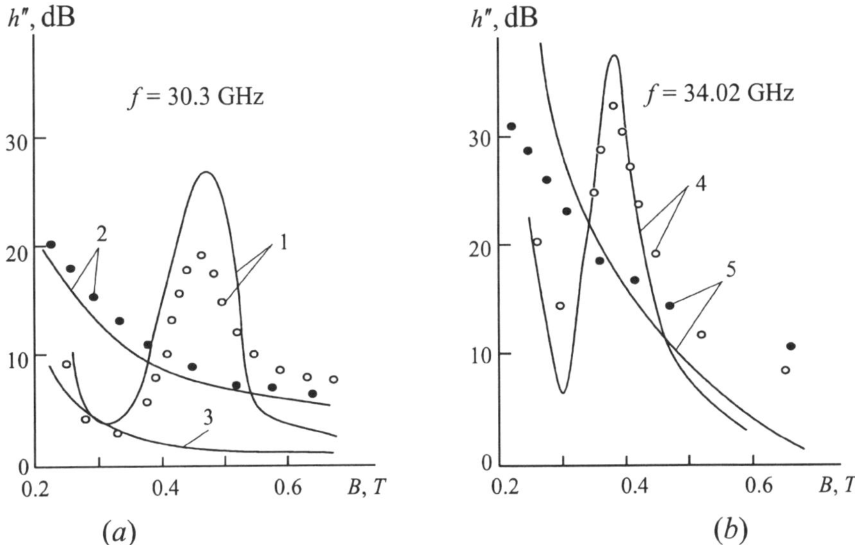

Fig. 7.35. *Dependence of the losses h'' on the magnetic induction B in the one metal strip MSL (Fig.7.29) (a) curve 1 corresponds $n = 2.0 \cdot 10^{20}\,\text{m}^{-3}$, $t = 14 \cdot 10^{-5}\,\text{m}$; curve 2 corresponds $n = 2.0 \cdot 10^{20}\,\text{m}^{-3}$, $t = 2 \cdot 10^{-5}\,m$ curve 3 corresponds $n = 1.0 \cdot 10^{20}\,\text{m}^{-3}$, $t = 14 \cdot 10^{-5}\,\text{m}$; (b) curve 4 corresponds $n = 2.0 \cdot 10^{20}\,\text{m}^{-3}$, $t = 14 \cdot 10^{-5}\,\text{m}$; curve 5 corresponds $n = 2.0 \cdot 10^{20}\,\text{m}^{-3}$, $t = 2 \cdot 10^{-5}\,\text{m}$. (Calculations are the solid lines and the experiment results are the circles and points).*

Fig.7.36 shows the electric field in the fixed cross–section of the MSL with the semiconductor–dielectric substrate at different moments of time within the one period of the electromagnetic wave. The parameters correspond to the MSL of the previous case with the concentration of the charge carriers $n = 2 \cdot 10^{20}\,\text{m}^{-3}$ at $B = 0.7\,\text{T}$.

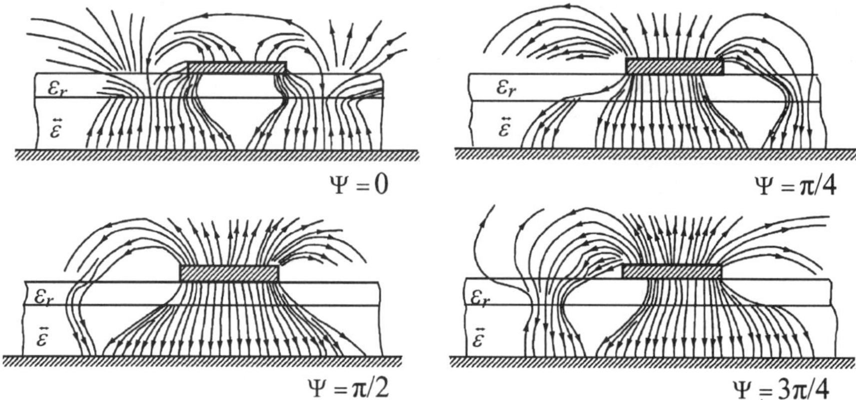

Fig. 7.36. *Distribution of the MSL electric field strength $\vec{E}$ force lines at different moments of time for the one–strip line (Fig.7.29) with parameters: $h = 0.5 \cdot 10^{-3}\,m$, $d = 0.25 \cdot 10^{-3}\,m$, $w = 10^{-2}\,m$, $t = 14 \cdot 10^{-5}\,m$, $\varepsilon_r = 9.8$ at $f = 34.02\ GHz$.*

7.4.4. Investigation of coupled MSLs with semiconductor – magneto-dielectric substrates. A number of devices (directional couplers, filters and phase shifters) have been designed using the interaction of the metal strips in the coupled MSL.

Here we will consider the influence of the dielectric ($\varepsilon_r \geq 1, \mu_r = 1$) and magnetodielectric ($\varepsilon_r \geq 1$, $\mu_r \succ 1$) layers on the characteristics of the coupled MSL (Fig.7.37).

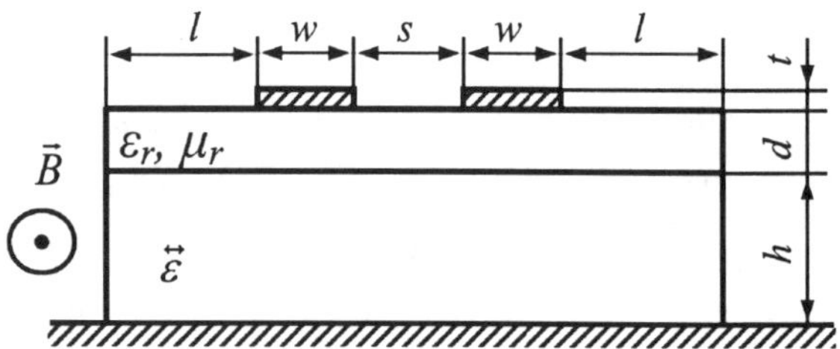

Fig. 7.37. *Cross–section of two metal strips MSL with the longitudinally magnetized semiconductor–magnetodielectric substrate.*

Our calculations were carried out for the two metal strips MSL with sizes: $w/h = 0.1$, $t/h = 0.002$, $l/h = 1$, $s/h = 0.1$, $d/h = 0.2$ at the charge carrier concentration $n = 10^{20}\,m^{-3}$, magnetic induction $B = 0.4\,T$ and frequency $f = 70\,GHz$. In Fig.7.38 we present the dependence of the propagation constants $\dot{h} = h' - ih''$ of the eigenwaves, their relative phase velocities $\beta_{1,2}$ and the current coupling coefficient $\dot{p}_{1,2} = p'_{1,2} - ip''_{1,2}$ on the relative permittivity ε_r of the dielectric layer [7.27]–[7.29].

One can see from Fig.7.38a that with ε_r increasing the losses of both eigenwaves increases. The first eigenwaves is the analogue of the even wave in the geometrically symmetric line with the dielectric substrate and the second eigenwaves is the analogue of the odd wave in the same MSL. Also in Fig.7.38b we show the corresponding dependence of the eigenwave phase velocities on the value ε_r. It is clear that with changing ε_r of the dielectric layer one can achieve the equality of the velocities. After changing of any parameter (for example frequency f), the eigenwaves velocity becomes inequality. We achieve the equality of the velocities again by the proper change of external constant magnetic field. Fig.7.38c shows the dependence of the eigenwave current coupling coefficient $\dot{p}_{1,2} = p'_{1,2} - ip''_{1,2}$ on the value ε_r. The imaginary parts $\mathrm{Im}\,\dot{p}_{1,2}$ show that the current in one metal strip has a phase shift when compared to the adjacent metal strip.

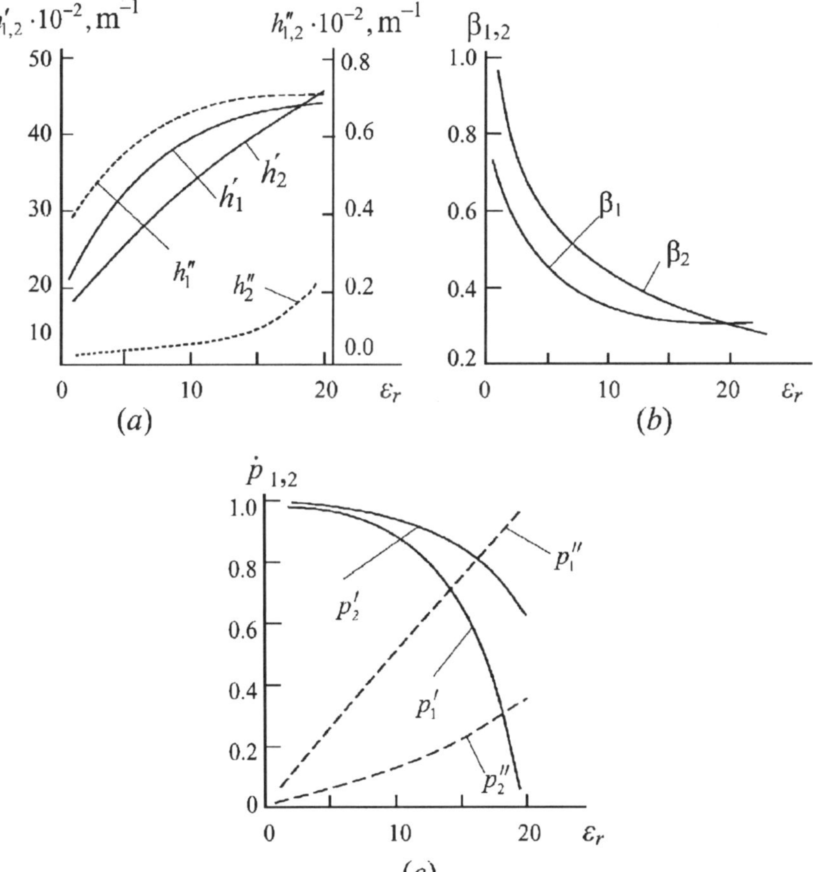

Fig. 7.38. *Dependence of (a) the propagation constants $\dot{h}_{1,2} = h'_{1,2} - ih''_{1,2}$; (b) the relative phase velocities $\beta_{1,2}$ and (c) the current coupling coefficients $\dot{p}_{1,2} = p'_{1,2} - ip''_{1,2}$ of the MSL eigenwaves on the relative permittivity ε_r of the upper substrate layer.*

Fig.7.39 presents the electric field vector polarization pictures in a cross–section of the MSL for different values of the value ε_r. One can see that when the permittivity increases the amplitude of the electric field in the substrate semiconductor layer increases. That explains the increase of the losses h'' when the value ε_r increases.

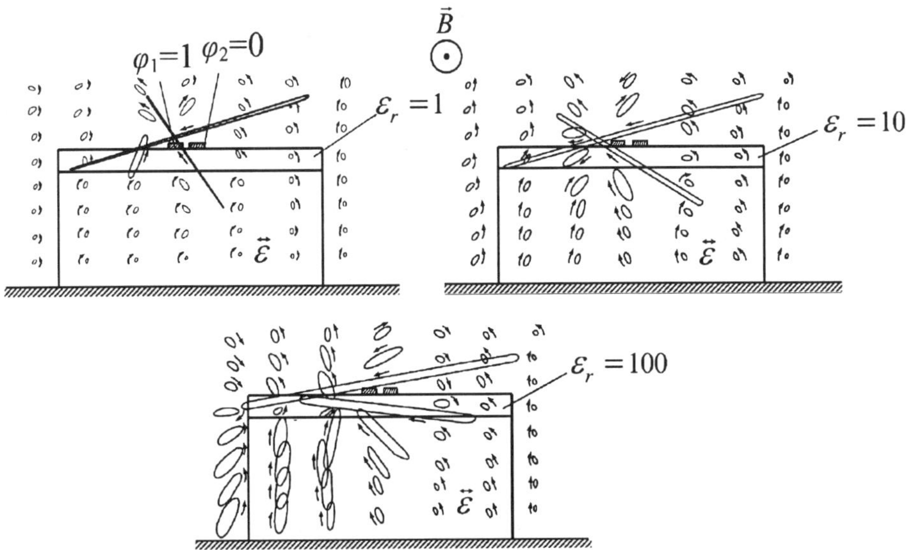

Fig. 7.39. Distribution of the electric field strength $\vec{E}$ polarization ellipses in the fixed cross–section of the MSL (the parameters correspond to Fig.7.38) for the different values of the permittivity ε_r of the upper substrate layer.

Fig.7.40 presents the dependence of the propagation constants $\dot{h}_{1,2} = h'_{1,2} - ih''_{1,2}$ and the current coupling coefficient $\dot{p}_{1,2} = p'_{1,2} - ip''_{1,2}$ on the permeability μ_r. Here we calculated the electrodynamical characteristics of two metal strips MSL with the magnetodielectric substrate when $\varepsilon_r = 10$ and $\mu_r \geq 1$. The MSL parameters are: $w/h = 0.1$, $t/h = 0.002$, $l/h = 1$, $s/h = 0.1$, $d/h = 0.2$, $n = 10^{20}$ m^{-3}, $B = 0.4$ T and $f = 70$ GHz. One can see that the dependence of the MSL losses h'' and the coupling coefficient $\dot{p}_{1,2} = p'_{1,2} - ip''_{1,2}$ on the permeability μ_r at $\varepsilon_r = 10$ is weaker than the same dependences on ε_r at $\mu_r = 1$ (Fig.7.38a,c).

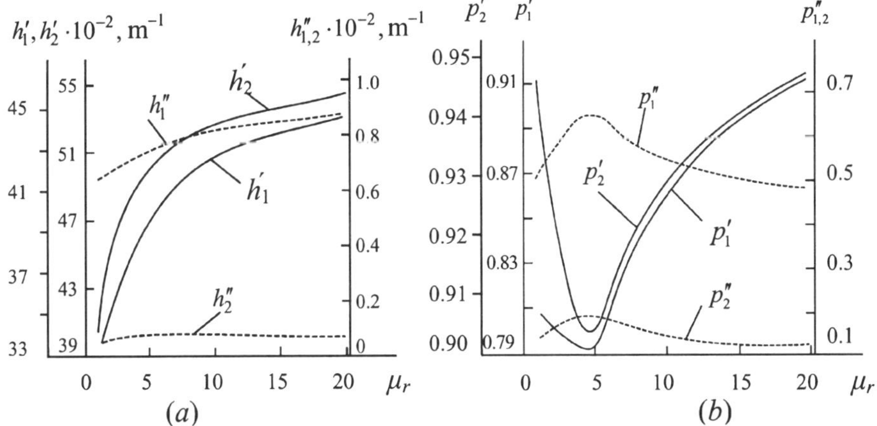

Fig. 7.40. *Dependence of (a) the propagation constants $\dot{h}_{1,2} = h'_{1,2} - i h''_{1,2}$; (b) the current coupling coefficients $\dot{p}_{1,2} = p'_{1,2} - i p''_{1,2}$ of the MSL eigenwaves on the relative permeability μ_r of the upper substrate layer.*

7.4.5. Coupled MSLs with layer semiconductor substrates. Figs. 7.41–7.43 show the electric field distribution in a cross–section of the MSL with the longitudinally magnetized two–layer semiconductor substrate at $B = 0.3\,\text{T}$ and $f = 70\,\text{GHz}$. The lower substrate layer has the concentration of the charge carriers $n_\ell = 2.0 \cdot 10^{20}\,\text{m}^{-3}$, the upper layer has the concentration of the charge carriers $n_u = 0.5 \cdot 10^{20}\,\text{m}^{-3}$. The normalized MSL sizes are: $w/h = 0.2$, $s/h = 0.2$, $d/h = 0.2$, $t/h = 0.02$ and $l/h = 1$.

Fig.7.41 shows the pictures of the electric field vector polarization for the opposite in phase excitation (Fig.7.41*a*) and the in phase excitation (Fig.7.41*b*) of the MSL metal strips. The field distribution is asymmetric in both cases. With the charge carrier concentration n and the magnetic induction B increase the electric field distribution asymmetry increases [7.30].

Figs.7.42 and 7.43 show the electric field correspondingly at the opposite in phase and the in phase excitation of the metal strips in a fixed MSL cross–section.

Fig.7.44 presents the electric field strength polarization ellipses when one of the metal strips has the zero potential. We can see that depending on which strip has the zero potential that the electric field is concentrated on either one or the other substrate side (the direction of the external constant magnetic field has not been changed).

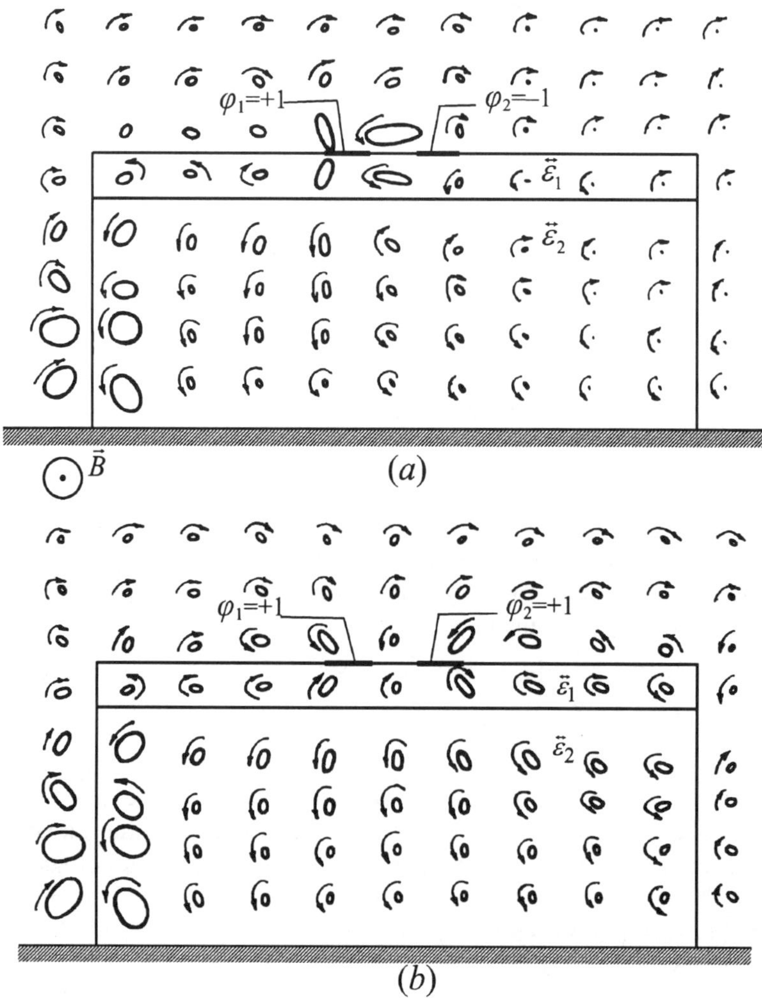

Fig. 7. 41. *Distribution of the electric field strength $\vec{E}$ polarization ellipses in a cross–section of the coupled MSL with the layer longitudinally magnetized semiconductor substrate for (a) the opposite excitation in phase of the metal strips and (b) the excitation of the metal strips in the same phase.*

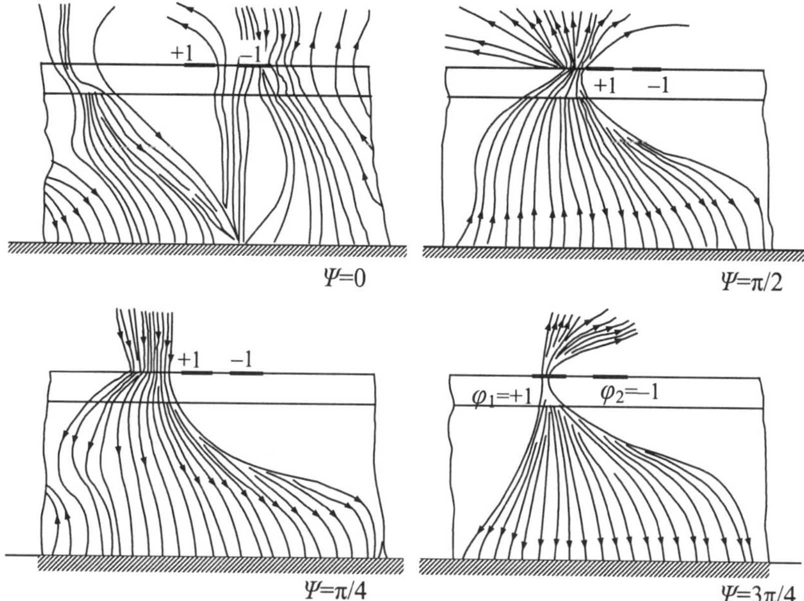

Fig. 7.42. *Distribution of the electric field strength $\vec{E}$ force lines at fixed time moments $\Psi = \omega t$ for the opposite in phase excitation of the MSL metal strips.*

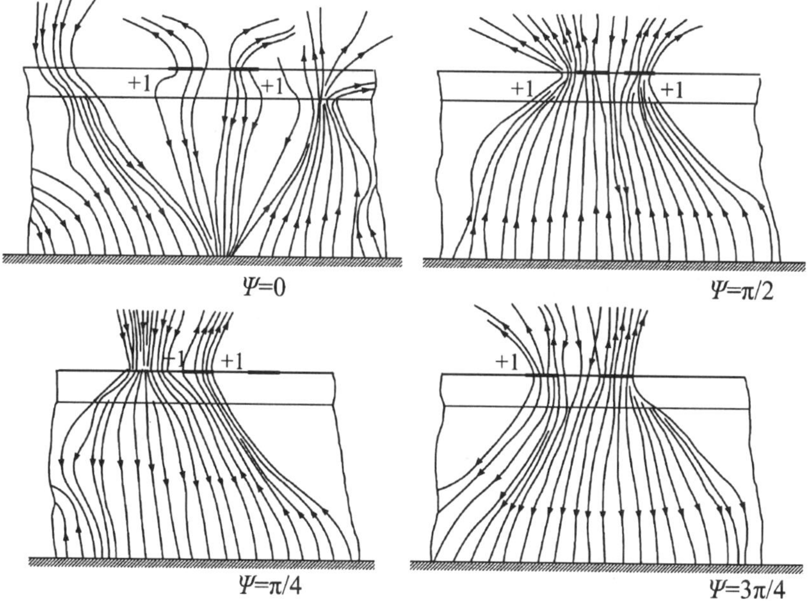

Fig. 7.43. *Distribution of the electric field strength $\vec{E}$ force lines at fixed time moments $\Psi = \omega t$ for the in phase excitation of the MSL metal strips.*

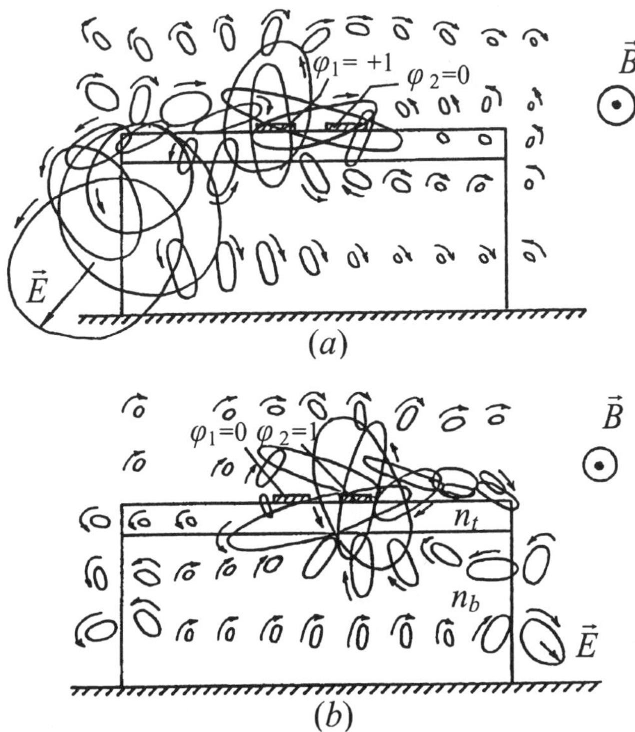

Fig. 7.44. *Distribution of the electric field strength $\vec{E}$ polarization ellipses in a cross–section of the coupled MSL with the layer semiconductor substrate. The two metal strips MSL potentials are: (a) $\varphi_1 =1$ and $\varphi_2 =0$; (b) $\varphi_1 =0$ and $\varphi_2 =1$.*

7.5. Microstrip Lines with Longitudinally Magnetized Substrates Having Double $(\bar{\varepsilon}, \bar{\mu})$ – Gyrotropy

We will now consider a one metal strip and a two metal strips MSL with the layer semiconductor–ferrite substrate and the two metal strips MSL with the substrate of the $(\bar{\varepsilon}, \bar{\mu})$– gyrotropic material, which was placed in an external constant magnetic field [7.30] and [7.31].

The permeability tensor components for the substrate material are found according to the formulae in chapter 4 section 4.1. The half–width ΔH of the resonance curve (which describes the magnetic losses) was equal to 0.05 A/m (A/m= $4\pi \cdot 10^{-7} T$) and the relative permittivity $\varepsilon_r = 10$. The permittivity tensor components are determined according to chapter 4 section 4.3. The values characterizing the semiconductor properties of material were as follows: $m^* = 0.014 m_{el}$, $\mu_{el} = 40$ m$^2/(\text{V} \cdot \text{s})$ and $\varepsilon_L = 17.8$.

7.5.1. MSLs with layer semiconductor – ferrite substrates. Certain layer semiconductor–ferrite structures were studied in [7.32]. The essential advantage of the semiconductor–ferrite structures is the ability to model a new artificial material with magnetic and semiconductor properties. The physical substrate layer parameters can be chosen in such a way that the ferrite is working more effectively in one frequency band and the semiconductor is working more effectively in another frequency band.

Here we will consider the open MSLs (Fig.7.45) with the layer longitudinally magnetized semiconductor–ferrite substrate. The less dissipative ferrite material is placed under the metal strip that the losses of the MSLs would be less [7.31].

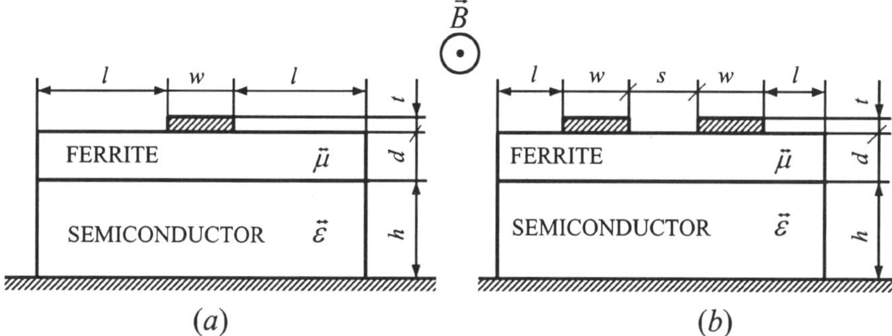

Fig. 7.45. *Cross–section of (a) one metal strip MSL and (b) two metal strips MSL with the longitudinally magnetized semiconductor–ferrite substrate.*

Fig.7.46 presents the dependence of the complex wave impedance $\dot{Z} = Z' - iZ''$ and of the propagation constant $\dot{h} = h' - ih''$ on the frequency at the magnetic induction $B = 0.4$ T for the one metal strip MSL. The normalized MSL sizes are: $w/h = 0.2$, $l/h = 1$, $t/h = 0.002$ and $d/h = 0.2$. The electrophysical parameters of the substrate layers are: the charge carrier concentration of the semiconductor $n = 10^{20}$ m^{-3} and the ferrite saturation magnetization $4\pi M_s = 396.8$ kA/m. One can see from this figure that when the frequency increases the complex wave impedance and losses in the MSL decrease.

Fig.7.47 shows the dependence of the values $\dot{Z} = Z' - iZ''$ and $\dot{h} = h' - ih''$ on the width of the metal strip for the MSL at $f = 70$ GHz and the magnetic induction $B = 0.4$ T. We can see that when we change the width of the metal strip only by a narrow interval ($w/h \le 2$) then the values of the complex wave impedance $\dot{Z}$ and the losses in MSL h'' are changed at wide intervals. The wider the metal strip the larger part of the electromagnetic wave energy is concentrated between the metal strip and the metal plate of the MSL. This leads to the more intensive interaction of the TEM-wave with the dissipative substrate material where the MSL complex wave impedance $\dot{Z}$ greatly decreases. It is important to match the MSL with the feeder lines of the external waveguide section.

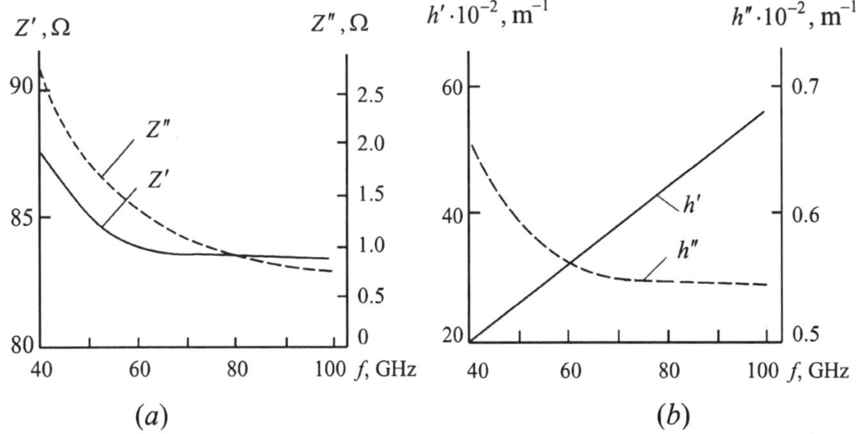

Fig. 7.46. *Dependence of (a) the complex wave impedance $\dot{Z}$ and (b) the propagation constant $\dot{h}$ on the signal frequency f.*

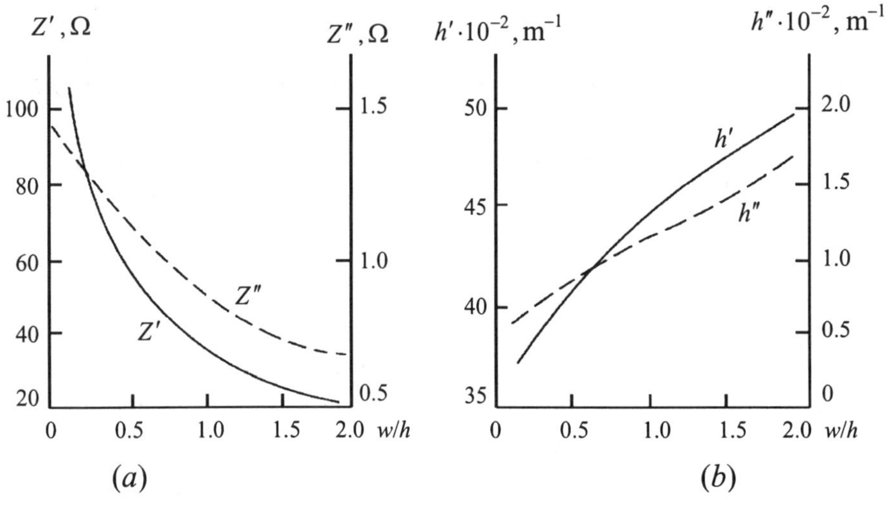

Fig. 7.47. *Dependence of (a) the complex wave impedance $\dot{Z}$ and (b) the propagation constant $\dot{h}$ on the normalized width w/h of the metal strip MSL.*

Fig.7.48 presents the dependence of the values $\dot{Z}$ and $\dot{h}$ on the charge carrier concentration n in the semiconductor substrate layer at $f = 70\,\text{GHz}$ and $B = 0.4\,\text{T}$. The normalized MSL sizes are: $w/h = 0.2$, $l/h = 1$, $t/h = 0.002$ and $d/h = 0.2$. The ferrite saturation magnetization is $4\pi M_s = 396.8$ kA/m. Despite the fact that the semiconductor does not directly contact the metal strip (there is a ferrite layer between them), the dependence of the real part of the complex wave impedance on the charge carrier concentration n increases very suddenly starting from some n value).

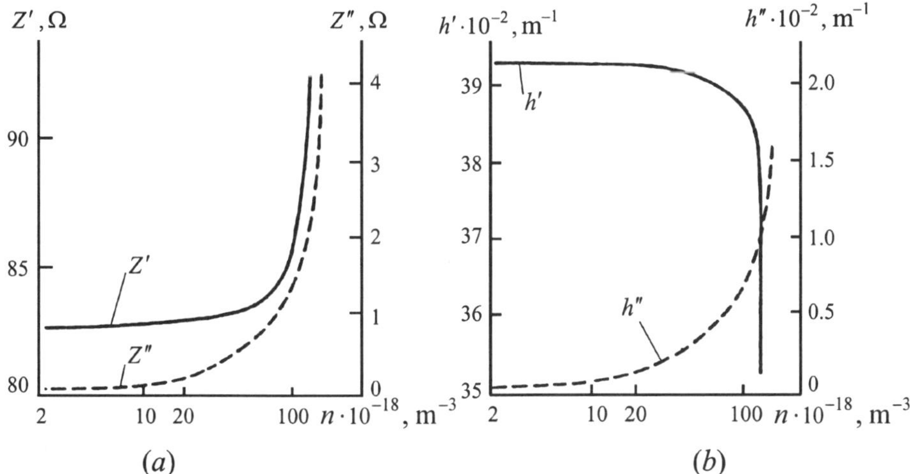

Fig. 7.48. *Dependence of (a) the complex wave impedance* $\dot{Z}$ *and (b) the propagation constant* $\dot{h}$ *on the concentration of charge carriers in the MSL substrate semiconductor material.*

We see that when the value n increases the reactive wave impedance part Z'' and the losses h'' increase. We point out here that the charge carrier concentration n is one of the most important parameters characterizing the properties of the semiconductor material. When the charge carrier concentration n diminishes the nondiagonal elements of the permittivity tensor (which describes the material gyrotropy) goes to zero and the semiconductor properties become nearer to the properties of a uniaxial crystal. From the TEM–approximation point of view the semiconductor transforms into the isotropic material.

Here we will consider the influence of the ferrite layer on the characteristics of the coupled MSL (Fig.7.45b) at $B=0.4\,\mathrm{T}$, $f=70\,\mathrm{GHz}$ and $n=10^{20}\,\mathrm{m}^{-3}$. The MSL sizes are: $w/h=0.1$, $t/h=0.01$, $l/h=1$, $s/h=0.1$ and $d/h=0.2$. Fig.7.49 shows the dependence of the propagation constant $\dot{h}=h'-ih''$ and the coupling coefficient $\dot{p}_{1,2}=p'_{1,2}-ip''_{1,2}$ of the eigenwaves on the saturation magnetization M_s of the substrate ferrite layer. We see that when the ferrite saturation magnetization increases then one of the eigenwaves propagates with low losses. And the other eigenwave attenuates greatly.

Fig.7.50 presents the dependence of the complex wave impedance $\dot{Z}=Z'-iZ''$ and of the voltage coupling coefficient $\dot{q}_{1,2}=q'_{1,2}-iq''_{1,2}$ on the ferrite layer thickness d/h at $4\pi M_s=396.8\,kA/m$. The sizes and other parameters of this MSL were as in the previous MSL only the value d/h is different.

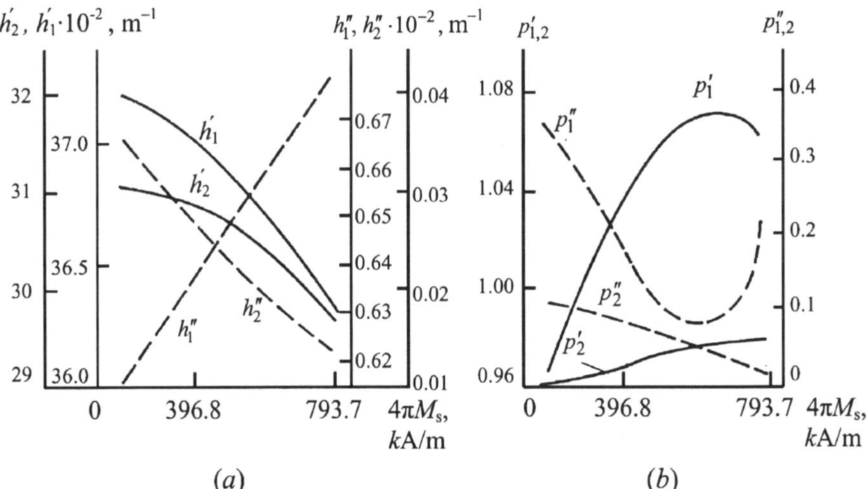

(a)   (b)

Fig. 7.49. *Dependence of (a) propagation constants $\dot{h}_{1,2} = h'_{1,2} - ih''_{1,2}$; (b) current coupling coefficients $\dot{p}_{1,2} = p'_{1,2} - ip''_{1,2}$ of eigenwaves on the saturation magnetization $4\pi M_s$ of the MSL substrate ferrite layer.*

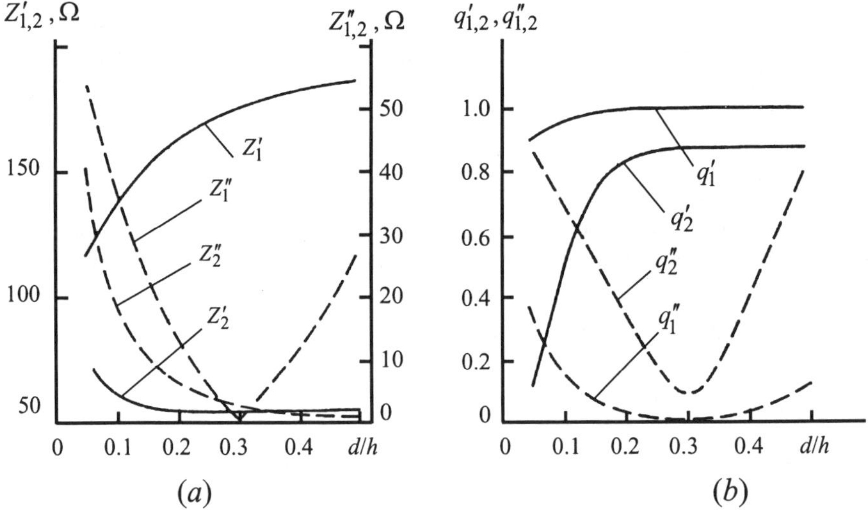

(a)   (b)

Fig. 7.50. *Dependence of (a) the complex wave impedance $\dot{Z}_{1,2} = Z'_{1,2} - iZ''_{1,2}$ and (b) the voltage coupling coefficient $\dot{q}_{1,2} = q'_{1,2} - iq''_{1,2}$ on the normalized ferrite layer thickness of the MSL.*

Fig.7.51 shows the dependence of the eigenwave propagation constant $\dot{h}_{1,2} = h'_{1,2} - ih''_{1,2}$ on the electron mobility μ_{el} at $4\pi M_s = 396.8$ *kA/m* for the MSL with parameters corresponding to Fig.7.49. In Fig.7.51 the losses decrease when the charge mobility μ_{el} increases.

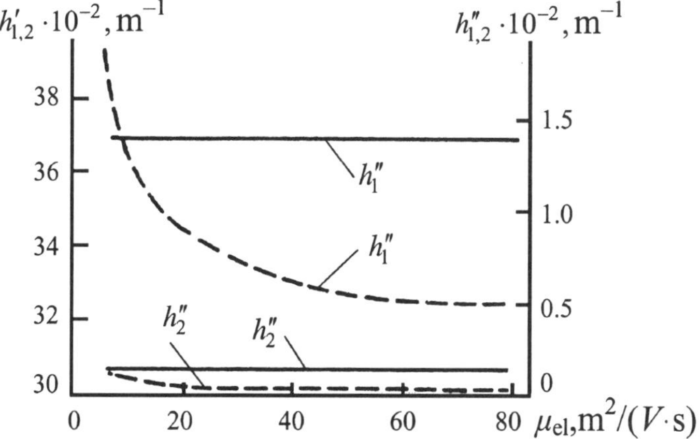

Fig. 7.51. *Dependence of the eigenwave propagation constant* $\dot{h}$ *on the electron mobility* μ_{el} *of the MSL substrate semiconductor layer.*

For the MSL with the mobility $\mu_{el} = 40 \text{ m}^2/(\text{V·s})$ Fig.7.52 presents the dependence of the current coupling coefficient $\dot{p}_{1,2} = p'_{1,2} - ip''_{1,2}$ on the magnetic induction. Note that some characteristics such as the complex wave impedance and coupling coefficient do change, but only slightly. Here the MSL losses $h''_{1,2}$ decrease when the magnetic induction B increases (this also happens in all others MSLs in this chapter).

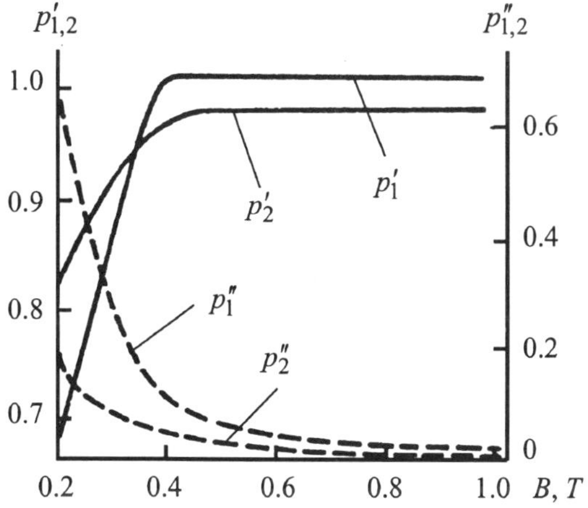

Fig. 7.52. *Dependence of the current coupling coefficient* $\dot{p}_{1,2} = p'_{1,2} - ip''_{1,2}$ *of the MSL (Fig.7.45b) eigenwaves on the magnetic induction* B *value.*

Fig.7.53 presents distribution of the electric field strength $\vec{E}$ in a cross–section of the above– mentioned MSL at some time moments ($\omega t = \Psi$).

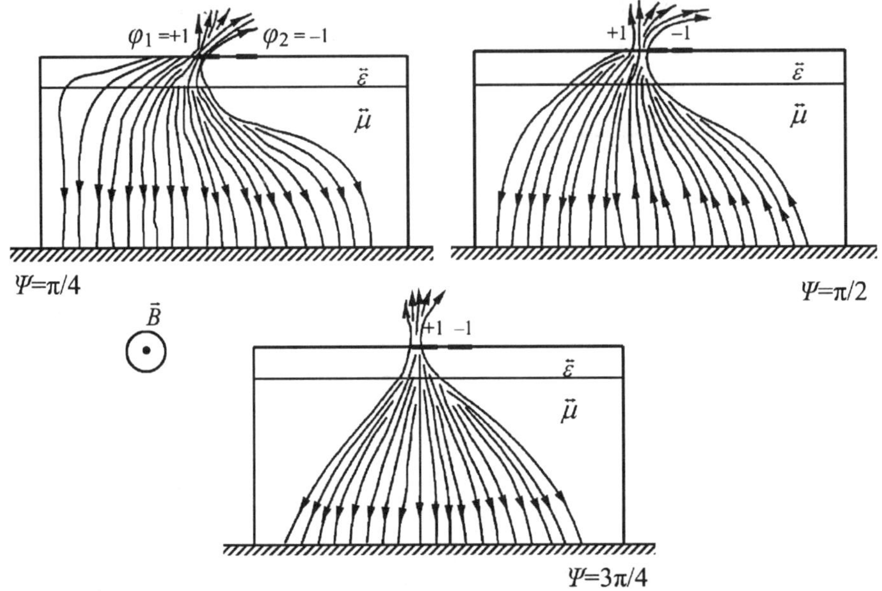

Fig. 7.53. *Distribution of the electric field strength* $\vec{E}$ *force lines at different time moments* $\Psi = \omega t$ *in a cross-section of the MSL.*

Fig.7.54 presents diagrams of the electric field strength $\vec{E}$ polarization ellipses and the magnetic field strength $\vec{H}$ polarization ellipses when one of the metal strips has a zero potential. In these calculations the parameters are: magnetic induction $B = 0.4\,\text{T}$, frequency $f = 70\,\text{GHz}$, charge carrier concentration in the semiconductor material $n = 10^{20}\,\text{m}^{-3}$ and the saturation magnetization of the substrate ferrite layer $4\pi M_s = 396.8\,\text{kA/m}$. The normalized MSL sizes are: $w/h = 0.5$, $t/h = 0.01$, $l/h = 1$, $s/h = 0.1$ and $d/h = 0.2$.

Fig.7.55 presents the diagram of the electric field strength $\vec{E}$ polarization and the pictures of their force line distributions at certain time moments $\omega t = \Psi$ for the MSL. This MSL has different sizes of metal strips, which are located on the edge of the semiconductor–ferrite substrate. The tensor components of the upper layer are: $\mu_{xx} = 0.97 + i2.7 \cdot 10^{-3}$, $\mu_{xy} = -0.21 - i8.4 \cdot 10^{-4}$, those of the lower layer are: $\varepsilon_{xx} = 1.96 + i1.3$ and $\varepsilon_{xy} = 20.7 - i0.2$. Other parameters of these MSLs are: $w_1/h = 0.5$, $w_2/h = 0.1$, $t_1/h = t_2/h = 0.05$, $l_1/h = 0.01$, $l_2/h = 2$, $s/h = 0.1$, $d/h = 0.2$, $B = 0.4\,\text{T}$, $n = 2 \cdot 10^{20}\,\text{m}^{-3}$, $\mu_{el} = 60\,\text{m}^2/(\text{V}\cdot\text{s})$, $f = 70\,\text{GHz}$, $4\pi M_s = 396.8\,\text{kA/m}$ and $\varepsilon_{fer} = 10$.

7.5.2. Coupled MSLs with the substrate having double $(\vec{\varepsilon},\vec{\mu})-$
gyrotropy. Here we present some calculated results for the coupled MSL with
the substrate having the double $(\vec{\varepsilon},\vec{\mu})-$ gyrotropy. Here the magnetic-
semiconductor material with electrophysical parameters, we interpret as a model
of artificial material. Our investigations were carried out at rather wide intervals
of the semiconductor and ferrite properties of this hypothetical magnetic-
semiconductor material with the charge carrier concentration n varying from
$2\cdot10^{18}\,\mathrm{m}^{-3}$ to $2\cdot20^{20}\,\mathrm{m}^{-3}$. The MSL under consideration (Fig.7.56) has the sizes:
$w/h = 0.1$, $t/h = 0.01$, $l/h = 1$ and $s/h = 0.5$.

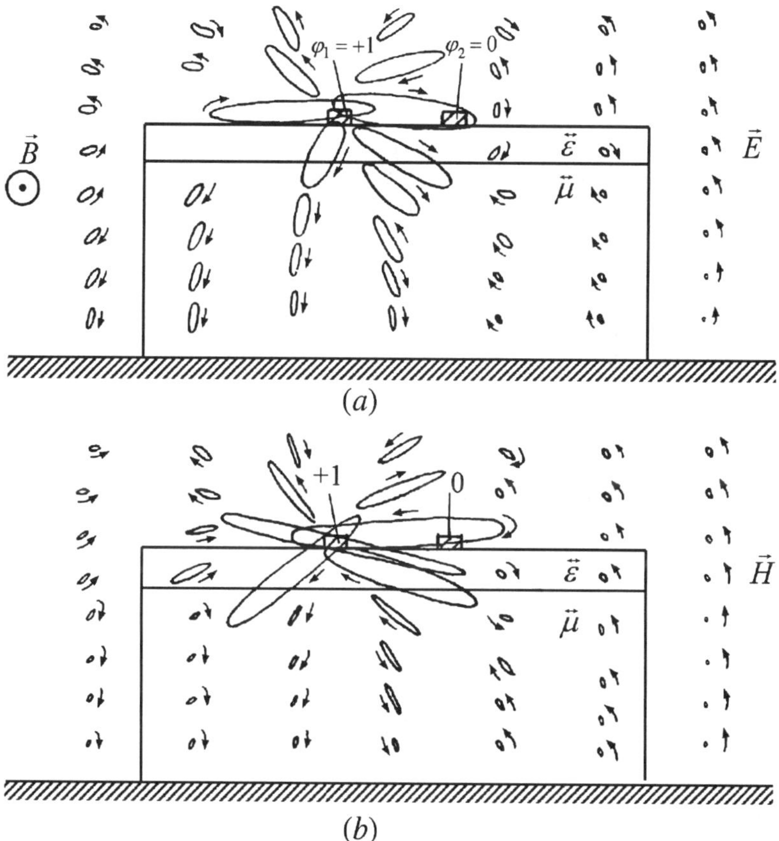

Fig. 7.54. *Diagrams of polarization ellipses in a cross–section of the coupled*
MSL (a) the electric field strength $\vec{E}$ *; (b) the magnetic field strength*
$\vec{H}$ at the saturation magnetization of the ferrite $4\pi M_s = 396.8\,kA/m.$
The two metal strips potentials are: $\varphi_1 = 1$ *and* $\varphi_2 = 0$.

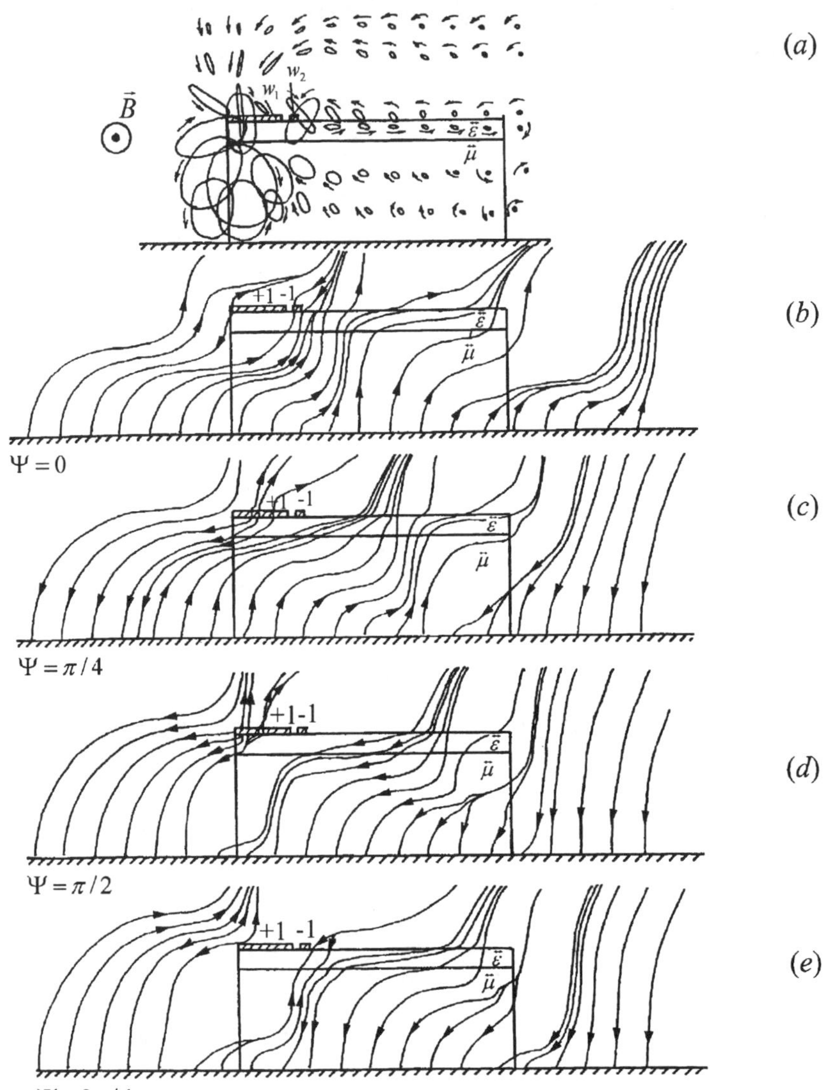

Fig. 7.55. *Distributions of the electric field strength $\vec{E}$ (a) polarization ellipses and (b,c,d,e) the force lines at certain time moments $\Psi = \omega t$ of the MSL.*

Fig.7.57 presents the dependence of the propagation constants $\dot{h}_{1,2} = h'_{1,2} - ih''_{1,2}$ and the current coupling coefficients $\dot{p}_{1,2} = p'_{1,2} - ip''_{1,2}$ of the eigenwaves on the charge carrier concentration n in the material of the MSL substrate at $B = 0.4\,\text{T}$, $f = 70\,\text{GHz}$ and $4\pi M_s = 396.8\ kA/\text{m}$. Fig.7.57$a$ shows that when the concentration is $n \geq 10^{16}\,\text{m}^{-3}$ then the losses of one eigenwave increases and the other eigenwave decreases. This MSL can be used as a one mode waveguide.

We see in Fig.7.57b that the imaginary parts of the current coupling coefficients depend on the value n in a very complicated way. Such complicated dependencies like these were not observed in the previous MSLs with the semiconductor and the layer semiconductor–ferrite substrates. It is clear from the Fig.7.57b that changing n we can obtain small values of $p''_{1,2}$ and that means the small phase shift between the eigenwave currents in the strips.

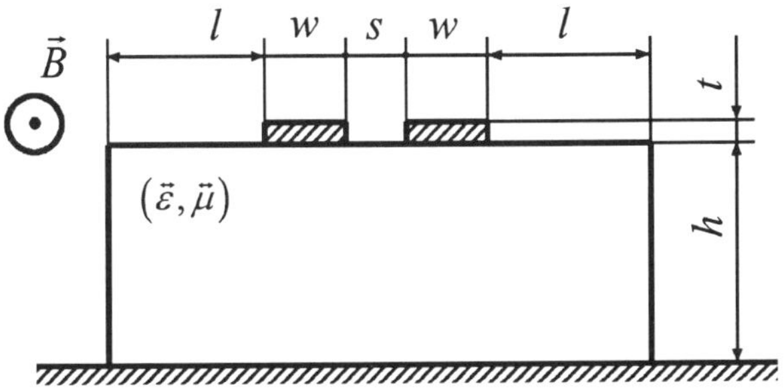

Fig. 7.56. *Cross– section of the coupled MSL with the longitudinally magnetized substrate having double* $(\vec{\vec{\varepsilon}}, \vec{\vec{\mu}})$ *– gyrotropy.*

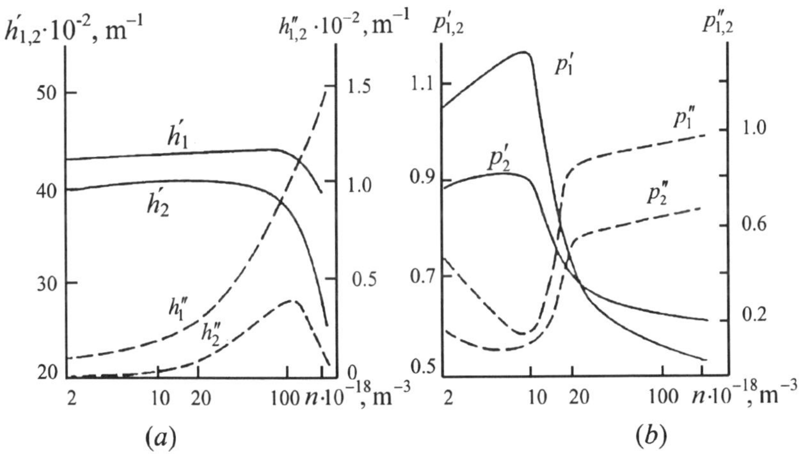

(a)   (b)

Fig. 7. 57. *Dependence of (a) the propagation constant* $\dot{h}_{1,2} = h'_{1,2} - ih''_{1,2}$ *and (b) the current coupling coefficient* $\dot{p}_{1,2} = p'_{1,2} - ip''_{1,2}$ *of the MSL eigenwaves on the charge carrier concentration* n *in the substrate material.*

Fig.7.58 shows the dependence of $\dot{h}_{1,2} = h'_{1,2} - ih''_{1,2}$ and $\dot{p}_{1,2} = p'_{1,2} - ip''_{1,2}$ on the saturation magnetization M_s value. Here we see that the eigenwave losses depend little on M_s and the dependence of the current coupling coefficients is monotonic.

The dependence of $\dot{h}_{1,2} = h'_{1,2} - ih''_{1,2}$ and $\dot{p}_{1,2} = p'_{1,2} - ip''_{1,2}$ on the magnetic induction B is shown in Fig.7.59 for the same MSL at $n = 10^{20}$ m^{-3} and

$4\pi M_s = 396.8\ kA/m$. Fig.7.59 shows that by changing the magnetic induction B of the external constant magnetic field it is possible to effectively control the characteristics of the MSL. For instance, it is possible to diminish the losses, to vary the value of $h'_{1,2}$ as well as regulate the strip coupling of the MSL.

Fig.7.60 shows the distribution of the polarization ellipses of the electric and magnetic fields in a cross–section of the MSL.

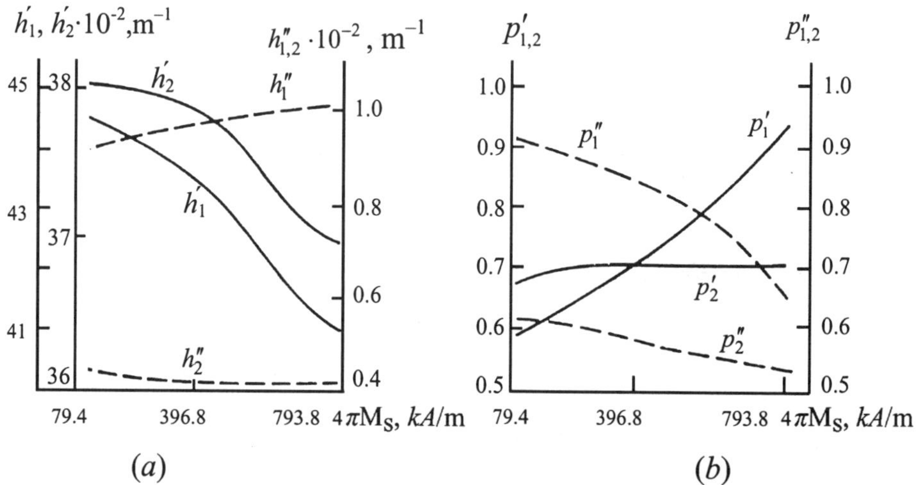

(a)   (b)

Fig. 7.58. *Dependence of (a) the propagation constant* $\dot{h}_{1,2} = h'_{1,2} - ih''_{1,2}$ *and (b) the current coupling coefficient* $\dot{p}_{1,2} = p'_{1,2} - ip''_{1,2}$ *of the MSL eigenwaves on the saturation magnetization* M_s *of the substrate material.*

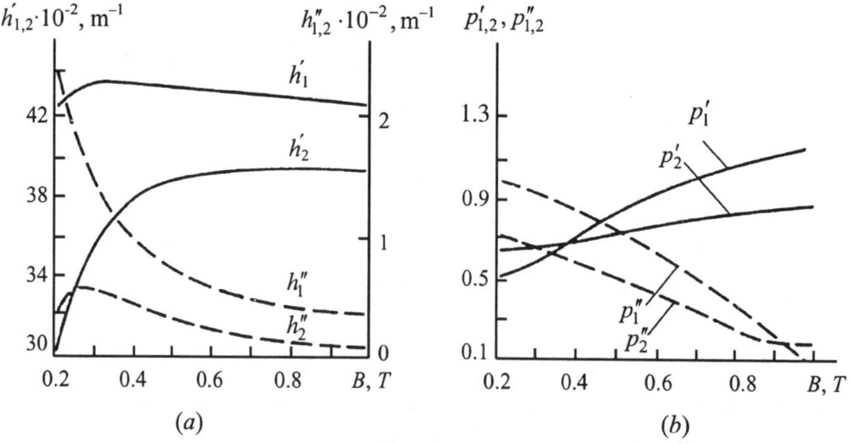

(a)   (b)

Fig. 7.59. *Dependence of (a) the propagation constant* $\dot{h}_{1,2} = h'_{1,2} - ih''_{1,2}$ *and (b) the current coupling coefficient* $\dot{p}_{1,2} = p'_{1,2} - ip''_{1,2}$ *of the MSL eigenwaves on the magnetic induction B .*

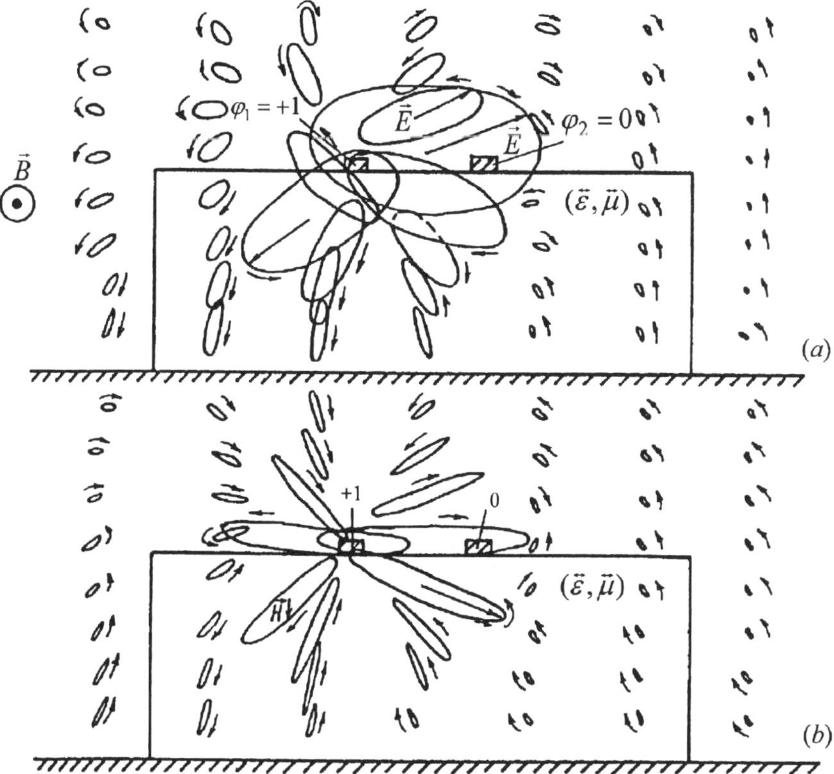

Fig. 7.60. *Distributions of polarization ellipses in a MSL cross– section: (a) the electric field strength $\vec{E}$; (b) the magnetic field strength $\vec{H}$.*

In Fig.7.61 we present the electric field strength $\vec{E}$ polarization ellipses and the distribution of the $\vec{E}$ force lines at certain time moments for the wide metal strips, which are placed at the substrate edges. The MSL substrate consists of an upper semiconductor layer and a lower magnetic– semiconductor layer. The tensor components of the lower layer are: $\varepsilon_{xx} = 19.6 - i1.3$, $\varepsilon_{xy} = -20.7 - i0.2$, $\mu_{xx} = 0.97 - i2.7 \cdot 10^{-3}$ and $\mu_{xy} - 0.21 - i8.4 \cdot 10^{-4}$. The other MSL parameters are: $w_1/h = w_2/h = 0.8$, $t_1/h = t_2/h = 0.05$, $s/h = 0.5$, $l_1/h = l_2/h = l/h = 0.01$, $d/h = 0.1$, $B = 0.4\,\text{T}$, $n_u = 0.5 \cdot 10^{20}\,\text{m}^{-3}$, $n_\ell = 2 \cdot 10^{20}\,\text{m}^{-3}$, $\mu_{el} = 60\,\text{m}^2/(\text{V} \cdot \text{s})$, $f = 70\,\text{GHz}$, $4\pi M_s = 396.8\,\text{kA/m}$, $\varepsilon_{fer} = 10$.

In this chapter we analyzed the electrodynamical characteristics of MSLs in TEM–approximation. To accomplish this we had to solve Laplace's equation. The EM–waves (TEM–waves) that propagated on these MSLs had only transversal components, because the longitudinal components were negligible when compared to the transversal ones.

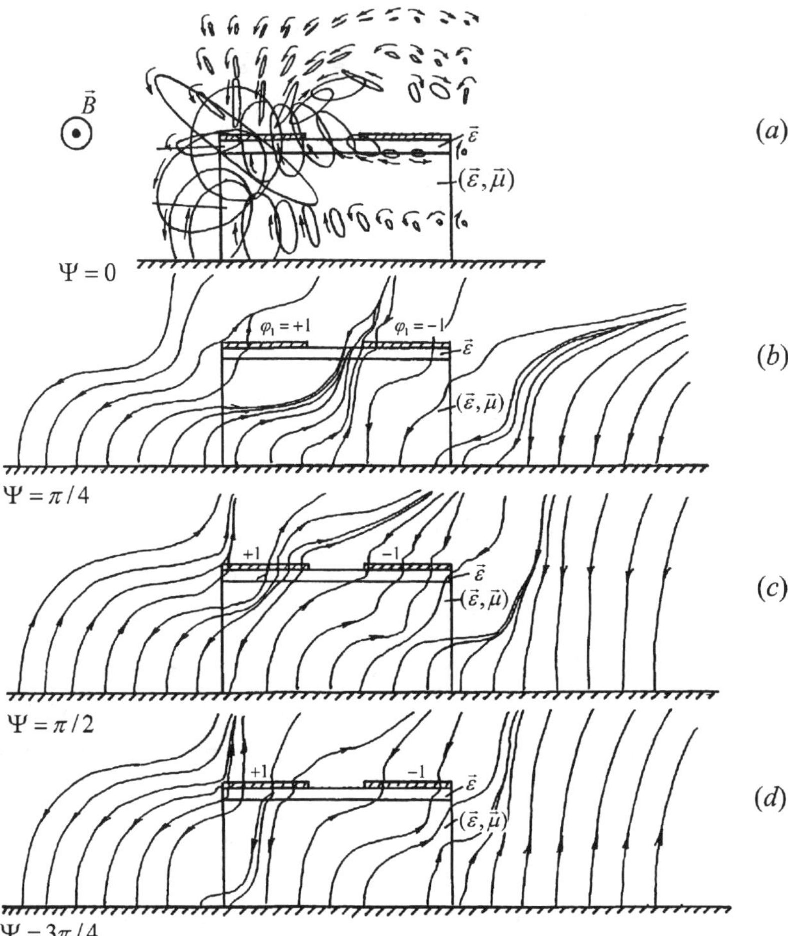

Fig. 7.61. *Distributions the electric field strength $\vec{E}$ (a) polarization ellipses and (b,c,d) its force lines at certain time moments $\Psi = \omega t$ in the MSL with the layer $\vec{\vec{\varepsilon}} - (\vec{\varepsilon}, \vec{\mu})$ substrate.*

However the criterion of TEM-approximation is not being satisfied for electrodynamical problems about propagating of EM waves on slot lines (SLs), dielectric waveguides and MSLs with "thick" substrates (that mean thicknesses of substrates are comparable with the MSL wavelength). For these electrodynamical problems it is necessary to solve Maxwell's equations in the rigorous problem formulation. In the next chapter we will describe a new version of the SIE method for solving Maxwell's equations and we will analyse the electrodynamical characteristics of MSLs, SLs and dielectric waveguides.

8. SOLUTION OF MAXWELL'S EQUATIONS BY THE SIE METHOD FOR ISOTROPIC WAVEGUIDES

8.1. The Integral Representation for the Solution of Maxwell's Equations

In this chapter, we describe the SIE method for solving Maxwell's equations in the rigorous problem formulation. When using the SIE method we make it possible to investigate the dispersion characteristics of main and higher modes in regular waveguides of arbitrary cross–section geometry containing piece–wise homogeneous material.

Our proposed method consists in finding the solution of differential equations with a point–source. Then the fundamental solution of differential equations is used in the integral representation of the general solution for each particular boundary problem. The integral representation automatically satisfies Maxwell's differential equations and has the unknown density functions μ_h and μ_e, which are found using the proper boundary conditions. Our SIE method is based on the SIE theory [3.3], [3.4]. Here we solve the problem using Maxwell's equations:

$$\mathrm{rot}\vec{H} = \varepsilon_0\varepsilon_r \frac{\partial\vec{E}}{\partial t},$$

$$\mathrm{rot}\vec{E} = -\mu_0\mu_r \frac{\partial\vec{H}}{\partial t}, \tag{8.1}$$

where $\vec{E}$ is the electric field strength vector and $\vec{H}$ is the magnetic field strength vector. Also ε_r, is the relative permittivity and μ_r is the relative permeability of medium. The electric and magnetic constants ε_0, μ_0 are called the permittivity and permeability of a vacuum.

The dependence on time t and on the longitudinal coordinate z are assumed in the form $e^{i(\omega t - hz)}$, where h is the longitudinal propagation constant. The magnitude $\omega = 2\pi f$ is the cyclic signal frequency and i is the imaginary unit $\left(i^2 = -1\right)$. The transversal components E_x, E_y, H_x, H_y of the electromagnetic (EM) field are being expressed through the longitudinal components E_z, H_z from Maxwell's equations as follows:

$$E_x = \frac{\mu_0\mu_r i\omega \dfrac{\partial H_z}{\partial y} + ih\dfrac{\partial E_z}{\partial x}}{\Delta},$$

$$E_y = \frac{-\mu_0\mu_r i\omega \dfrac{\partial H_z}{\partial x} + ih\dfrac{\partial E_z}{\partial y}}{\Delta},$$

$$H_x = \frac{-\varepsilon_0\varepsilon_r i\omega \dfrac{\partial E_z}{\partial y} + ih\dfrac{\partial H_z}{\partial x}}{\Delta},$$

$$H_y = \frac{\varepsilon_0\varepsilon_r i\omega \dfrac{\partial E_z}{\partial x} + ih\dfrac{\partial H_z}{\partial y}}{\Delta}, \qquad \Delta = h^2 - k^2\varepsilon_r\mu_r. \tag{8.2}$$

The longitudinal components E_z, H_z satisfy scalar wave equations, which are Helmholtz's equations:

$$\left[\Delta_\perp + k_\perp^2\right]\left\{\begin{matrix} H_z \\ E_z \end{matrix}\right\} = 0. \tag{8.3}$$

Here $\Delta_\perp = \dfrac{\partial^2}{\partial x^2} + \dfrac{\partial^2}{\partial y^2}$ is the transversal Laplacian. The magnitudes are:

$k_\perp^2 = -\Delta = k^2\varepsilon_r\mu_r - h^2$, $k = \omega/c$ and c is the light velocity in a vacuum. The fundamental solution of the second order differential equations (8.3) in the cylindrical coordinates (or in the polar coordinates, since the dependence on the longitudinal coordinate has already been determined) is the Hankel function of the zeroth order.

    The problem is formulated in this way. We have in the complex plane a piece–wise smooth contour L (Fig.8.1). This contour subdivides the plane into two regions; the inner S^+ and the outer S^- region. These regions according to the physical problem are characterized by different electrophysical parameters: the region S^+ has constitutive parameters ε_1, μ_1 and S^- has constitutive parameters ε_0, μ_0. The positive direction of going round the contour is when the region S^+ is on the left side. One has to determine in region $S^\pm$ solutions of Helmholtz's equation (8.3), which satisfies the boundary conditions for the tangent components of the magnetic and electric fields:

$$H^+\big|_L = H^-\big|_L,$$

$$E^+\big|_L = E^-\big|_L. \tag{8.4}$$

If there is perfect metal the tangent components of the electric and magnetic fields are equalized to zero on the metal surface. We write the solution to the equations (8.3) in the form:

$$H_z(\vec{r}) = \int\limits_{L_m} \mu_h(\vec{r}_s) H_0^{(2)}(k_\perp r')ds, \tag{8.5}$$

$$E_z(\vec{r}) = \int\limits_{L_m} \mu_e(\vec{r}_s) H_0^{(2)}(k_\perp r')ds, \tag{8.6}$$

where $H_z(\vec{r})$, $E_z(\vec{r})$ are the longitudinal components of the magnetic and electric fields. Here $\vec{r} = \vec{i}x + \vec{j}y$ is the radius vector of the point, where the fields are determined (see Fig.8.1b), where $\vec{i},\vec{j}$ are the unit vectors. Here ds is an element of the contour L_m. The magnitudes $\mu_h(\vec{r}_s)$ and $\mu_e(\vec{r}_s)$ are the unknown functions satisfying Hoelder's condition [3.4]. Here $H_0^{(2)}(k_\perp r')$ is the Hankel function of the zeroth order and the second kind, where $r' = \left| \vec{r}_s - \vec{r} \right|$ and $k_\perp = \sqrt{k^2 \varepsilon_r \mu_r - h^2}$ is the transversal wave–number. The magnitude $\vec{r}_s = \vec{i}x_s + \vec{j}y_s$ is the radius vector of the contour point (at this point we will determine the functions $\mu_e(\vec{r}_s)$ and $\mu_h(\vec{r}_s)$).

Fig.8.2 shows the points of the contours L_m (m=1,2), where we satisfy the boundary conditions on the boundary line DD', dividing the media with the constitutive parameters ε_1, μ_1 and ε_0, μ_0.

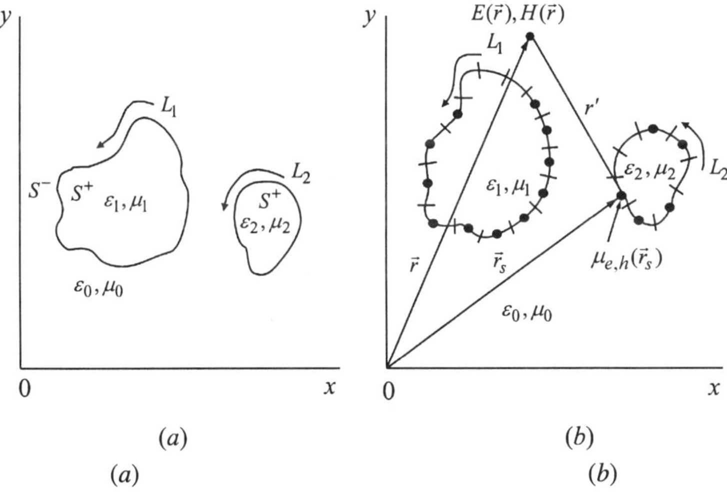

(a)   (b)

Fig. 8.1. *Geometry of a waveguide cross–section.*

We say that the solutions (8.5) and (8.6) are electrodynamically rigorous because they satisfy Maxwell's equations, the condition in infinity and the boundary conditions.

The slot line (SL) with the cross–section geometry shown in Fig.8.3 where the upper contours are filled with perfect metal and the lower is magnetodielectric.

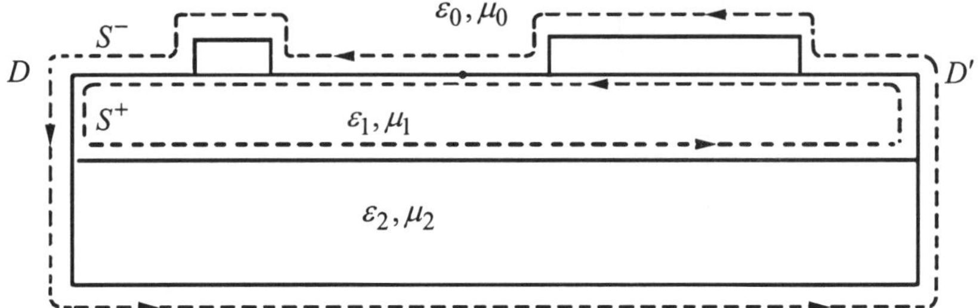

Fig. 8.2. *The integration contours.*

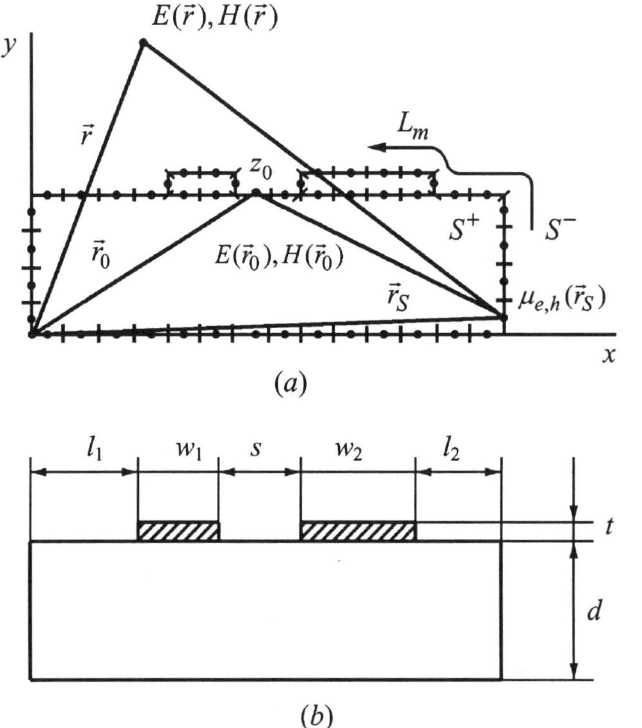

(a)

(b)

Fig. 8.3. *Geometry of a SL cross–section.*

When solving electrodynamic problems we satisfy boundary conditions. On the boundary dividing the air and the magnetodielectric we require the equality of the tangent components of magnetic and electric fields:

$$H_z^+ = H_z^-,$$

$$H_s^+ = H_s^-,$$

$$E_z^+ = E_z^-,$$

$$E_s^+ = E_s^-. \tag{8.7}$$

On the boundary of perfect metal:

$$E_z^+ = 0 \,,$$

$$E_s^+ = 0 \,, \tag{8.8}$$

where s is one of the transversal coordinates x or y and z is the longitudinal coordinate. Here E_z^+ and E_s^+ are tangent components of the electric field.

The singularities in the expressions (8.7) and (8.8) for the transversal components E_s and H_s appear from the term $\ln r'$ of the Hankel function $H_0^{(2)}(k_\perp r')$ in the representations (8.5) and (8.6). Therefore, we first consider the function:

$$\varphi(\vec{r}) = \int\limits_{L_m} \mu(\vec{r}_s)\ln r' \, ds \ . \tag{8.9}$$

The partial derivatives of the function (8.9) are:

$$\frac{\partial\varphi}{\partial x} = \operatorname{Re}\, \int\limits_{L_m} \mu(\vec{r}_s)\frac{ds}{z-t} \,,$$

$$\frac{\partial\varphi}{\partial y} = -\operatorname{Im}\, \int\limits_{L_m} \mu(\vec{r}_s)\frac{ds}{z-t} \,. \tag{8.10}$$

These partial derivatives are the real and imaginary parts of the Cauchy type integral. The quantities $z = x + jy$, $t = x_s + jy_s$ where j is the imaginary unit. Before we used another imaginary unit i, which was being used in time–factor $e^{i\omega t}$ and that described the field component phase shift in time.

Here, with Sokhotsky–Plemelj formulae (3.4) we write the limit values of the derivatives (8.10) with respect to the tangent τ and the normal n of the contour:

$$\frac{\partial\varphi^\pm}{\partial\tau} = \operatorname{Re}\, e^{j\theta} \int\limits_{L_m} \frac{\mu(r_s)ds}{z_0-t} \,, \tag{8.11}$$

$$\frac{\partial\varphi^\pm}{\partial n} = \mp\pi\mu(r_0) - \operatorname{Im}\, e^{j\theta_o} \int\limits_{L_m} \frac{\mu(r_s)ds}{z_0-t} \,. \tag{8.12}$$

In these formulae the contour point z_0 (with the radius vector $\vec{r}_0$), can be reach from the left side or the right. The upper plus sign in the formulae (8.11) and (8.12) corresponds the reaching the point z_0 from the left side. The lower minus sign in these formulae corresponds the reaching the point z_0 from the right. The integrals in the formulae (8.11) and (8.12) are to understand as the Cauchy principal value integrals. Here θ_0 is the angle between the x axis and the tangent to the contour (in the integration direction) at the point $z_0 = x_0 + jy_0$. When applying the Krylov–Bogoliubov method we obtained for transversal field components at the contour points:

$$
\begin{pmatrix} H_x \\ H_y \end{pmatrix}^{\pm} = \pm \frac{2\varepsilon_0 \varepsilon_r \omega \, \cos\theta_0}{k_{\perp}^{\pm 2}} \mu_e^{\pm}(s_j) + \begin{bmatrix} \chi & 0 \\ 0 & -Q \end{bmatrix} \times
$$

$$
\times \left[k_{\perp}^{\pm} \sum_{j=1}^{n_x} \begin{pmatrix} \mu_e(s_j) \\ \mu_h(s_j) \end{pmatrix}^{\pm} \int\limits_{\Delta L_x} H_1^{(2)}(k_{\perp}^{\pm} r') \frac{y_s - y_0}{r'} dx_s - \right.
$$

$$
\left. - \sum_{j=1}^{n_y} \begin{pmatrix} \mu_e(s_j) \\ \mu_h(s_j) \end{pmatrix}^{\pm} H_0^{(2)}(k_{\perp}^{\pm} r') \big| \Delta L_y \right] +
$$

$$
+ \begin{bmatrix} Q & 0 \\ 0 & \chi \end{bmatrix} \left[\sum_{j=1}^{n_x} \begin{pmatrix} \mu_h(s_j) \\ \mu_e(s_j) \end{pmatrix}^{\pm} H_0^{(2)}(k_{\perp}^{\pm} r') \big| \Delta L_x - \right.
$$

$$
\left. - k_{\perp}^{\pm} \sum_{j=1}^{n_y} \begin{pmatrix} \mu_h(s_j) \\ \mu_e(s_j) \end{pmatrix}^{\pm} \int\limits_{\Delta L_y} H_1^{(2)}(k_{\perp}^{\pm} r') \frac{x_s - x_0}{r'} dy_s \right] \tag{8.13}
$$

$$
\begin{pmatrix} E_x \\ E_y \end{pmatrix}^{\pm} = \mp \frac{2\mu_0 \mu_r \omega \, \cos\theta}{k_{\perp}^{\pm 2}} \mu_h^{\pm}(s_j) - \begin{bmatrix} V & 0 \\ 0 & Q \end{bmatrix} \times
$$

$$
\times \left[k_{\perp}^{\pm} \sum_{j=1}^{n_x} \begin{pmatrix} \mu_h(s_j) \\ \mu_e(s_j) \end{pmatrix}^{\pm} \int\limits_{\Delta L_x} H_1^{(2)}(k_{\perp}^{\pm} r') \frac{y_s - y_0}{r'} dx_s - \right.
$$

$$
\left. - \sum_{j=1}^{n_y} \begin{pmatrix} \mu_h(s_j) \\ \mu_e(s_j) \end{pmatrix}^{\pm} H_0^{(2)}(k_{\perp}^{\pm} r') \big| \Delta L_y \right] +
$$

$$
+ \begin{bmatrix} Q & 0 \\ 0 & -V \end{bmatrix} \left[\sum_{j=1}^{n_x} \begin{pmatrix} \mu_e(s_j) \\ \mu_h(s_j) \end{pmatrix}^{\pm} H_0^{(2)}(k_{\perp}^{\pm} r') \big| \Delta L_x - \right.
$$

$$
\left. - k_{\perp}^{\pm} \sum_{j=1}^{n_y} \begin{pmatrix} \mu_e(s_j) \\ \mu_h(s_j) \end{pmatrix}^{\pm} \int\limits_{\Delta L_y} H_1^{(2)}(k_{\perp}^{\pm} r') \frac{x_s - x_0}{r'} dy_s \right], \tag{8.14}
$$

where $\chi = i\varepsilon_0 \varepsilon_r \omega / k_{\perp}^{\pm 2}$, $V = i\mu_0 \mu_r \omega / k_{\perp}^{\pm 2}$, $Q = ih / k_{\perp}^{\pm 2}$. The magnitudes n_x and n_y are the numbers of sections into which the horizontal and vertical parts of the contour L_m ($m = 1,2,3,...$) are divided. The magnitudes ΔL_x and ΔL_y are the horizontal and vertical sections of the contour L_m correspondingly. The whole number of the sections is $n_x + n_y$ and s is the *arc* abscissa. The field components and the values of the functions $\mu_h(s_j)$ and $\mu_e(s_j)$ are noted in the upper–right corner with the sign corresponding to different waveguide regions, for instance, the functions $\mu_h^{+}(s_j)$, $\mu_e^{+}(s_j)$ or $\mu_h^{-}(s_j)$, $\mu_e^{-}(s_j)$. These functions at the same contour point are different for the field components in the regions S^{+} and

S^- and $\mu_h^+(s_j) \neq \mu_h^-(s_j)$. Here the subscript j indicates the ordinal number of the sections ΔL_x and ΔL_y that varies in the range from 1 to $(n_x + n_y)$. The integrals in (8.13) and (8.14), having the kernel:

$$H_1^2(k_\perp r') = RH_1^{(2)}(k_\perp r') + \frac{2i}{\pi}\frac{1}{k_\perp r'}, \tag{8.15}$$

have been calculated numerically singling out the singularity analytically. The first term in the expression (8.15) is the regular part of the function and the second is the singular part. Note that the first terms in the expressions (8.13) and (8.14) contribute only to the diagonal elements of the system as follows from the Sokhotsky–Plemelj formulae. The computations of electrodynamical problems show that absolute values of diagonal elements by one–two orders exceed the absolute values of nondiagonal. Because the system is well conditioned this leads to the stability of the solution algorithms. The longitudinal field components after applying the Krylov–Bogoliubov method take the form

$$E_z^\pm = \sum_{j=1}^{n_x} \mu_e^\pm(s_j) \int_{\Delta L_x} H_0^{(2)}(k_\perp^\pm r')ds + \sum_{j=1}^{n_y} \mu_e^\pm(s_j) \int_{\Delta L_y} H_0^{(2)}(k_\perp^\pm r')ds ,$$

$$H_z^\pm = \sum_{j=1}^{n_x} \mu_h^\pm(s_j) \int_{\Delta L_x} H_0^{(2)}(k_\perp^\pm r')ds + \sum_{j=1}^{n_y} \mu_h^\pm(s_j) \int_{\Delta L_y} H_0^{(2)}(k_\perp^\pm r')ds .$$

The system of the algebraic equations for the unknowns $\mu_h^+(s_j)$, $\mu_h^-(s_j)$, $\mu_e^+(s_j)$, $\mu_e^-(s_j)$ obtained from the boundary conditions is homogeneous. The condition of solvability is obtained equalizing the determinant of the system to zero. The root of the equation obtained determines the propagation constant. After obtaining the propagation constant the determination of the electric field of the waveguide is not difficult. For the correct formulated problem [3.4] the solution is one–valued and stable with respect to small changes of the coefficients and the contour form.

8.2. Examples of Testing the Algorithm

In our numerical study we investigated the algorithm stability by changing slightly sizes of waveguides and other parameters. Here are some examples of our calculations after some slight changes of the contour forms and the signal frequencies:

1. Table 8.1 presents the results for SLs with the sizes (Fig.8.3b): $l_1 = l_2 = 0$, $w_2 = 10^{-3}\,\mathrm{m}$, $s = 10^{-3}\,\mathrm{m}$, $d = 10^{-3}\,\mathrm{m}$, $t = 2.5 \cdot 10^{-6}\,\mathrm{m}$ at $f = 10\,\mathrm{GHz}$ and $\varepsilon_r = 11.8$.

2. Table 8.2 presents the results for SLs with the sizes (Fig.8.3b): $l_2 = 2 \cdot 10^{-3}$ m, $s = 10^{-3}$ m, $w_1 = w_2 = 2 \cdot 10^{-3}$ m, $t = 2.5 \cdot 10^{-6}$ m at $f = 10$ GHz and $\varepsilon_r = 11.8$.

3. Table 8.3 presents the results for the SLs, which differs from the previous ones only in the substrate sizes $l_1 = l_2 = 0$ as a function of the slight frequency changes.

Table 8.1. *Dependence of the longitudinal propagation constant h when we change the metal strip w_1 slightly.*

Mode number	h, m^{-1}				
	Strip width, $w_1 \cdot 10^{-3}$, m				
	$w_1 = 3.0$	2.9	2.7	2.5	2.0
The main mode h	633	633	632	632	631
The first higher mode h	565	565	564	564	563
The second higher h	522	522	522	521	520

Table 8.2. *Dependence of the longitudinal propagation constant h when we change the substrate length l_1 slightly.*

Mode number	h, m^{-1}				
	The substrate length, $l_1 \cdot 10^{-3}$ (m)				
	$l_1 \cdot = 0$	0.05	0.1	0.5	1.0
The main mode h	633	621	600	597	591
The first higher mode h	560	552	550	549	546

Table 8.3. *Dependence of the longitudinal propagation constant h when we change the frequency slightly.*

Mode number	h, m^{-1}					
	Frequency f, (GHz)					
	$f = 9.9$	10.0	10.1	10.2	10.4	10.5
The main mode h	623	633	644	653	664	674
The first higher mode h	558	560	567	574	586	591
The second higher h	523	528	534	539	545	556

It is worth mentioning that when computing the waveguide dispersion characteristics, the smooth change of the electrophysical parameters and the waveguide sizes it always leads to the smooth variation of the computation results. Here we studied the accuracy and stability of our calculations, which depend on the system determinant order.

Eigenmodes of open waveguides. Our calculation algorithm and our computer programs were first applied to MSLs and then after that to SLs. We investigated the planar and nonplanar slot lines (NSL) and compared the theoretical and experimental results. We optimized the SLs in the frequency range from 37.5 to 78 GHz as a function of the width of the slot between metal strips and the distance to the substrate lateral edges from both metal strips. The influence of the asymmetry in the SL geometry on the dispersion characteristics was investigated and the new nontrivial results for the SLs with the micron slot size were obtained.

8.3. Microstrip Line Dispersion Dependences and Comparison of our Calculated Results with Data from References

In the paper [7.32] the open one metal strip MSL on the dielectric substrate was studied and the sizes were: $l_1/d = l_2/d = \infty$, $t/d = 0$. The same magnitudes in our calculation (Fig.8.4) had other values of sizes: $l_1/d = l_2/d = 1.5$, $t/d = 0.005$. Here in our calculations the thicknesses of the metal strip and the metal plate were equal. We designated the geometrically symmetric microstrip line as MSL–S. Fig.8.4 presents the dependence of the propagation constant h of the main mode and two higher modes on the signal frequency for the MSL–S. In this Fig.8.4 the dependence of the wavelength λ_{MSL} on the signal frequency is shown for only the main mode. Our calculations are solid lines and the data from [7.32] are circles. We noted here, that the largest deviation in the results is observed when the wavelength is compared to the substrate width, as we see here. So when λ_{MSL} was compared to the substrate width we could not assume that the substrate was limitless in its lateral directions. Our last conclusion is especially important when someone would like create microstrip devices.

Fig.8.5 presents the dispersion dependences of the main mode and two higher modes for the geometrically asymmetric microstrip line (MSL–A).In [8.1] and [8.2] we describe in more detail the characteristics of the MSL–S and MSL–A.

Fig.8.6 presents the dispersion dependences of the main made and two higher modes for the transmission line resembles as a MSL with dielectric substrate and without the bottom metal plate (we designate as DML). Fig.8.7 presents the dispersion dependences of the main mode for the three above mentioned transmission lines: MSL–S, MSL–A, DML and for the coplanar line (CPL) with the sizes: $w_1 = w_2 = 2 \cdot 10^{-3}$ m, $d = 10^{-3}$ m, $l_1 = l_2 = 2 \cdot 10^{-3}$ m, $t = 3 \cdot 10^{-6}$ m at $\varepsilon_r = 11.7$. The MSL–A appears to have least frequency dependence when compared to the other three waveguides.

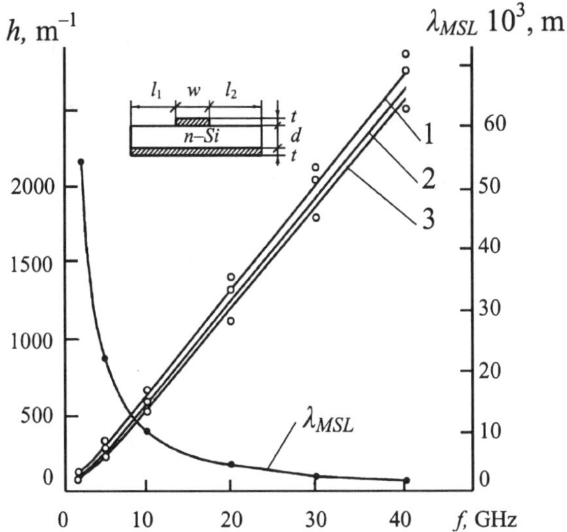

Fig. 8.4. *Dispersion dependences of the longitudinal propagation constant* h *of the main mode (curve 1) and two higher modes (curves 2, 3) and the wavelength* λ_{MSL} *in the geometrically symmetric MSL with parameters:*

$d = 3.17 \cdot 10^{-3}$ m, $w/d = 0.96$, $\varepsilon_r = 11.7$. *Here the solid lines are the calculated results and the circles are from [7.32].*

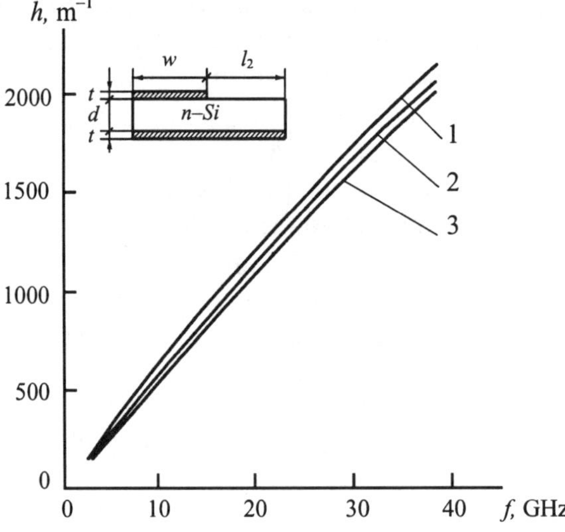

Fig. 8.5. *Dispersion dependences of the longitudinal propagation constant* h *of the main mode (curve 1) and two higher modes (curves 2, 3) for the MSL–A. The sizes are:* $d = 10^{-3}$ m, $w = 2 \cdot 10^{-3}$ m, $l_1 = 0$, $l_2 = 2 \cdot 10^{-3}$ m, $t = 3 \cdot 10^{-6}$ m.

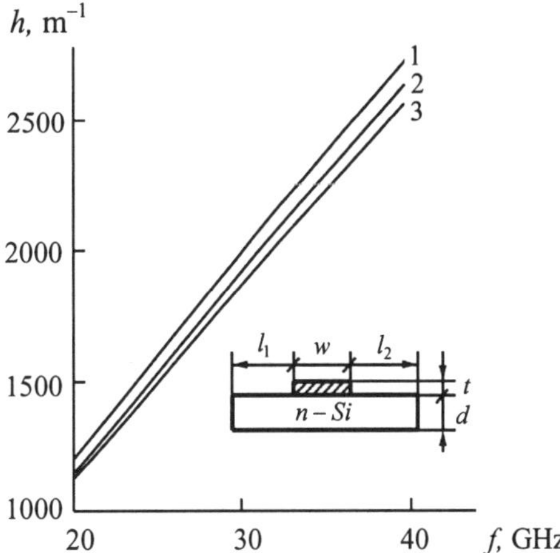

Fig. 8.6. *Dispersion dependences of the longitudinal propagation constant* h *for the main mode (curve 1) and two higher modes (curves 2, 3) of the DML. The line sizes are:* $d = 3.17 \cdot 10^{-3}$ m, $w = 3.043 \cdot 10^{-3}$ m, $l_1 = l_2 = 5 \cdot 10^{-3}$ m, $t = 3 \cdot 10^{-6}$ m.

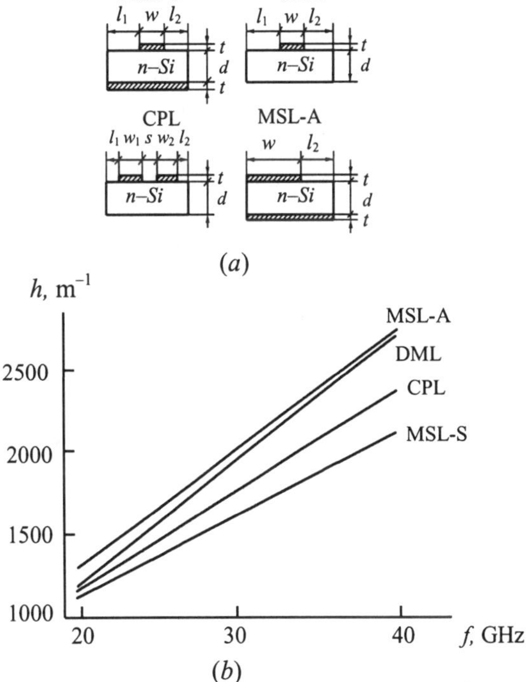

Fig. 8.7. *(a) the geometry of waveguides, (b) dispersion dependences of the main mode for waveguides.*

8.4. The Dispersion Dependences of Planar Slot Lines and the Comparison of our Calculated Results with Experimental Data

Here our calculations assume the thickness of the metal strips and the substrate size to be finite (Fig.8.3b). Fig.8.8a presents our calculated results and those from the paper [8.3]. This paper shows the SL parameters as: $d = 0.64 \cdot 10^{-3}$ m, $s = 4 \cdot 10^{-3}$ m, $l_1 = l_2 = \infty$, $t = 0$, $\varepsilon_r = 9.7$, $\mu_r = 1.0$. In our calculations we have used the same parameters with the exception of the sizes $l_1 = l_2 = 7 \cdot 10^{-3}$ m and $t = 10^{-6}$ m. Fig.8.8b presents our calculations as solid lines and circles [8.4]. This paper shows the parameters of the SL as: $d = 10^{-3}$ m, $s = 2 \cdot 10^{-3}$ m, $l_1 = l_2 = \infty$, $t = 0$, $\varepsilon_r = 9.0$, $\mu_r = 1.0$. In our calculations we used the same parameters except the sizes $l_1 = l_2 = 16 \cdot 10^{-3}$ m, $t = 2.5 \cdot 10^{-3}$ m.

Fig.8.8c presents the calculated results for the SL with parameters similar to the Fig.8.8b except the slot size are $s = 0.5 \cdot 10^{-3}$ m. At the frequencies below 20GHz the wavelength in the SL becomes comparable with the width of the metal strip and the width of the substrate. Therefore, the calculated characteristics of the SL with the limited substrate are noticeably different then the characteristics of the SL with the infinite substrate. One can see from Fig.8.8c that when the λ_{SL} is five times less than the substrate width, then the influence of the side facets is weak, and the results of our calculations and of [8.4] coincide well. Now if the wavelength is comparable to the SL substrate width, the boundary conditions on the substrate side facets have be taken considered in the calculations.

Theoretical and experimental studies of the planar SL were described in [8.5] and [8.6]. Where the SL substrate was made of electronic silicon (n–Si) with the specific resistivity of 2.45 to 3.36 $(\Omega \cdot m)$ and had the thickness $d = 10^{-3}$ m and the length of the SL was $45 \cdot 10^{-3}$ m. The metal strip thickness was $t = 3 \cdot 10^{-6}$ m. The SL was excited by the H_{10} wave of the rectangular metal waveguide. The coupling element of the SL with the waveguide was made of a wire frame, which was connected with two thin knife–like metal strips which extended past the SL edge. The adjustment of the SL excitation was reached by moving the SL end a little into the waveguide.

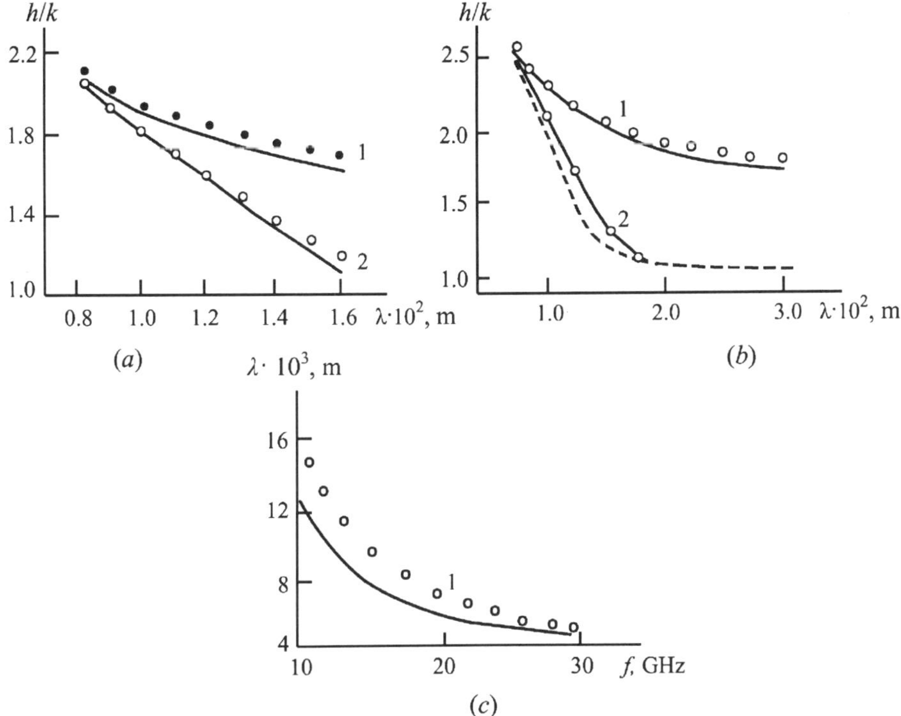

Fig. 8.8. *Dispersion dependences of the longitudinal propagation constant h of the main mode (curve 1) and the first higher mode (curve 2). The solid lines are our calculation and the circles are the results from [8.3], [8.4].*

It was not possible in this experiment to control the type of the propagating waves. The value of the wavelength λ_{SL} measured in the experiment was the most different compared to our theoretical results, when our calculations shown that in this SL could propagate several (the main and higher) modes. In Figs.8.9–8.12 we present the calculated dispersion characteristics for the SL mode number, which was detected by two frequencies 8 and 10 GHz. The dielectric permittivity $\varepsilon_r = 11.8$ was used in our calculations. Every circle in Figs.8.9–8.12 shows the average mean over several measurements of the longitudinal propagation constant. The SL size choices in Figs.8.9–8.12 were used only to illustrate the ability of our SIE method. This is why we not only study ones–mode SLs but also multi–mode SLs.

Fig.8.9a,b presents the dependence of the longitudinal propagation constant on the signal frequency for the asymmetric SL with the slot sizes $s = 0.2 \cdot 10^{-3}$ m and $s = 10^{-3}$ m. We see the measured results given in Fig.8.9a only for the SL with the slot width $0.2 \cdot 10^{-3}$ m. Analyzing the theoretical dispersion dependences for these SLs, we see that with the slot width increasing there are changes, but not only with the wave length in the SL but also with an increased number of propagating eigenwaves. Comparing the theoretical and experimental results of Fig.8.9*a* we see that with the frequency increasing the experimental data moves from the theoretical dispersion curve for the main mode (curve 1) nearer to the curve 2 of the higher mode. The sensor reacts to the superposition of the main mode and higher modes. When the frequency increases, the influence of the higher mode becomes stronger. As a result, the contribution of the higher mode in the experimental signal increased and the wavelength registered value became nearer to the higher mode value.

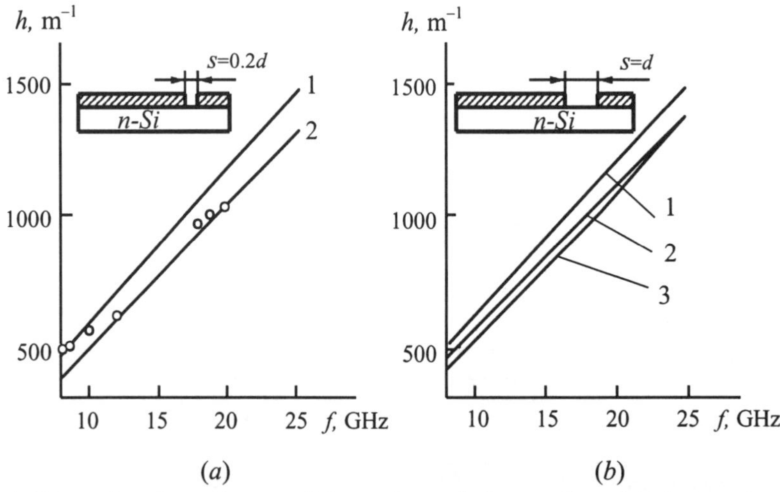

Fig. 8.9. *Dispersion dependences of the main mode (curve 1) and higher modes (curves 2, 3) for the asymmetric SL. The line parameters are:* $d = 10^{-3}$ *m,* $w_1 = 3 \cdot 10^{-3}$ *m,* $w_2 = 10^{-3}$ *m,* $l_1 = l_2 = 0$, $t = 3 \cdot 10^{-6}$ *m. The solid lines are our calculations and the circles are the experiment [8.5], [8.6].*

Fig.8.10*a,b* presents similar dependences for the main mode and three higher modes of the symmetric SL with substrates protruding out of the metal strip limits. When we compared Fig.8.9*b* with Fig.8.10*b*, we saw that the last SL is more multi–mode and has other dispersion dependence characteristics. The last SL also has a denser spectrum. Here the symmetric SL has the same sizes $l_1 = l_2 = 2 \cdot 10^{-3}$ m as in the previous SL but without the protruding substrate edges $l_1 = l_2 = 0$. Fig.8.11*a* presents the dispersion dependences of the first three eigenmodes (the main mode and two higher modes). Comparing Fig.8.10*b* with Fig.8.11*b* we see that the dispersion dependence characters are different, despite the fact that the SL differs only in the protruding substrate edges.

In Fig.8.11*b* there is not a high frequency cutoff in the frequency range that we considered. But when we investigated at a higher frequency range we saw that in the SL with parameters of Fig.8.11*b* that the dispersion curves of the first and the second higher modes (curves 1 and 2) have a high frequency cutoff at 34GHz and 40GHz respectively. We do not show these in Fig.8.11*b*.

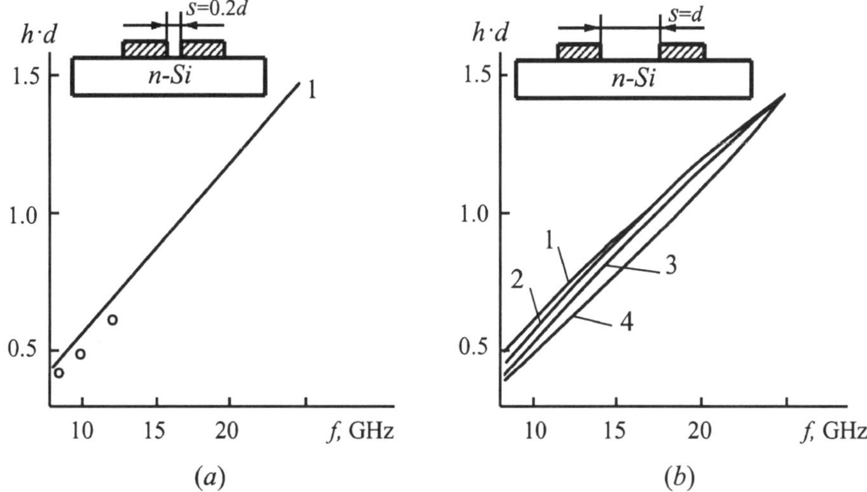

Fig. 8.10. *Dispersion dependences of the main mode (curve 1) and three higher modes (curves 2, 3, 4) for the SL with narrow metal strips. The SL parameters are:* $d = 10^{-3}$ m, $w_1 = w_2 = 2 \cdot 10^{-3}$ m, $l_1 = l_2 = 2 \cdot 10^{-3}$ m, $t = 3 \cdot 10^{-6}$ m. *The solid lines are our calculations and the circles are the experimental results.*

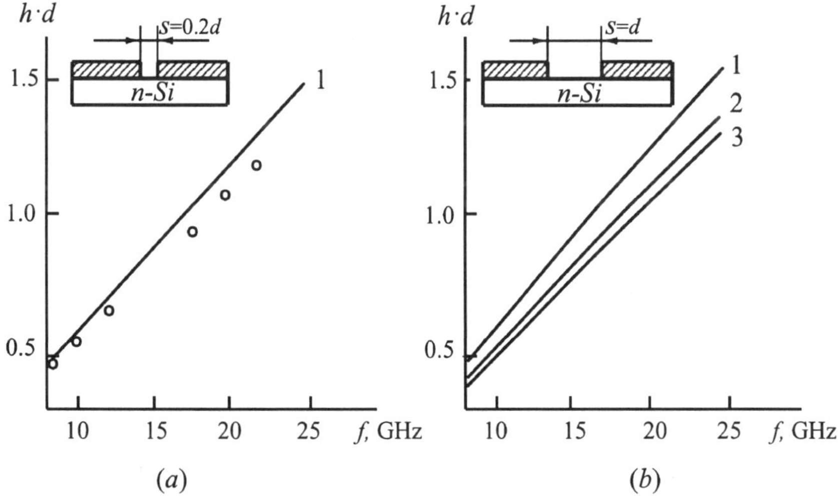

Fig. 8.11. *Dispersion dependences of the SL with metal strips placed at the substrate edges for the main mode (curve 1) and two higher modes (curves 2, 3). The SL parameters are:* $d = 10^{-3}$ m, $w_1 = w_2 = 2 \cdot 10^{-3}$ m, $l_1 = l_2 = 0$, $t = 3 \cdot 10^{-6}$ m. *The solid lines are our calculations and the circles are experimental results.*

Fig.8.12 presents the dispersion dependences of the main mode and the first higher mode. All four SL types we considered are united by the total width of the metal strips $w_1 + w_2$, which is $4 \cdot 10^{-3}$ m. One SL type differs from another in the redistribution of the metal over the substrate surface and as a result the boundary conditions change slightly for these SL.

If we change the boundary conditions the SL characteristics will change noticeably. This is demonstrated by the calculated characteristics of the multi–mode SL (Figs.8.9b–8.11b). It is difficult to measure each mode of the multi–mode SL with high accuracy. Because the measurement of multi–mode SL electrodynamical parameters is difficult, it is necessity to develop rigorous methods for these SL calculations.

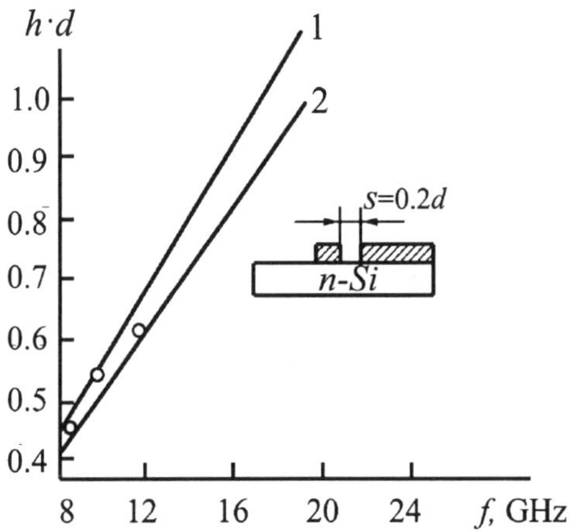

Fig. 8.12. Dispersion dependences of the asymmetric SL with the protruding substrate edge for the main mode (curve 1) and the first higher (curve 2) mode. The SL parameters are: $d = 10^{-3}$ m, $w_1 = 10^{-3}$ m, $w_2 = 3 \cdot 10^{-3}$ m, $l_1 = 10^{-3}$ m, $l_2 = 0$, $t = 3 \cdot 10^{-6}$ m. *The solid lines are our calculations and the circles are the experimental results.*

8.5. The Numerical Investigation of Planar Slot Lines with Asymmetrically Placed Metal Strips and with Micron Slots Between them

Here we will present some interesting results of our calculations. We investigated regular open SLs with substrates made of n–Si ($\varepsilon_r = 11.8$) at $s \ll \lambda_{SL}$ and studied the dispersion characteristics for different sizes of metal strips, slots, substrate widths and thicknesses. We now show the results in the normalized form multiplying all magnitudes by the substrate thickness

$d = 10^{-4}$ m. The bulk of our calculations were carried out for the SL with the slot width $s = (0.05 \div 0.2) \cdot d$, when the wavelength of the main mode and higher modes are within the interval $\lambda \sim (10 \div 30) \cdot d$. The metal strip thickness is $t = 3 \cdot 10^{-6}$ m. The most interesting result of the theoretical SL investigation is the discovery of the nontrivial dependence of the waveguide eigenmode number on the slot width s. (Contrary to the generally accepted opinion; the narrower the slot the less number of propagating SL modes.) We have found that at $s \prec\prec \lambda_{SL}$, when slot size decreases then the number of modes increase.

Fig.8.13 presents the dependence of the SL mode number on the slot width s for two kinds of SLs. One can see that the mode number increases in the SL in two different intervals of s. The first interval of s when λ_{SL} is comparable to the slot width s ($\lambda_{SL} = \phi(s)$). In this first interval of s when the slot width increases the mode number increases. The second interval of s is when $s \prec\prec \lambda_{SL}$. Here when the slot width decreases the mode number increases. It is possible to explain it in this way: when the SL has a very small slot width ($s \prec\prec \lambda_{SL}$) the EM wave does not "feel" this slot. The SL electrodynamic characteristics become similar to the mirror waveguide, which is known as the multi–mode waveguide. In designing integral circuits it is necessary to take into account the feature of open SL become multi–mode waveguide when the slot width decreases.

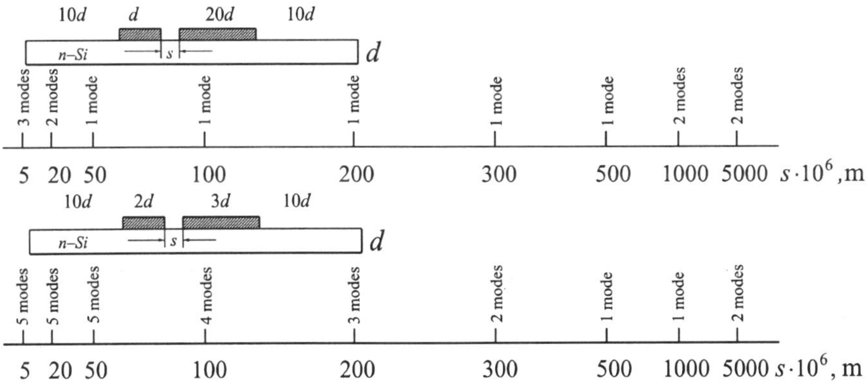

Fig. 8.13. *Dependence of the SL eigenmode number on the slot width s at $d = 100 \cdot 10^{-6}$ m and $f = 37.5$ GHz.*

The investigation of the dispersion characteristics of the SL with a micron size slot as a function of the substrate thickness shows that the dependence do not have particularities as compared to what was known from the references for such lines. For example, the thicker the substrate the larger the number of the eigenmodes can exist in the SL. Fig.8.14 presents the dispersion characteristics for the SL on the thick substrate (the thickness is $3d$). Here we see six eigenmodes can propagate in the indicated frequency range.

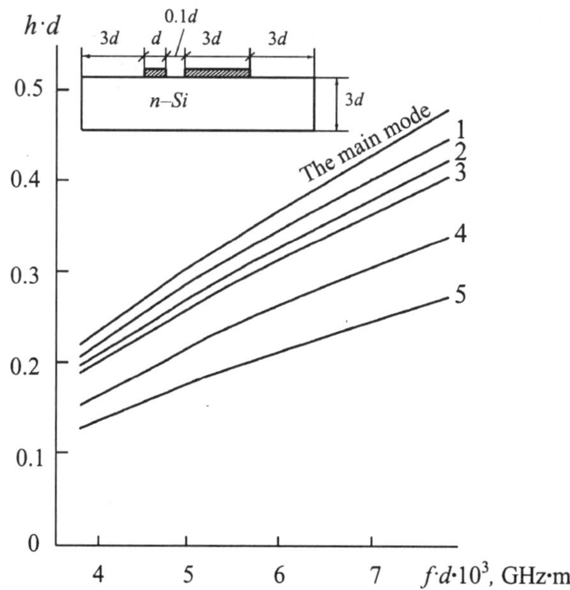

Fig. 8.14. Dependence of the SL longitudinal propagation constants on the frequency at $d = 100 \cdot 10^{-6}$ m *for the main mode and five higher modes.*

Fig.8.15a,b presents the dispersion dependences of the SL eigenmodes (the main mode and three higher modes, curves 1–3) for the different slot widths at $d = 100 \cdot 10^{-6}$ m. In Fig.8.15a the width is $s = 5 \cdot 10^{-6}$ m and in Fig.8.15b the width is $s = 20 \cdot 10^{-6}$ m. We see that even a little change of the slot width and the character of the dispersion dependences of the SL are significantly different.

Fig.8.16a,b presents a similar dependence for the SL differing only in the substrate width. Here in Fig.8.16a the SL size is $l_2 = d$. In Fig.8.16b there is no distance between the metal strips and the substrate lateral edges on the left side $(l_2 = 0)$.

Comparison of these two figures (Figs.8.16a and 8.16b) shows that it is possible to change the bandwidth of a SL device by changing the substrate width. The size of the substrate arm (the distance between the metal strips and the substrate lateral edges) has more influence on the SL mode spectrum when the metal strip is closest to the substrate side facet.

In Fig.8.17 we see the dependences of the longitudinal propagation constant of the main mode for three SLs with different substrate widths. The wider the substrate the less the SL wavelength λ_{SL} and so a larger energy portion can propagate in the substrate material. We understand these features as we compared the dispersion dependences 1 with 2 and 1 with 3 in Fig.8.17. When we restrict the substrate from the right side (where the metal is at the very edge of the substrate) the dispersion dependences of SL is easy to change. The electrical field is more concentrated in the part of the substrate between its right facet and the right rib of the metal strip and we could change the SL electrodynamical characteristics by polishing this right facet (of the substrate edge).

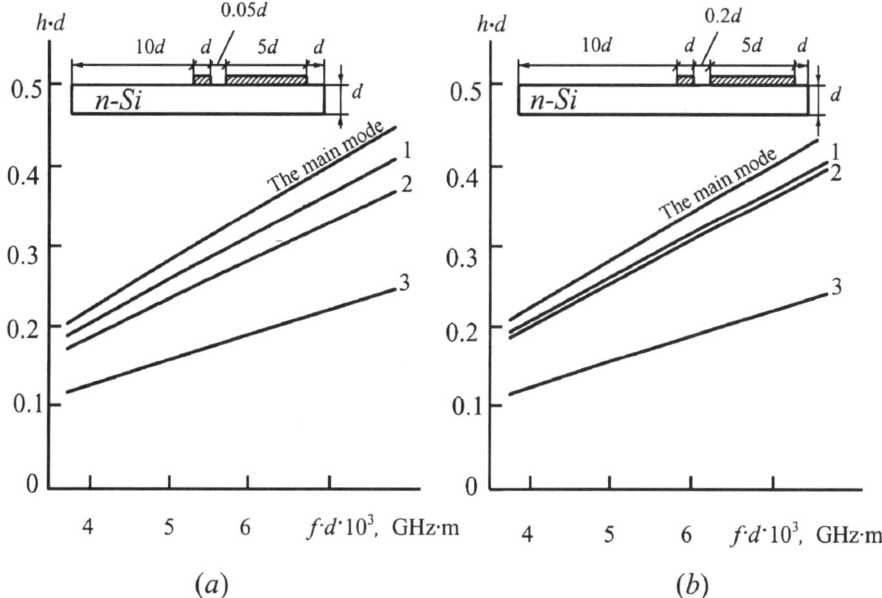

Fig. 8.15. Dependence of the SL longitudinal propagation constants of the main mode and three higher modes (curves 1–3) on the frequency.

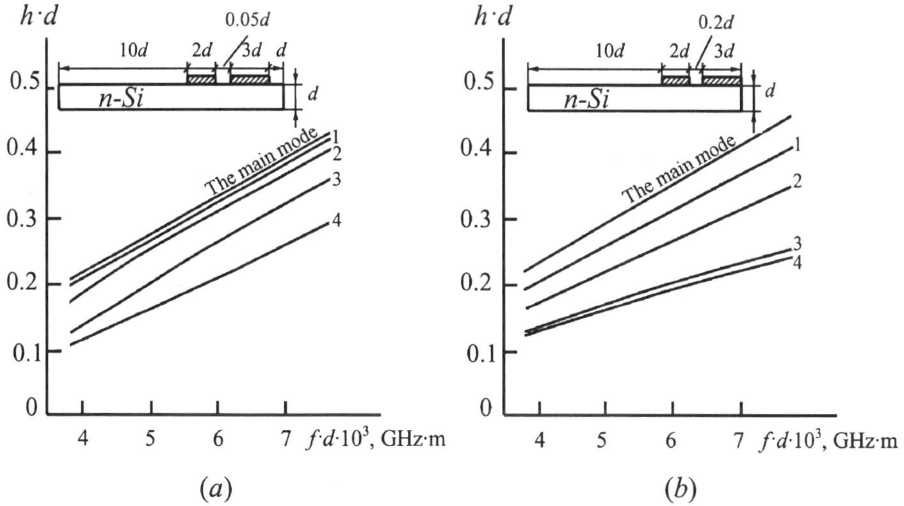

Fig. 8.16. Dependence of the SL longitudinal propagation constants of the main mode and four higher modes (curves 1–4) on the frequency.

The dispersion characteristics are mostly dependent on the sizes of this part of the substrate. We see that there is an additional possibility in receiving the needed electrodynamical characteristics of the SL, which are needed (see the dispersion dependences 2 and 3 in Fig.8.17).

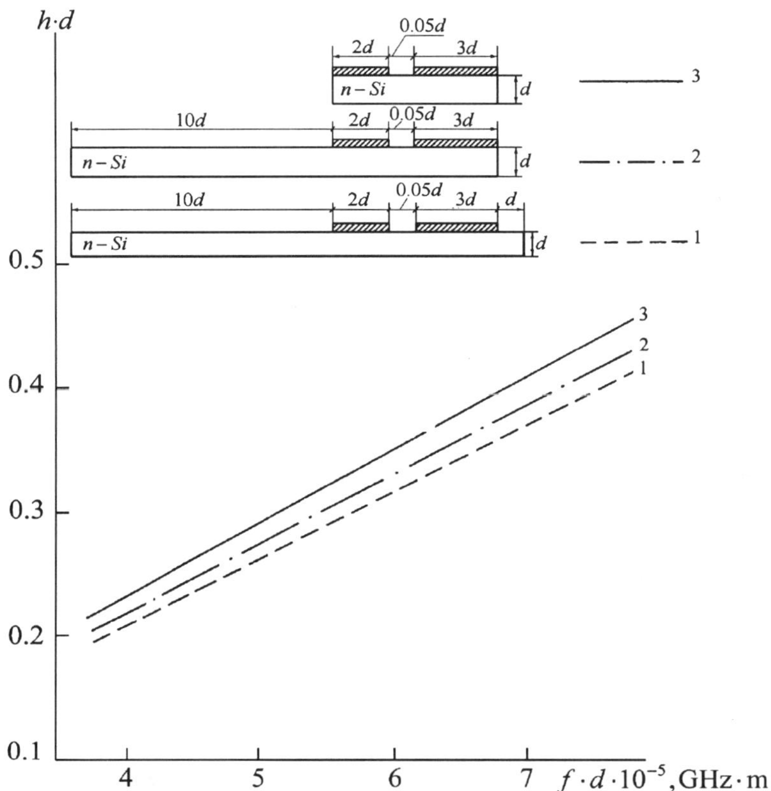

Fig. 8.17. Dispersion dependences for the main modes in the SL having different substrate widths at $d = 100 \cdot 10^{-6}$ m.

Fig.8.18 shows how the dispersion dependences of the main mode differs depending on the left metal strip sizes for two SLs with $w = d$ and $w = 2d$. Fig.8.19 shows how the dispersion dependence of the main mode differs depending on the right metal strip sizes for two SLs with $w = 3d$ and $w = 5d$.

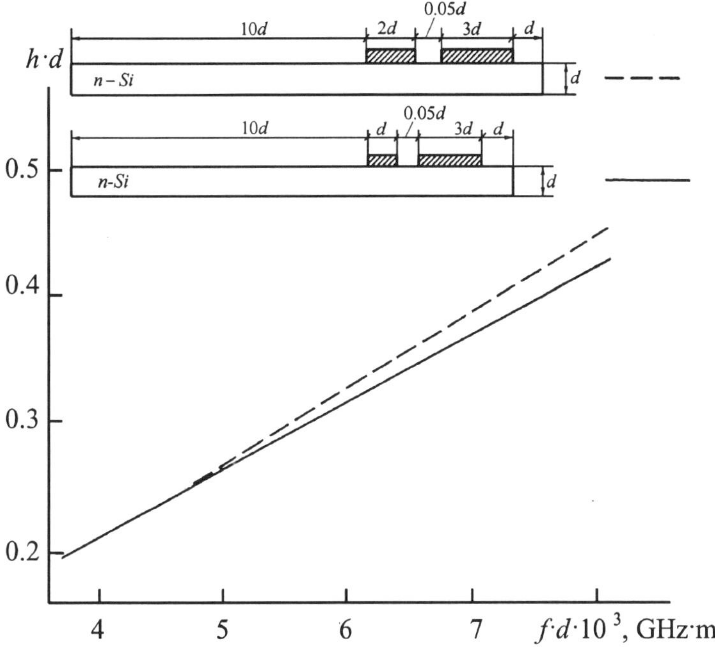

Fig. **8.18**. *Dispersion dependences of the main modes for the SLs having different left metal strip widths.*

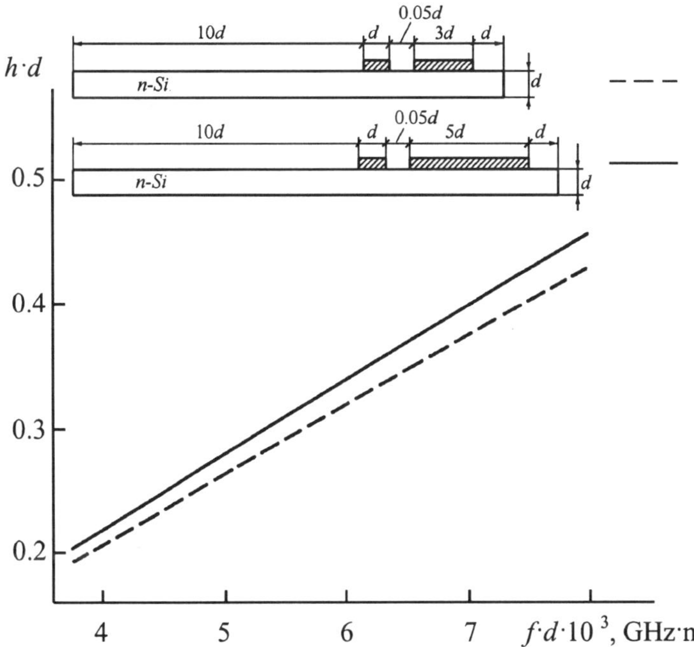

Fig. **8.19**. *Dispersion dependences of the main modes of the SLs having different right metal strip widths.*

8.6. The Dispersion Dependences of the Nonplanar Slot Lines and a Comparison of our Calculated Results with the Experimental Data

There are few papers on the numerical investigation of open nonplanar SL (NSL) in a rigorous electrodynamic problem formulation and among them is [8.4]. It theoretically dealt with a SL having limitless sizes of metal strips and substrates also it did not take into account the metal strips thickness.

We have not found any other articles on the investigation of the NSL in the rigorous electrodynamic problem formulation for the line geometry considered in this section (Fig.8.20). The SL substrates were made of electronic silicon (n–Si) with specific resistivity from 2.45 to 3.36 $(\Omega \cdot m)$ having the thickness $d = 10^{-3}$ m and the length of the NSL was $L = 45 \cdot 10^{-3}$ m (in experiment). The SL metal strips have thicknesses of $t_1 = t_2 = 30 \cdot 10^{-6}$ m. The NSL was exited by the rectangular waveguide H_{10} mode by the method of [8.7] and [8.8]. In our calculations the permittivity of the substrate material was $\varepsilon_r = 11.8$.

In our theoretical computations all sizes of SLs were taken as in experiments. But here we did not take into account the losses in the substrate material (the n-Si specific resistivity was ≥ 4.5 $(\Omega \cdot m)$ in these experiments). And we did not try to optimize the SL geometry to get the propagation of only one–mode. In [8.5] and [8.6] we show that by changing the width of the metal strips and the substrate one can obtain the propagation of only one–mode in the SL (Fig.8.20). We also will see that in the SLs under consideration there maybe multi–mode propagations EM waves. Thus, it is of interest to compare the theoretical calculations with the experimental data for such SL. The calculations were carried out in the frequency range of 8 to 30 GHz and showed that the NSL under consideration was multi–mode [8.9].

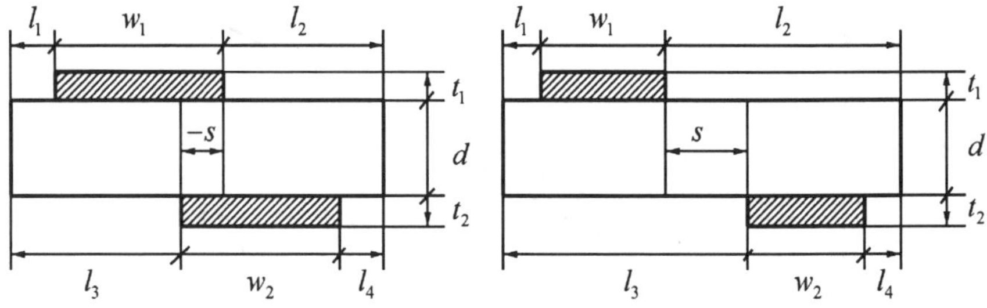

Fig. 8.20. *Cross–section geometry of the NSL.*

Fig.8.21 presents the dependence of the longitudinal propagation constants h on the signal frequency f for the main and four higher modes propagating in the NSL. Fig.8.22 presents similar dispersion dependences for the main and higher modes in the NSL differ in the metal strips and substrate sizes. Fig.8.23 presents similar dispersion dependences for the main and higher modes in the NSL with a different distribution of the metal strips on the substrate surfaces of the SL.

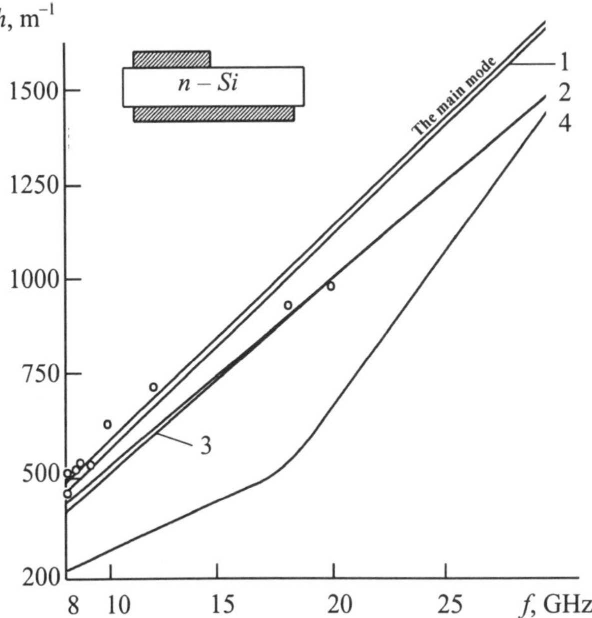

Fig. 8.21. *Dispersion dependences of the main and four higher modes (curves 1–4) for the NSL. The SL sizes are:* $d = 10^{-3}$ m, $w_1 = 2 \cdot 10^{-3}$ m, $w_2 = 4 \cdot 10^{-3}$ m, $l_1 = l_3 = l_4 = 10^{-3}$ m, $l_2 = 2.1 \cdot 10^{-3}$ m *and* $s = -2 \cdot 10^{-3}$ m. *The solid lines are our calculations and the circles are the experimental results* [8.5], [8.6].

The circles in Figs.8.21–8.23 are the average between several measured values of longitudinal propagation constants. The propagation constant $h = 2\pi/\lambda_{SL}$ is determined from the wavelength λ_{SL} measured in the SL. It was not possible to control the type of the propagating modes in these experiments. We saw that at lower frequencies for the SL with two overlapping metal strips (Figs.8.21 and 8.22) that there were propagation constants of an experimentally exited mode and a main mode in our calculations, which were approximately the same.

We can see in Fig.8.23 there was considerable scattering of the experimental results for the SL with the nonoverlapping metal strips. This scattering can be explained by exciting not only the main mode but higher modes at lower frequencies too.

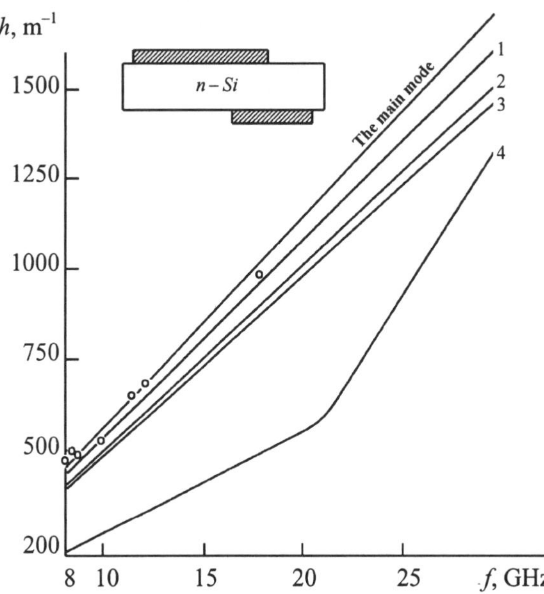

Fig. 8.22*. Dispersion dependences of the main and four higher modes for the NSL. The SL sizes are:* $d = 10^{-3}$ *m,* $w_1 = 3 \cdot 10^{-3}$ *m,* $w_2 = 2 \cdot 10^{-3}$ *m,* $l_1 = l_4 = 0.1 \cdot 10^{-3}$ *m,* $l_2 = 1.1 \cdot 10^{-3}$ *m,* $l_3 = 2.1 \cdot 10^{-3}$ *m and* $s = -10^{-3}$ *m. The solid lines are our calculations and the circles are experimental results* [8.5], [8.6].

The scattering of the experimental results (for three NSLs) when the frequency increases can be explained in a similar way to that of the previous case. In Figs.8.21 and 8.23, we observed the high frequency cutoff in the NSL. The propagation constants of the second and third higher modes become equal at $f = 18$ GHz for the NSL (Fig.8.21) and that of the main mode and first higher mode become equal at $f = 20$GHz for the NSL (Fig.8.23). Comparing the dispersion dependences of Figs.8.21 and 8.23 NSL, we see that they differ greatly when the frequency increases. This happened because the wavelengths of the corresponding NSL modes become comparable with the width of the metal strips in the high frequency band of the frequency that we considered. And NSLs sizes begin to influence the dispersion curve stronger. When comparing (Figs.8.21–8.23) we see that when the frequency increases the dispersion dependences character varies. This is the consequence of another mutual displacement of metal strips on the substrate.

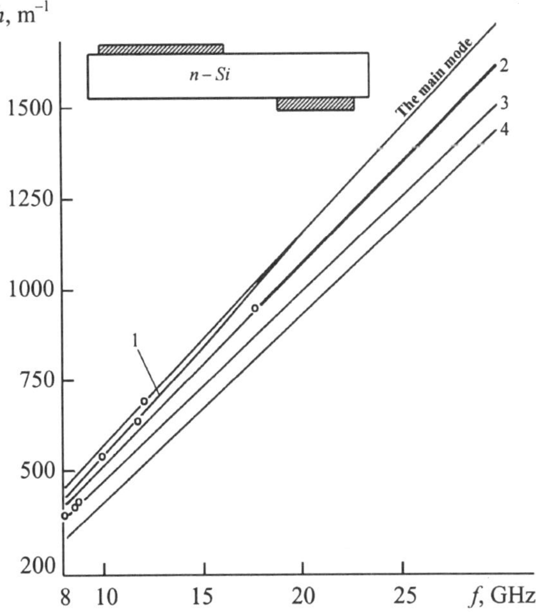

Fig. 8.23. *Dispersion dependences of the main and four higher modes for the NSL with the nonoverlapping metal strip. The SL sizes are:* $d = 10^{-3}$ m, $w_1 = 3 \cdot 10^{-3}$ m, $w_2 = 2 \cdot 10^{-3}$ m, $l_1 = l_4 = 0.1 \cdot 10^{-3}$ m, $l_2 = 3.1 \cdot 10^{-3}$ m, $l_3 = 4.1 \cdot 10^{-3}$ m *and* $s = 10^{-3}$ m. *The solid lines are our calculations and the circles are the experimental results* [8.5], [8.6].

8.7. The Numerical Investigation of Nonplanar Slot Lines with Asymmetrically Placed Strips and Micron Thicknesses of Substrates

Here the object of our theoretical investigation is the regular open NSL with metal strips placed on opposite *n–Si* substrate sides and displaced with respect to each other in the transversal direction (Fig.8.20*a*). The NSL dispersion characteristics were investigated for different sizes of metal strips, distance between a metal strip and the substrate lateral edge as well as different displacements between the metal strips. The relative permittivity of the substrate material (*n–Si*) was taken as $\varepsilon_r = 11.8$. The NSL sizes and all values in Figs.8.24 and 8.25 were multiplied by the substrate thickness $d = 5 \cdot 10^{-6}$ m. The metal strip thickness in all cases was taken as $t = 0.6d$. The calculations were obtained in the frequency range of 37.5 to 78 GHz. When we investigated NSLs there were usually one or two propagating modes. Figs.8.24 and 8.25 present the dispersion dependences for the NSL with metal strips of different sizes.

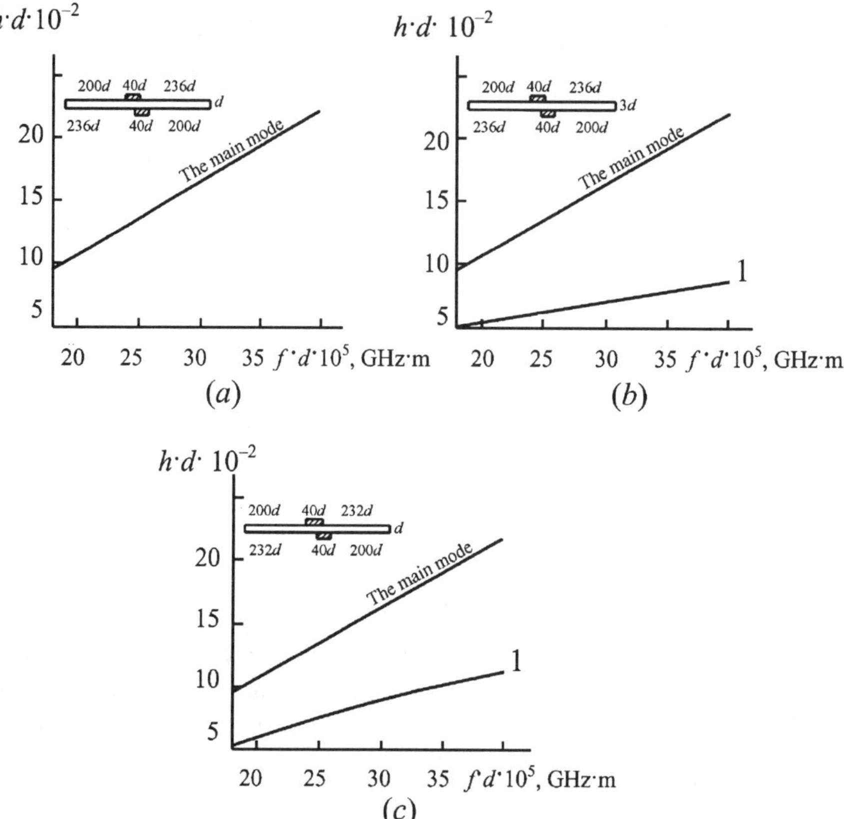

Fig. 8.24. *Dependence of the longitudinal propagation constants of the NSL eigenmodes on the frequency for different substrate thicknesses and overlapping of metal strips: (a) $d = 5 \cdot 10^{-6}$ m, $s = -20 \cdot 10^{-6}$ m; (b) $3d = 15 \cdot 10^{-6}$ m, $s = -20 \cdot 10^{-6}$ m; (c) $d = 5 \cdot 10^{-6}$ m, $s = -40 \cdot 10^{-6}$ m.*

Comparing Fig.8.24a and Fig.8.25a ($d = 5 \cdot 10^{-6}$ m) with Fig.8.24b and Fig.8.25b ($3d = 15 \cdot 10^{-6}$ m) we see that when the substrate thickness increases the main mode propagation constant is practically unchanged. This is because the change of the substrate thickness was insignificant when compared to the wavelength in the NSL. But comparing Figs.8.24a and 8.24b, we see that the higher mode starts to propagate when we change the substrate thickness from size d to size $3d$. Comparing Figs.8.24a and 8.25a with Figs.8.24c and 8.25c, we see that when the value s (the overlapping) changes from $s = -4d$ to $s = -8d$ then the values of main mode propagation constants do not change. This happens because the value s changes was insignificant compared to the wavelength change in the NSL. We observed a similar situation when we compared Figs.8.24a and 8.24c with Figs.8.25a and 8.25c.

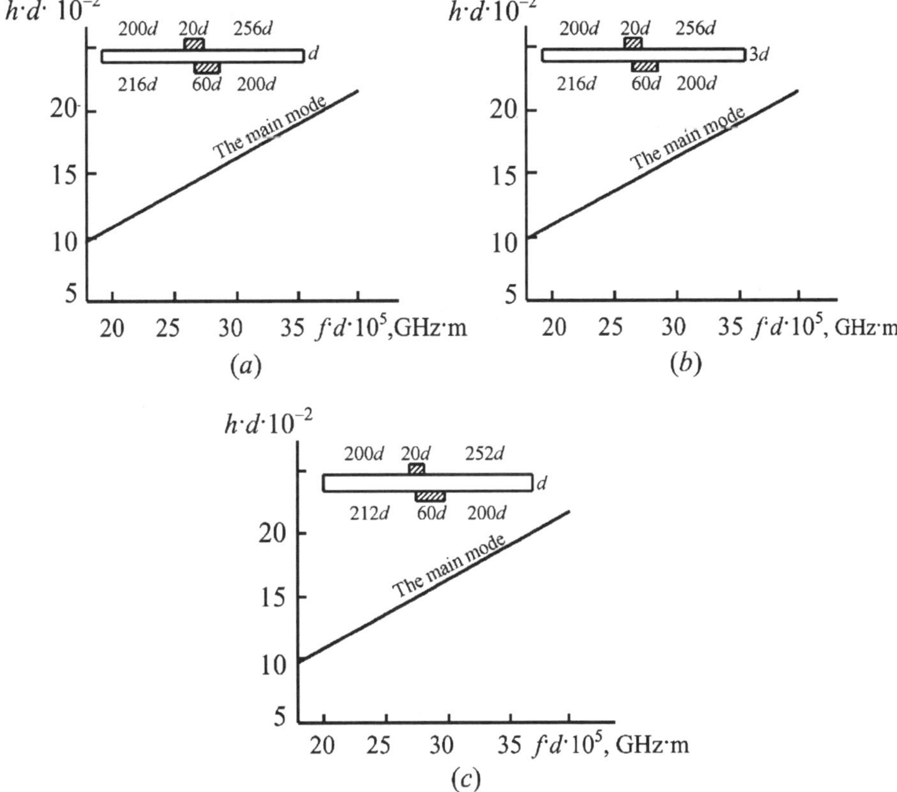

Fig. 8.25. *Dependence of the longitudinal propagation constants of the NSL eigenmodes on the frequency for different substrate thickness and overlapping of metal strips: (a) $d = 5 \cdot 10^{-6}\,m$, $s = -20 \cdot 10^{-6}\,m$; (b) $d = 15 \cdot 10^{-6}\,m$, $s = -20 \cdot 10^{-6}\,m$; (c) $d = 5 \cdot 10^{-6}\,m$ and $s = -40 \cdot 10^{-6}\,m$.*

We see that if the total width $w_1 + w_2$ is a fixed quantity for both $w_1 = w_2$ and $w_1 \neq w_2$, then the dispersion curve of the main mode remains practically unchanged. Here we compare Figs. 8.24b,c with 8.25b,c for two NSL that have the same substrates and the same total width $w_1 + w_2$. For the first NSL Fig.8.24b,c the metal strips width $w_1 = w_2$ is equal. For the second NSL Fig.8.25b,c the metal strips width $w_1 \neq w_2$ is not equal. It is important to notice that for the NSL with $w_1 \neq w_2$ the higher mode is cutoff and this NSL becomes a one–mode transmission waveguide.

8.8. The Dispersion Dependences of Dielectric Waveguides and the Comparison of our Calculated Results with Reference Data

Fig.8.26 presents the dispersion dependences of the main and higher modes for the rectangular dielectric waveguide (DW). One can see for the DW made of the material with $\varepsilon_{r_2} = 2.06$ that the discrepancy of the calculated results and the experimental data given in [8.10] does not exceed 2%. Fig.8.26 presents the dispersion dependences of three lowest modes for the DW of the same geometry and sizes with the constitutive parameters $\varepsilon_{r_2} = 13.5$ and $\mu_{r_2} = 0.6$ (the DW geometry is placed in the lower part of this Fig.).

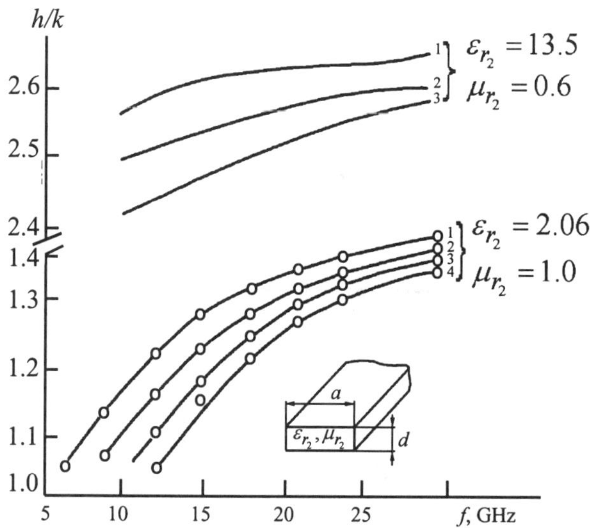

Fig. 8.26. *Dispersion dependences of the longitudinal propagation constant of the main mode (curve 1) and three higher modes (curves 2, 3, 4) for the rectangular DW. The waveguide parameters are: $a = 15 \cdot 10^{-3}$ m and $d = 5 \cdot 10^{-3}$ m. The solid lines are our calculations and the circles are results of [8.10].*

Fig.8.27 presents the dispersion dependences for the four modes of the strip DW [8.11].The deviation in our results and that of [8.11] doesn't exceed 4%. The dispersion dependences of the mirror DW is presented in Fig.8.28. The comparisons of our calculation results and that of [8.12] shows that when the frequency increases then the deviation increases. It can be explained like this, because the metal plate width is infinite in [8.12] and this metal plate width was $a = 0.7 \cdot 10^{-3}$ m in our calculations. When the frequency increases then the DW wavelength becomes closer as compare to the DW width and the influence of the plate ends in our calculations increases. The deviation of our calculations and that of [8.12] and [8.13] does not exceed 1.2%.

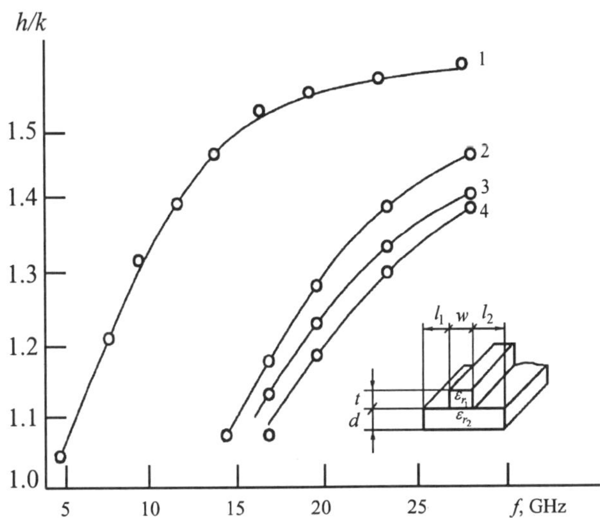

Fig. 8.27. *Dispersion dependences of the longitudinal propagation constant for the main (curve 1) and three higher modes (curves 2, 3, 4) of the strip DW. The waveguide parameters are: $d = t = l_1 = l_2 = 5 \cdot 10^{-3}$ m, $w = 10^{-2}$ m, $\varepsilon_{r_1} = 2.62$, $\varepsilon_{r_2} = 2.66$ and $\mu_r = 1$. The solid lines are our calculations and the circles are the results of [8.11].*

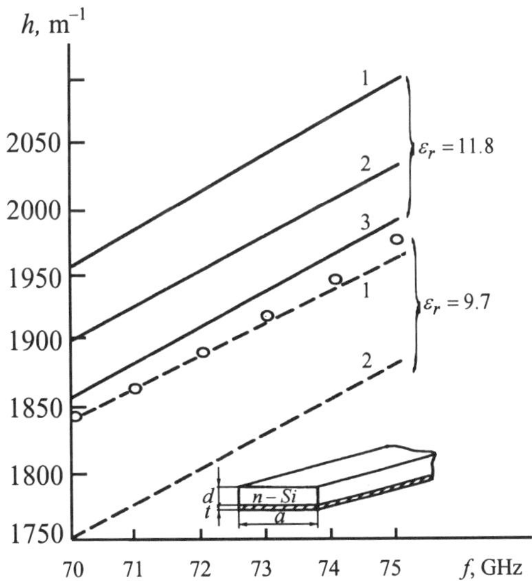

Fig. 8.28. *Dispersion dependences of the longitudinal propagation constant of the main mode (curve 1) and two higher modes (curve 2, 3) for the mirror DW. The waveguide parameters are: $d = 0.35 \cdot 10^{-3}$ m, $a = 0.7 \cdot 10^{-3}$ m and $t = 10^{-6}$ m. The solid ($\varepsilon_r = 11.8$) and the dash lines ($\varepsilon_r = 9.7$) are our calculations and the circles are the results from [8.12].*

Here we did not calculate the EM field components for the mirror DW because the classification of the hybrid modes *HE* and *EH* was not executed. So we numbered the DW modes in an increasing order. But we agreed with the classification of [8.14] for the smaller ε_r values and [8.15] for the larger ε_r values.

Fig.8.29 presents the dispersion dependences of the main and higher modes (the solid lines) of the open mirror demagnetized ferrite waveguide and the waveguide material was ferrite ISCh4. According to the classification [8.15] the main mode is EH_{11}, two higher modes are HE_{21} and the second higher mode is EH_{21}. In the Fig.8.29 we also presented our calculated dispersion dependences of the main mode HE_{11} and the first higher mode EH_{11} (depicted by the dash lines) for the open circular demagnetized ferrite waveguide. The radius of the circular waveguide was chosen so that its cross–section area was equal to that of the investigated mirror DW. Such comparisons can give only qualitative conclusions.

The characteristics of the circular and square DW with equal cross–section areas differed insignificantly. It is known, for instance in [8.14], that the larger ratio of the rectangular DW sides, one can obtain a higher propagation constant value.

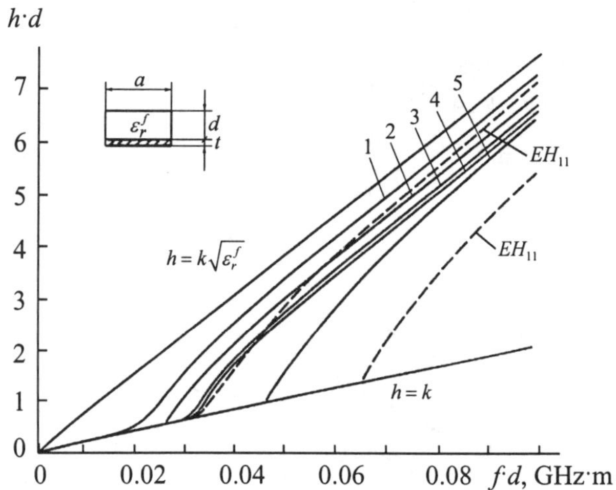

Fig. 8.29. *Dispersion dependences of the longitudinal propagation constant of the main mode (curve 1) and four higher (curves 2–5) modes for the mirror demagnetized ferrite waveguide. The waveguide parameters are: $d = 10^{-3}$ m, $a = 2 \cdot 10^{-3}$ m, $t = 2 \cdot 10^{-5}$ m, $\varepsilon_r^f = 13.5$ and $\mu_r = 1$. The solid lines are dispersion curves for the open ferrite waveguide (mirror waveguide), the dash lines are for the cylindrical demagnetized ferrite waveguide with the radius d .*

When we investigated the waveguide with the side ratio $a : d = 2 : 1$ we found that the longitudinal propagation constant value of the mirror DW (curve 1) is larger than that of the circular waveguide mode HE_{11} (the dash line). When the frequency increases the influence of the magnitude of ratio a/d on quantity h value decreases as seen in Fig.8.29. The dispersion curves for the main modes of the rectangular and circular DW are nearing each other as the frequency increases. For the circular waveguide, there are the axially symmetric modes H_{01}, E_{01} and the hybrid mode HE_{21} in the interval between the modes HE_{11} and EH_{11} [8.16] – [8.18].

In this chapter we described the use of our SIE method to make it possible to investigate the dispersion characteristics of waveguides of arbitrary cross–section geometry, containing piece–wise homogeneous isotropic material. In the next chapter we will be describing a new version of our SIE method for solving Maxwell's equations, which will enable one to investigate the dispersion characteristics of waveguides of arbitrary cross – sections, containing piecewise $\bar{\bar{\varepsilon}}$ –, $\bar{\bar{\mu}}$ – or $\bar{\bar{\varepsilon}} \, \& \, \bar{\bar{\mu}}$ – gyrotropic materials.

9. SOLUTION OF MAXWELL'S EQUATIONS BY THE SIE METHOD FOR LONGITUDINALLY MAGNETUZED $\vec{\varepsilon}\,-,\ \ \vec{\mu}\,-$ AND $\vec{\varepsilon}\,\&\,\vec{\mu}-$ GYROTROPIC WAVEGUIDES

In this chapter we will describe our SIE method for solving Maxwell's equations, which will enable one to investigate the dispersion characteristics of the main mode and higher modes of regular waveguides. These waveguides have an arbitrary cross–section with piecewise uniform longitudinally magnetized $\vec{\varepsilon}\,-,\ \vec{\mu}\,-$ and $\vec{\varepsilon}\,\&\,\vec{\mu}$–gyrotropic materials. The idea of our method is the same as was described in previous chapters except that here the problems are more complicated because the waveguides are gyrotropic.

In solving these gyrotropic problems we must first find the solution of differential equations with point sources. Then this (fundamental) solution is used in the integral representation of the general solution for each particular boundary problem. This integral representation automatically satisfies Maxwell's equations and has the unknown density functions. The density functions obtained from the boundary conditions that are the equality of the tangential components on the contour dividing different media. From these boundary conditions we obtained the singular integral equations (SIE). These SIE are solved by use of Bogoliubov–Krylov method to determine the unknown density functions.

Here we will start to construct a SIE method (which we will be using in this chapter) from the solution of Maxwell's equations for a gyrotropic circular waveguide. Then the formulae we obtained for the gyrotropic circular waveguide we will use in a solution for waveguides having any complicated cross – sections.

9.1. The Solution for an Open Circular Longitudinally Magnetized Waveguide

Here we will briefly give the solution of Maxwell's equations for the circular waveguides. We will consider the medium described by the permittivity tensor [4.11].

$$\ddot{\varepsilon} = \begin{vmatrix} \varepsilon_{xx} & i\varepsilon_{xy} & 0 \\ -i\varepsilon_{xy} & \varepsilon_{xx} & 0 \\ 0 & 0 & \varepsilon_{zz} \end{vmatrix} \qquad (9.1)$$

and the permeability tensor [9.1]

$$\ddot{\mu} = \begin{vmatrix} \mu_{xx} & i\mu_{xy} & 0 \\ -i\mu_{xy} & \mu_{xx} & 0 \\ 0 & 0 & \mu_{zz} \end{vmatrix}. \qquad (9.2)$$

We will assume that the wave is propagating along z–axis and so the dependence on time t and the coordinate z is represented by the factor $e^{i\omega t - ihz}$.

Here we used Maxwell's equations for the $\ddot{\varepsilon}$ – and $\ddot{\mu}$ –gyrotropic medium in this form:

$$rot\vec{H} = i\omega\ddot{\varepsilon}\vec{E} \ ,$$
$$rot\vec{E} = -i\omega\ddot{\mu}\vec{H} \ . \qquad (9.3)$$

Then we express the transversal components through the longitudinal ones and we obtained (such as [9.2]) the system of coupled wave equations for the longitudinal components of the electric E_z and magnetic H_z fields:

$$\left[k^2\mu_{xx}\left(\varepsilon_{xx}^2 - \varepsilon_{xy}^2\right) - h^2\varepsilon_{xx} \right]\Delta_\perp E_z + ikhZ_0\left(\varepsilon_{xx}\mu_{xy} + \varepsilon_{xy}\mu_{xx}\right)\ \Delta_\perp H_z +$$

$$+ \varepsilon_{zz}E_z\Delta_{PF} = 0 \ ,$$

$$-\frac{ikh}{Z_0}\left(\varepsilon_{xx}\mu_{xy} + \varepsilon_{xy}\mu_{xx}\right)\ \Delta_\perp E_z + \left[k^2\varepsilon_{xx}\left(\mu_{xx}^2 - \mu_{xy}^2\right) - h^2\mu_{xx} \right]\Delta_\perp H_z +$$

$$+ \mu_{zz}H_z\Delta_{PF} = 0. \qquad (9.4)$$

where $k = \omega/c$ is the wave number in a vacuum, h is the longitudinal propagation constant and i is the imaginary unit $\left(i^2 = -1\right)$. The value $Z_0 = \sqrt{\mu_0/\varepsilon_0} = 120\pi\ \Omega$ is the wave impedance (characteristic impedance) of a vacuum. The value $\Delta_\perp = \partial^2/\partial x^2 + \partial^2/\partial y^2$ is the transversal Laplace's operator. The value Δ_{PF} equals: $\Delta_{PF} = k^4(\mu_{xx}^2 - \mu_{xy}^2)(\varepsilon_{xx}^2 - \varepsilon_{xy}^2) - 2h^2k^2(\varepsilon_{xx}\mu_{xx} + \varepsilon_{xy}\mu_{xy}) + h^4$.

The solution of the equation (9.4) for the region I (Fig.9.1) can be written in this form:

$$E_z = \left(aJ_m\left(k_{\perp 1}r\right) + J_m\left(k_{\perp 2}r\right)\right) A\ e^{\pm im\varphi}), \qquad (9.5)$$

$$H_z = \left(J_m(k_{\perp 1}r) + bJ_m(k_{\perp 2}r)\right) B\ e^{\pm im\varphi}), \qquad (9.6)$$

where $J_m(k_{\perp 1,2}r)$ is the Bessel function of the m–th order for the complex argument $k_{\perp 1,2}r$ and $m = 0, 1, 2....$. The values A and B are free amplitudes and values $k_{\perp 1}$, $k_{\perp 2}$ are the transversal propagation numbers.

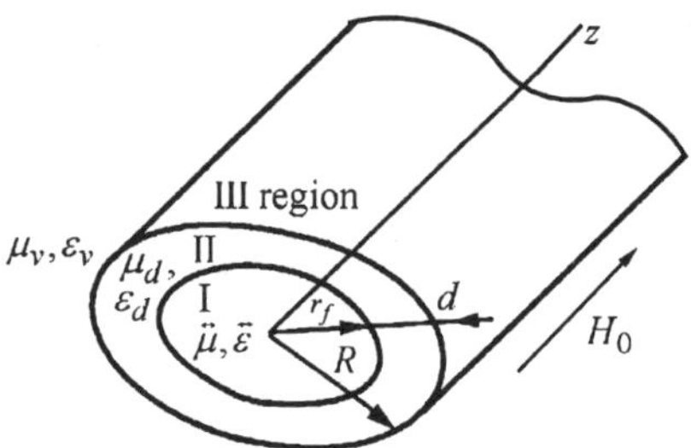

Fig. 9.1. *The geometry of a layer gyrotropic waveguide.*

The coefficients a and b point out the $H-$ and $E-$ modes mixing ratio in a hybrid wave and are expressed by the formulae:

$$a = \frac{ikhZ_0(\varepsilon_{xx}\mu_{xy} + \varepsilon_{xy}\mu_{xx})k_{\perp 1}^2}{\varepsilon_{zz}\Delta_{PF} - k_{\perp 1}^2 \varepsilon_{xx}\left[k^2 \dfrac{\mu_{xx}}{\mu_{zz}} \varepsilon^{ef} - h^2\right]}, \tag{9.7}$$

$$b = \frac{-ikh(\varepsilon_{xx}\mu_{xy} + \varepsilon_{xy}\mu_{xx})k_{\perp 2}^2}{Z_0\left\{\mu_{zz}\Delta_{PF} - k_{\perp 2}^2 \mu_{xx}\left[k^2 \dfrac{\varepsilon_{xx}}{\varepsilon_{zz}} \mu^{ef} - h^2\right]\right\}}, \tag{9.8}$$

where $\varepsilon^{ef} = \mu_{zz}\left(\varepsilon_{xx} - \varepsilon_{xy}^2/\varepsilon_{xx}\right)$ and $\mu^{ef} = \varepsilon_{zz}\left(\mu_{xx} - \mu_{xy}^2/\mu_{xx}\right)$.

The transversal wave numbers for the bigyrotropic medium are:

$$k_{\perp 1,2} = \frac{1}{2}\left\{\left[k^2(\varepsilon^{ef} + \mu^{ef}\right] - h^2\left(\frac{\varepsilon_{zz}}{\varepsilon_{xx}} - \frac{\mu_{zz}}{\mu_{xx}}\right) \pm \right.$$

$$\left. \pm\sqrt{\left[k^2\left(\varepsilon^{ef} - \mu^{ef}\right) + h^2\left(\frac{\varepsilon_{zz}}{\varepsilon_{xx}} - \frac{\mu_{zz}}{\mu_{xx}}\right)\right]^2 + 4h^2k^2\mu_{zz}\varepsilon_{zz}\left(\frac{\varepsilon_{xy}}{\varepsilon_{xx}} + \frac{\mu_{xy}}{\mu_{xx}}\right)^2}\right\}. \tag{9.9}$$

The tangential components are written in the form:

$$E_\varphi = \frac{1}{\Delta_{PF}}\omega\mu_0\left[\varepsilon_{xy}k^2\left(\mu_{xx}^2 - \mu_{xy}^2\right) + \mu_{xy}h^2\right]\frac{1}{r}\frac{\partial H_z}{\partial\varphi} -$$

$$-ih\left[k^2(\varepsilon_{xx}\mu_{xx} + \varepsilon_{xy}\mu_{xy}) - h^2\right]\frac{1}{r}\frac{\partial E_z}{\partial\varphi} +$$

$$+i\omega\mu_0\left[\varepsilon_{xx}k^2\left(\mu_{xx}^2 - \mu_{xy}^2\right) - \mu_{xx}h^2\right]\frac{\partial H_z}{\partial\varphi} + hk^2\left(\varepsilon_{xx}\mu_{xy} + \varepsilon_{xy}\mu_{xx}\right)\frac{\partial E_z}{\partial r}\right\}, \tag{9.10}$$

$$H_\varphi = -\frac{1}{\Delta_{PF}}\left\{\omega\varepsilon_0\left[k^2\mu_{xy}\left(\varepsilon_{xx}^2 - \varepsilon_{xy}^2\right) + \varepsilon_{xy}h^2\right]\frac{1}{r}\frac{\partial E_z}{\partial\varphi} +$$

$$+ih\left[k^2(\varepsilon_{xx}\mu_{xx} + \varepsilon_{xy}\mu_{xy}) - h^2\right]\frac{1}{r}\frac{\partial H_z}{\partial\varphi} +$$

$$+i\omega\varepsilon_0\left[k^2\mu_{xx}\left(\varepsilon_{xx}^2-\varepsilon_{xy}^2\right)-\varepsilon_{xx}h^2\right]\frac{\partial E_z}{\partial r}-hk^2\left(\varepsilon_{xy}\mu_{xx}+\varepsilon_{xx}\mu_{xy}\right)\frac{\partial H_z}{\partial r}\Bigg\}. \tag{9.11}$$

The wave equations for dielectric regions become simple as compared to gyrotropic medium, when we use these electrophysical material parameters $\varepsilon_{xy}=0$, $\mu_{xy}=0$, $\varepsilon_{xx}=\varepsilon_f$ and $\mu_{xx}=1$.

The solution for region II has this form:

$$H_z=\left[A_2 J_m(k_\perp^d r)+D_2 N_m(k_\perp^d r)\right]e^{\pm im\varphi}, \tag{9.12}$$

$$E_z=\left[B_2 J_m(k_\perp^d r)+C_2 N_m(k_\perp^d r)\right]e^{\pm im\varphi}. \tag{9.13}$$

And the solution for region III has this form [9.3]:

$$H_z=A_3 H_m^{(2)}(\xi_\perp r)e^{\pm im\varphi}, \tag{9.14}$$

$$E_z=B_3 H_m^{(2)}(\xi_\perp r)e^{\pm im\varphi}. \tag{9.15}$$

Where $k_\perp^{d2}=k^2\varepsilon_d\mu_d-h^2$, $\xi_\perp^2=h^2-k^2\varepsilon_v\mu_v$. Magnitudes A_2, B_2, C_2 and D_2 are the coefficients which must be determined. $N_m(k_\perp^d r)$ is the Neumann function of the m–th order and $H_m^{(2)}(\xi_\perp r)$ is the Hankel function of the m–th order and the second kind. The value $\xi_\perp$ is the outside wave number (in the outer region III) and the value $k_\perp^d$ is the transversal propagation number in the dielectric layer (region II). The value k is the wave number in a vacuum and h is the longitudinal propagation constant.

When equalizing the tangential components of electric and magnetic fields on the boundaries we obtained a system of equations. The determinant of the system yields the dispersion equation for the waveguide under our consideration. This determinant has the eight order and some of its elements are:

$$\det\left(a_{ik}\right)=0, \qquad\qquad i,k=1\div8,$$

$$a_{11}=J_m(k_{\perp1}r_f), \qquad\qquad a_{12}=bJ_m(k_{\perp2}r_f),\ldots,$$

$$a_{45}=-\frac{hm}{\left(k_\perp^d\right)^2 r_f}N_m(k_\perp^d r_f), \qquad a_{46}=-\frac{i\omega\mu_0\mu_d}{k_\perp^d}N_m'(k_\perp^d r_f),\ldots,$$

$$a_{87}=-\frac{i\omega\mu_0\mu_v}{\xi_\perp}H_m^{(2)\prime}(\xi_\perp R), \qquad a_{88}=-\frac{hm}{\xi_\perp^2 R}H_m^{(2)}(\xi_\perp R). \tag{9.16}$$

9.1.1. Comparison of our calculated results with experimental data. Here we will mostly consider waveguides having ferrites in weak magnetic fields. And so according to [9.4] the nondiagonal tensor element is expressed like this: $\mu_{xy}=\mu_{xy}^s(4\pi M/4\pi M_s)$, where μ_{xy}^s is the nondiagonal tensor element of the saturated ferrite [9.1] and $4\pi M_s$ is the saturated ferrite magnetization. In this chapter we will use the notation δ for the half–width of the resonance curve.

In the Figs.9.2–9.6 we present the electrodynamic characteristics of the longitudinally magnetized circular layer ferrite–dielectric waveguide (Fig.9.1).

The inner waveguide rod is made of ferrite *ISCh4* with the relative permittivity $\varepsilon_f = 13.5 \, (1 - i5 \cdot 10^{-4})$, magnetic losses $\mu_f'' = 5 \cdot 10^{-3}$ and the magnetization $4\pi M_s = 377$ kA/m. Our calculations were executed assuming $\varepsilon_{xx} = \varepsilon_{zz} = \varepsilon_f$, $\varepsilon_{xy} = 0$ and the outside dielectric layer thickness was $d/r_f = 0.3$ with $\varepsilon_r = \varepsilon_r' - i\varepsilon_r''$ for the materials taken from [9.5] at $\varepsilon_r' = 2 \div 20$.

In order to use our calculated results in different frequency ranges we multiply all values by the ferrite rod radius. The EM wave classification that we will use corresponds to that for the dielectric waveguide. The EM wave type of the ferrite waveguide is determined by the EM wave of the dielectric waveguide. We gradually take off the material gyrotropy then: $\mu_{xx} = \mu_{zz} = 1$ and $\mu_{xy} \to 0$.

The Figs.9.2 and 9.3 show the dispersion dependences for the waveguides with dielectric layers of different thickness having the relative permittivity $\varepsilon_d = 9.6$. The main mode HE_{11} dispersion curves are given for two normalized ferrite magnetizations. For the higher modes, we placed the results for $M/M_s = 0.7$ only, because of their weak dependence on the ferrite magnetization.

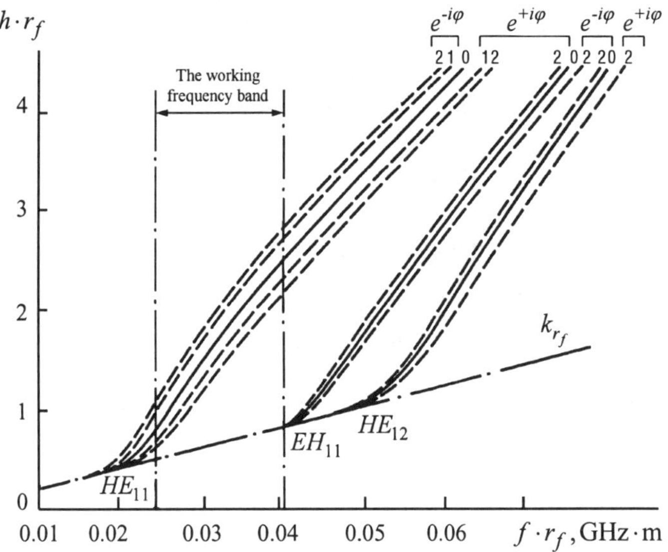

Fig. 9.2. *Dispersion dependences of the main mode HE_{11} and two higher modes EH_{11} and HE_{12} for the layer ferrite–dielectric waveguide (Fig.9.1) at the layer thickness $d/r_f = 0.3$ and several ferrite magnetizations. $0 - M = 0$ (solid lines), $1 - M/M_s = 0.5$ and $2 - M/M_s = 0.7$ (dash lines).*

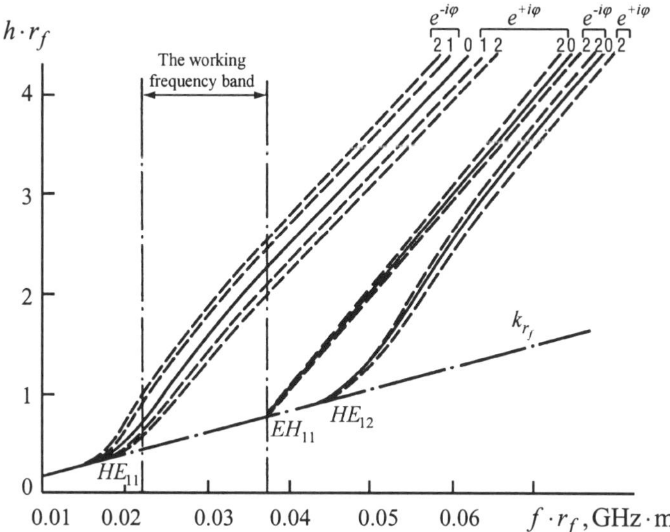

Fig. 9.3. *Dispersion dependences of the main mode* HE_{11} *and two higher modes* EH_{11} *and* HE_{12} *for the layer ferrite–dielectric waveguide (Fig.9.1) at the layer thickness* $d/r_f = 0.5$ *and several ferrite magnetizations.* $0 - M = 0$ *(solid lines),* $1 - M/M_s = 0.5$ *and* $2 - M/M_s = 0.7$ *(dash lines).*

Also between the vertical dash–dot lines there is a waveguide working frequency band. The cut off frequency of the higher mode EH_{11} with the angular dependence $e^{+i\varphi}$ decreases and the angular dependence $e^{-i\varphi}$ increases when the magnetization increases. Because of this, it lowers the bandwidth of the device working on the mode with polarization $e^{+i\varphi}$ and increases its bandwidth for the mode of opposite polarization.

Comparing the dispersion dependences (Figs.9.2 and 9.3) of the layer ferrite–dielectric waveguides, we see that the thicker the dielectric layer is the more the working frequency band is displacing into the low frequency region.

The Fig.9.4*a* presents the dependence of the differential phase shift $\Delta(hr_f)$ on the normalized ferrite magnetization for the main mode HE_{11} (of both polarizations) of the layer waveguide at its central frequency $(fr_f)_c = 0.033$ GHz·m. The differential phase shift is counted from the phase of the demagnetized waveguide material for the waves of both polarizations, so that: $\Delta(hr) = (hr)_M - (hr)_{M=0}$, where the first term corresponds to the magnetized ferrite and the second term corresponds to the demagnetized ferrite. The dependence Fig.9.4*a* of the layer waveguide is practically linear at the considered ferrite magnetizations. The Fig.9.4*b* presents the dependence of the differential phase shift for the mode HE_{11} with the polarization $e^{+i\varphi}$ on the normalized layer thickness d/r_f at three ferrite magnetizations $M/M_s = 0.3$, $M/M_s = 0.5$ and $M/M_s = 0.7$. For this mode HE_{11} with the other polarization $e^{-i\varphi}$ the dependence is very similar.

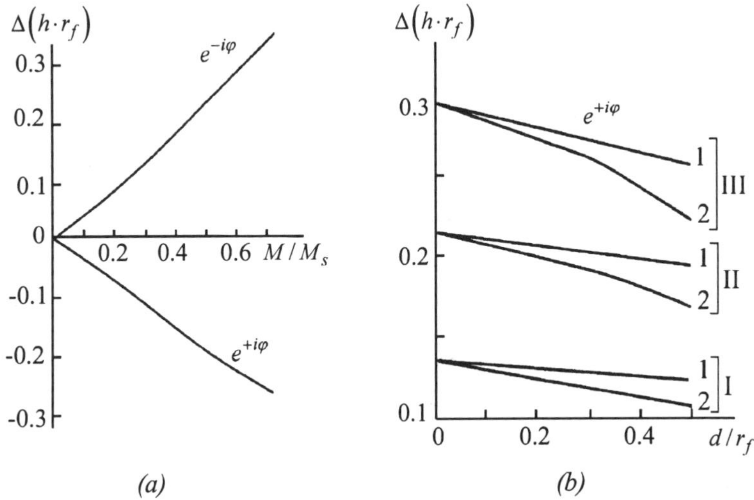

Fig. 9.4. *Dependence of the differential phase shift on: (a) the ferrite magnetization at the dielectric layer thickness $d/r_f = 0.3$ and $\varepsilon_d = 9.6$; (b) on the dielectric layer thickness where symbol 1 corresponds to $\varepsilon_d = 9.6$, symbol 2 corresponds to $\varepsilon_d = 15$ and symbols I, II, III equates to $M/M_s = 0.3, 0.5, 0.7$ accordingly.*

We draw your attention to the fact that all calculations of the magnitude $\Delta(hr_f)$ for every waveguide with d/r_f were fulfilled at its central frequency. The central frequency is $(fr_f)_c = 0.033\ GHz \cdot m$ for the waveguide with $\varepsilon_d = 9.6$ and $d/r_f = 0.3$. The central frequency is $(fr_f)_c = 0.031\ GHz \cdot m$ for the waveguide with $\varepsilon_d = 15$ and $d/r_f = 0.3$. One can see from Fig.9.4b that the dielectric layer thickness influence on the differential phase shift is stronger then the larger ferrite magnetization. It can be explained like this, when the ferrite magnetization increases the larger part of electromagnetic wave energy is drawn from the air into the waveguide. Therefore this part of the EM energy interacts with the dielectric layer and the EM wave reaction on the layer becomes more significant.

Fig.9.5 shows the losses of two ferrite waveguides one with a dielectric layer and the other without a dielectric layer for the main mode of polarization $e^{+i\varphi}$. Here the outside radiuses are equal for both ferrite waveguides and we see that the layer waveguide might have less of a loss than the waveguide without a dielectric layer.

Fig.9.6 presents the theoretical and experimental dependence of the differential phase shift modulus on the material magnetization for the open ferrite waveguides with the working length L. The differential phase shift was measured for three open ferrite waveguides of different sizes at two fixed frequencies (Table 9.1) and as a function of the ferrite magnetization. Our calculations and experimental results coincide well.

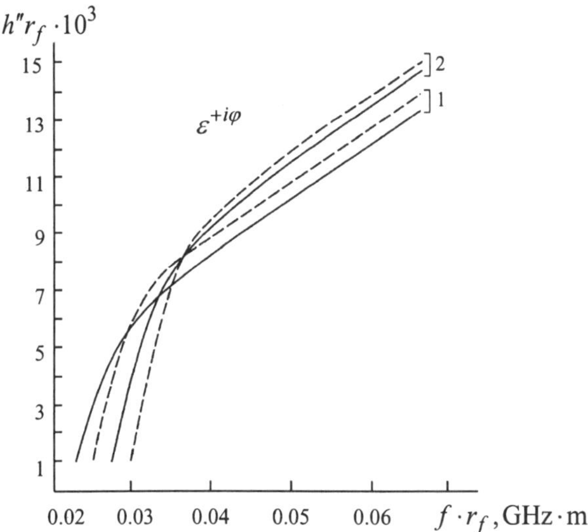

Fig. 9.5. *Dependence of losses on the frequency for the waveguide with and without a layer. The solid lines correspond to the demagnetized ferrite M=0 and the dash lines correspond to the ferrite magnetization M/M_s=0.7. Symbol 1 shows dependences for the ferrite waveguide with the dielectric layer d/r_f =0.3 and ε_d =9.6. Symbol 2 shows dependences for the ferrite waveguide without a dielectric layer.*

Table 9.1 *Parameters for three ferrite waveguides given in Fig.9.6.*

Fig.9.6	fr (GHz · m)	$4\pi M_s$ (kA/m)	L (m)
(a)	0.0319	312.3	0.008
(b)	0.0413	404.2	0.012
(c)	0.0553	320.4	0.016

In Figs.9.4–9.6 we see it is possible to choose a dielectric layer thickness with such a value that the differential phase shift will decrease insignificantly, when compared to the ferrite waveguide without a dielectric layer. As is known when with the frequencies increase the waveguide losses increase sharply and then there is a significant modulation of the signal. By choosing the electrophysical properties and the dielectric layer thickness we are able to diminish the change of losses in the working frequency band. This last dependence (if losses do not change) diminishes modulations of the signal.

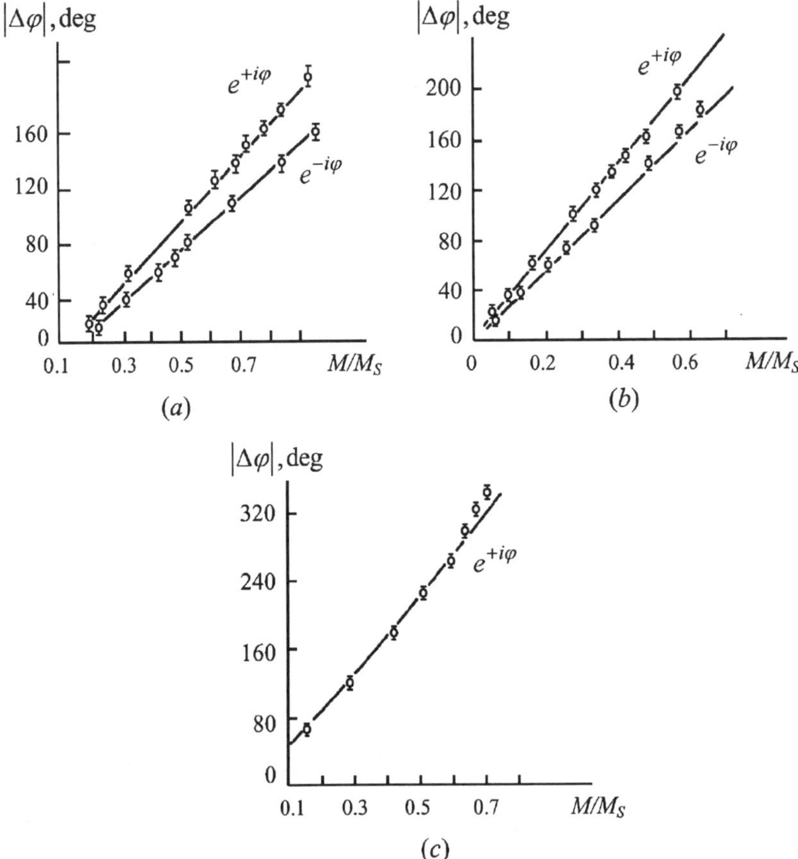

Fig. 9.6. Experimental and theoretical dependences of the differential phase shift $\Delta\varphi$ modules on the material magnetization M/M_s for three ferrite waveguides. We designate with the circle the experimental results and with the solid line our calculations.

9.2. Analysis of Longitudinally Magnetized $\ddot{\varepsilon}$– & $\ddot{\mu}$– Gyrotropic Waveguides

Here we have the solution for a circular gyrotropic waveguide (Section 9.1). The expressions (9.5) and (9.6) show that the eigenwaves are different then the transversal propagation constants $k_{\perp 1}$ and $k_{\perp 2}$. From the solution of the cylinder waveguide with uniform filling we see that both waves are a mixture of H–wave and E–wave. The first wave (corresponding $k_{\perp 1}$) degenerates into H–wave with gyrotropy diminishing and the second wave (corresponding $k_{\perp 2}$) degenerates into E–wave.

Maxwell's equations written in Cartesian coordinates for the harmonic wave, which propagates along the z–axis (with standard dependence on z and time t as $e^{i\omega t-ihz}$) in the gyrotropic medium described by tensors (9.1) and (9.2) are:

$$\frac{\partial H_z}{\partial y}+ihH_y=i\omega\varepsilon_0\left(\varepsilon_{xx}E_x+i\varepsilon_{xy}E_y\right),\tag{9.17}$$

$$-ihH_x-\frac{\partial H_z}{\partial x}=i\omega\varepsilon_0\left(-i\varepsilon_{xy}E_x+\varepsilon_{xx}E_y\right),\tag{9.18}$$

$$\frac{\partial H_y}{\partial x}-\frac{\partial H_x}{\partial y}=i\omega\varepsilon_0\varepsilon_{zz}E_z,\tag{9.19}$$

$$\frac{\partial E_z}{\partial y}+ihE_y=-i\omega\mu_0\left(\mu_{xx}H_x+i\mu_{xy}H_y\right),\tag{9.20}$$

$$-ihE_x-\frac{\partial E_x}{\partial x}=-i\omega\mu_0\left(-i\mu_{xy}H_x+\mu_{xx}H_y\right),\tag{9.21}$$

$$\frac{\partial E_y}{\partial x}-\frac{\partial E_x}{\partial y}=-i\omega\mu_0\mu_{zz}H_z.\tag{9.22}$$

From the equations (9.17), (9.18), (9.20) and (9.21) one can express the transversal components through the longitudinal components:

$$E_x=\frac{1}{\Delta_{PF}}\left\{-\frac{\partial E_z}{\partial x}ih\left[k^2(\varepsilon_{xx}\mu_{xx}+\varepsilon_{xy}\mu_{xy})-h^2\right]-\frac{\partial E_z}{\partial y}hk^2(\varepsilon_{xy}\mu_{xx}+\varepsilon_{xx}\mu_{xy})+\right.$$

$$\left.+\frac{\partial H_z}{\partial x}\omega\mu_0\left[k^2\mu_{xx}\varepsilon_{xy}\mu^{ef}+h^2\mu_{xy}\right]-\frac{\partial H_z}{\partial y}i\omega\mu_0\mu_{xx}\left[k^2\varepsilon_{xx}\mu^{ef}-h^2\right]\right\},\tag{9.23}$$

$$E_y=\frac{1}{\Delta_{PF}}\left\{\frac{\partial E_z}{\partial x}hk^2(\varepsilon_{xx}\mu_{xy}+\varepsilon_{xy}\mu_{xx})-\frac{\partial E_z}{\partial y}ih\left[k^2(\varepsilon_{xx}\mu_{xx}+\varepsilon_{xy}\mu_{xy})-h^2\right]+\right.$$

$$\left.+\frac{\partial H_z}{\partial x}i\omega\mu_0\mu_{xx}\left[k^2\varepsilon_{xx}\mu^{ef}-h^2\right]+\frac{\partial H_z}{\partial y}\omega\mu_0\left[k^2\mu_{xx}\varepsilon_{xy}\mu^{ef}+h^2\mu_{xy}\right]\right\},\tag{9.24}$$

$$H_x=\frac{1}{\Delta_{PF}}\left\{-\frac{\partial E_z}{\partial x}\omega\varepsilon_0\left[k^2\varepsilon_{xx}\mu_{xy}\varepsilon^{ef}+h^2\varepsilon_{xy}\right]+\frac{\partial E_z}{\partial y}i\omega\varepsilon_0\varepsilon_{xx}\left[k^2\mu_{xx}\varepsilon^{ef}-h^2\right]-\right.$$

$$\left.-\frac{\partial H_z}{\partial x}ih\left[k^2(\varepsilon_{xx}\mu_{xx}+\varepsilon_{xy}\mu_{xy})-h^2\right]-\frac{\partial H_z}{\partial y}hk^2(\varepsilon_{xx}\mu_{xy}+\varepsilon_{xy}\mu_{xx})\right\},\tag{9.25}$$

$$H_y=\frac{1}{\Delta_{PF}}\left\{-\frac{\partial E_z}{\partial x}i\omega\varepsilon_0\varepsilon_{xx}\left[k^2\mu_{xx}\varepsilon^{ef}-h^2\right]-\frac{\partial E_z}{\partial y}\omega\varepsilon_0\left[k^2\varepsilon_{xx}\mu_{xy}\varepsilon^{ef}+h^2\varepsilon_{xy}\right]+\right.$$

$$\left.+\frac{\partial H_z}{\partial x}hk^2(\varepsilon_{xy}\mu_{xx}+\varepsilon_{xx}\mu_{xy})-\frac{\partial H_z}{\partial y}ih\left[k^2(\varepsilon_{xx}\mu_{xx}+\varepsilon_{xy}\mu_{xy})-h^2\right]\right\}.\tag{9.26}$$

The designation we defined in Section 9.1. When putting the formulae (9.23–9.26) into (9.19) and (9.22) we obtained a system of coupled differential equations for the longitudinal components E_z and H_z, which coincide with the system (9.4). Here we look for the solution to this system in this form:

$$E_z = A\ Z_m(k_\perp r)e^{im\varphi},$$

$$H_z = B\ Z_m(k_\perp r)e^{im\varphi}. \tag{9.27}$$

Magnitude Z_m is the Bessel function of the order m. The amplitudes A, B and the transversal wave number $k_\perp$ are yet unknown. Putting (9.27) into the system (9.4) and taking into account the Bessel differential equation we obtain the homogeneous system of linear equations:

$$A\left\{\varepsilon_{zz}\Delta_{PF} - k_\perp^2\varepsilon_{xx}\left[k^2\mu_{xx}\varepsilon^{ef} - h^2\right]\right\} - iBk_\perp^2 Z_0 hk\left(\varepsilon_{xx}\mu_{xy} + \varepsilon_{xy}\mu_{xx}\right) = 0,$$

$$iAk_\perp^2\frac{hk}{Z_0}(\varepsilon_{xx}\mu_{xy} + \varepsilon_{xy}\mu_{xx}) + B\left\{\mu_{zz}\Delta_{PF} - k_\perp^2\mu_{xx}\left[k^2\varepsilon_{xx}\varepsilon^{ef} - h^2\right]\right\} = 0 \tag{9.28}$$

for the coefficients A and B. The solution of the system (9.28) exists when its determinant is equal to zero. That gives the algebraic equations of the fourth order for the $k_\perp$:

$$k_\perp^4 - k_\perp^2\cdot\left[k^2(\varepsilon^{ef} + \mu^{ef}) - h^2(\frac{\mu_{zz}}{\mu_{xx}} + \frac{\varepsilon_{zz}}{\varepsilon_{xx}})\right] + \frac{\varepsilon_{zz}}{\varepsilon_{xx}}\frac{\mu_{zz}}{\mu_{xx}}\Delta_{PF} = 0, \tag{9.29}$$

for every root one finds from (9.28) the ratio A/B.

Now we will consider more simple cases. For isotropic medium when $\varepsilon_{xy} = \mu_{xy} = 0$ the determinant of formula (9.28) becomes:

$$\Delta_{PF} = (k^2\varepsilon_{xx}\mu_{xx} - h^2)^2.$$

From the system (9.28) we find two solutions. The first is:

$$A = 0, \quad k_\perp^2 = \frac{(k^2\varepsilon_{xx}\mu_{xx} - h^2)\mu_{zz}}{\mu_{xx}}, \quad B \neq 0, \tag{9.30}$$

that corresponds to the first root (9.9) when one takes a plus sign before the radical. The second is:

$$A \neq 0, \quad k_\perp^2 = \frac{(k^2\varepsilon_{xx}\mu_{xx} - h^2)\varepsilon_{zz}}{\varepsilon_{xx}}, \quad B = 0, \tag{9.31}$$

that corresponds to the second root (9.9) when one takes a minus sign before the radical. The first solution (9.30) gives the H–wave and the second solution (9.31) gives the E–wave. Thus for the $\varepsilon-$ & $\mu-$ gyrotropic medium the solution (9.9) taken with the upper sign describes the hybrid wave. This hybrid wave, which is the superposition of H–wave and E–wave (9.27) degenerates into a pure H–wave with the gyrotropy decreasing. This wave is called a quasi–H–wave. Then one obtains the quasi–E–wave taking in (9.9) the lower sign.

The solutions of the system (9.4) for the point source are also a quasi–H–wave:

$$H_z(\vec{r}) = AH_0^{(2)}(k_{\perp 1}r'),$$

$$E_z(\vec{r}) = AaH_0^{(2)}(k_{\perp 1}r'),$$

and for a quasi–E–wave:

$$H_z(\vec{r}) = BbH_0^{(2)}(k_{\perp 2}r'),$$

$$E_z(\vec{r}) = BH_0^{(2)}(k_{\perp 2}r'),$$

where the coefficients a and b show the ratio of the H–wave and E–wave in the mixture.

Here we assumed the contour L_m is surrounding the gyrotropic medium and the contour point has the radius vector $r_s = ix_s + jy_s$ (Fig.9.7).

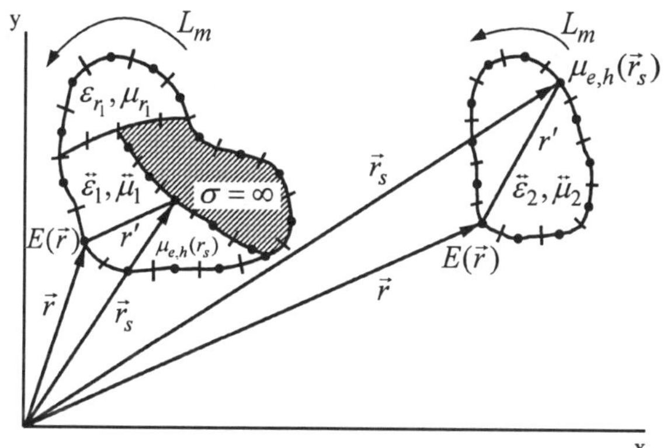

Fig. 9.7. *The topology of the piece–wise uniform gyrotropic waveguide.*

Then we write the integral representation for the quasi–H–wave:

$$H_z(\vec{r}) = \int_{L_m} \mu_h(\vec{r}_s)H_0^{(2)}(k_{\perp 1}r')ds, \tag{9.32}$$

$$E_z(\vec{r}) = a\int_{L_m} \mu_h(\vec{r}_s)H_0^{(2)}(k_{\perp 1}r')ds. \tag{9.33}$$

Similarly, for a quasi–E–wave:

$$H_z(\vec{r}) = b\int_{L_m} \mu_e(\vec{r}_s)H_0^{(2)}(k_{\perp 2}r')ds, \tag{9.34}$$

$$E_z(\vec{r}) = \int_{L_m} \mu_e(\vec{r}_s)H_0^{(2)}(k_{\perp 2}r')ds, \tag{9.35}$$

where $\mu_e(\vec{r}_s)$ and $\mu_h(\vec{r}_s)$ are the unknown functions of the contour point coordinate. Here it is clear that the representations (9.32–9.35) satisfy Maxwell's equations. The general solution of the wave equation system (9.4) is:

$$E_z(\vec{r}) = \int_{L_m} \mu_e(\vec{r}_s) H_0^{(2)}(k_{\perp 2} r')ds + a \int_{L_m} \mu_h(\vec{r}_s) H_0^{(2)}(k_{\perp 1} r')ds , \tag{9.36}$$

$$H_z(\vec{r}) = \int_{L_m} \mu_h(\vec{r}_s) H_0^{(2)}(k_{\perp 1} r')ds + b \int_{L_m} \mu_e(\vec{r}_s) H_0^{(2)}(k_{\perp 2} r')ds . \tag{9.37}$$

The coefficient $a = A/B$ is found from the first equations of (9.28) and is presented by the formula (9.7). The coefficient $b = B/A$ is also found from the equations of (9.28) and is shown by the formula (9.8).

Putting (9.36) and (9.37) into the formulae (9.23)–(9.26) we obtain the expressions for the transversal components E_x, E_y, H_x, H_y.

Using the formulae (9.23) and (9.24), we write the general expression for the tangential transversal components of the electric field:

$$E_S = \frac{1}{\Delta_{PF}}\left\{ -ih\left[k^2(\varepsilon_{xx}\mu_{xx} + \varepsilon_{xy}\mu_{xy}) - h^2\right]\frac{\partial E_z}{\partial s} - hk^2(\varepsilon_{xx}\mu_{xy} + \varepsilon_{xy}\mu_{xx})\frac{\partial E_z}{\partial n} + \right.$$
$$\left. +\omega\mu_0\left[k^2\varepsilon_{xy}(\mu_{xx}^2 - \mu_{xy}^2) + h^2\mu_{xy}\right]\frac{\partial H_z}{\partial s} - i\omega\mu_0\mu_{xx}(k^2\varepsilon_{xx}\mu^{ef} - h^2)\frac{\partial H_z}{\partial n}\right\}.$$

As for the points on the metal $E_z = 0$ then we have derivatives at the points on the metal surface $\partial E_z/\partial s = 0$. So then the general expression on the metal surface:

$$E_{SM} = \frac{1}{\Delta_{PF}}\left\{ -hk^2(\varepsilon_{xx}\mu_{xy} + \varepsilon_{xy}\mu_{xx})\frac{\partial E_z}{\partial n} + \right.$$
$$\left. +\omega\mu_0\left[k^2\varepsilon_{xy}(\mu_{xx}^2 - \mu_{xy}^2) + h^2\mu_{xy}\right]\frac{\partial H_z}{\partial s} - i\omega\mu_0\mu_{xx}(k^2\varepsilon_{xx}\mu^{ef} - h^2)\frac{\partial H_z}{\partial n}\right\}. \tag{9.38}$$

Similarly, from the formulae (9.25) and (9.26) we have the general expression for the tangential transversal components of the magnetic field:

$$H_S = \frac{1}{\Delta_{PF}}\left\{ -\omega\varepsilon_0\left[k^2\mu_{xy}\left(\varepsilon_{xx}^2 - \varepsilon_{xy}^2\right) + h^2\varepsilon_{xy}\right]\frac{\partial E_z}{\partial s} + \right.$$
$$+i\omega\varepsilon_0\varepsilon_{xx}(k^2\mu_{xx}\varepsilon^{ef} - h^2)\frac{\partial E_z}{\partial n} - hk^2(\varepsilon_{xx}\mu_{xy} + \varepsilon_{xy}\mu_{xx})\frac{\partial H_z}{\partial n} - $$
$$\left. -ih\left[k^2\left(\varepsilon_{xx}\mu_{xx} + \varepsilon_{xy}\mu_{xy}\right) - h^2\right]\frac{\partial H_z}{\partial s}\right\}. \tag{9.39}$$

The following steps to our solution are similar to chapter 8. The unknown densities functions $\mu_h(\vec{r}_s)$ and $\mu_e(\vec{r}_s)$ are determined from the boundary conditions, which present the equality of the tangential components on the contour dividing different media:

$$E_z^+ = E_z^- , \tag{9.40}$$

$$H_z^+ = H_z^- , \tag{9.41}$$

$$E_s^+ = E_s^- , \tag{9.42}$$

$$H_s^+ = H_s^- .$$

(9.43)

On the contour of any perfect metal the tangential components of the electric field are zeros:

$$E_z^- = 0 ,$$

(9.44)

$$E_s^- = 0 .$$

(9.45)

Here E_s, H_s are the transversal tangential components and E_z, H_z are the longitudinal tangential components of the EM field.

We present the Hankel function of the first order as a sum of the regular part (the first term) and of the singular part (the second term):

$$H_1^{(2)}(k_\perp r') = R H_1^{(2)}(k_\perp r) + i \frac{2}{\pi} \frac{1}{k_\perp r} .$$

(9.46)

The condition (9.44) in the contour points of the perfect metal yields the regular integral equation (Fig.9.8a):

$$a^{(-)} \int\limits_{L_m^{(-)}} \mu_h^{(-)}(\vec{r}_s) H_0^{(2)}(k_{\perp 1}^{(-)} r') ds + \int\limits_{L_m^{(-)}} \mu_e^{(-)}(\vec{r}_s) H_0^{(2)}(k_{\perp 2}^{(-)} r') ds = 0 .$$

(9.47)

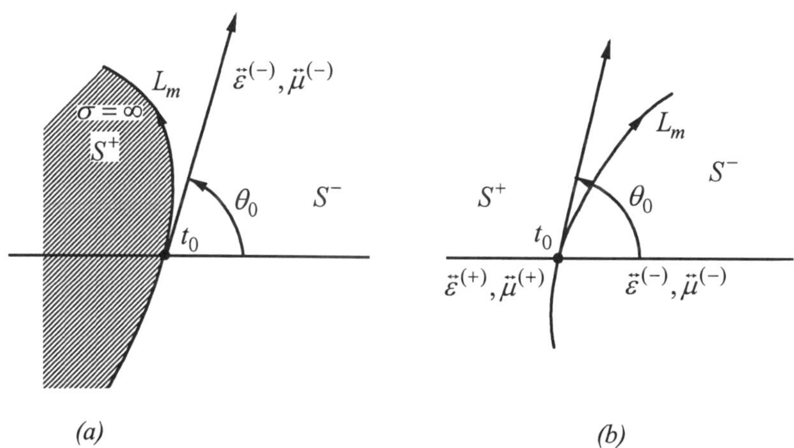

Fig. 9.8. *The integration contour L_m surrounding different media.*

The condition (9.45) at the contour points of the perfect metal (Fig.9.8a) are written in this form:

$$-hk^2 \left(\varepsilon_{xx}^{(-)} \mu_{xy}^{(-)} + \varepsilon_{xy}^{(-)} \mu_{xx}^{(-)} \right) \left\{ -a^{(-)} k_{\perp 1}^{(-)} \int\limits_{L_m^{(-)}} \mu_h^{(-)}(\vec{r}_s) R H_1^{(2)} \left(k_{\perp 1}^{(-)} r' \right) \frac{\partial r'}{\partial n} ds - \right.$$

$$-k_{\perp 2}^{(-)} \int\limits_{L_m^{(-)}} \mu_e^{(-)}(\vec{r}_s) R H_1^{(2)}(k_{\perp 2}^{(-)} r') \frac{\partial r'}{\partial n} ds - \frac{2i}{\pi} \left[-\pi (\mu_e^{(-)}(\vec{r}_0) + a^{(-)} \mu_h^{(-)}(\vec{r}_0)) - \right.$$

$$-\operatorname{Im} e^{j\theta_0} \int_{L_m^{(-)}} (\mu_e^{(-)}(\vec{r}_s) + a^{(-)}\mu_h^{(-)}(\vec{r}_s)) \frac{ds}{t_0 - t} \Bigg]\Bigg] +$$

$$+\omega\mu_0 \left[k^2 \varepsilon_{xy}^{(-)} \left(\mu_{xx}^{(-)2} - \mu_{xy}^{(-)2} \right) + h^2 \mu_{xy}^{(-)} \right] \Bigg\{ -k_{\perp 1}^{(-)} \int_{L_m^{(-)}} \mu_h^{(-)}(\vec{r}_s) \times$$

$$\times RH_1^{(2)}(k_{\perp 1}^{(-)}r')\frac{dr'}{\partial s} ds - b^{(-)}k_{\perp 2}^{(-)} \int_{L_m^{(-)}} \mu_e^{(-)}(\vec{r}_s) RH_1^{(2)}(k_{\perp 2}^{(-)}r')\frac{dr'}{\partial s} ds -$$

$$-\frac{2i}{\pi}\operatorname{Re} e^{j\theta_0} \int_{L_m^{(-)}} \left(\mu_h^{(-)}(\vec{r}_s) + b^{(-)}\mu_e^{(-)}(\vec{r}_s) \right)\frac{ds}{t_0 - t} \Bigg\} -$$

$$-i\omega\mu_0\mu_{xx}^{(-)}(k^2\varepsilon_{xx}^{(-)}\mu_{(-)}^{ef} - h^2)\Bigg\{ -k_{\perp 1}^{(-)} \int_{L_m^{(-)}} \mu_h^{(-)}(\vec{r}_s) RH_1^{(2)}(k_{\perp 1}^{(-)}r')\frac{\partial r'}{\partial n} ds -$$

$$-b^{(-)}k_{\perp 2}^{(-)} \int_{L_m^{(-)}} \mu_e^{(-)}(\vec{r}_s) RH_1^{(2)}(k_{\perp 2}^{(-)}r')\frac{\partial r'}{\partial n} ds + \frac{2i}{\pi}\Big[-\pi(\mu_h^{(-)}(\vec{r}_0) +$$

$$+b^{(-)}\mu_e^{(-)}(\vec{r}_0)) + \operatorname{Im} e^{j\theta_0} \int_{L_m^{(-)}} (\mu_h^{(-)}(\vec{r}_s) + b^{(-)}\mu_e^{(-)}(\vec{r}_s))\frac{ds}{t_0 - t} \Bigg]\Bigg\} = 0 . \tag{9.48}$$

Similarly the conditions (9.40) and (9.41) are written in this form (Fig.9.8*b*):

$$a^{(+)} \int_{L_m^{(+)}} \mu_h^{(+)}(\vec{r}_s) H_0^{(2)}(k_{\perp 1}^{(+)}r')ds - a^{(-)} \int_{L_m^{(-)}} \mu_h^{(-)}(\vec{r}_s) H_0^{(2)}(k_{\perp 1}^{(-)}r')ds +$$

$$+ \int_{L_m^{(+)}} \mu_e^{(+)}(\vec{r}_s) H_0^{(2)}(k_{\perp 2}^{(+)}r')ds - \int_{L_m^{(-)}} \mu_e^{(-)}(\vec{r}_s) H_0^{(2)}(k_{\perp 2}^{(-)}r')ds = 0 , \tag{9.49}$$

$$\int_{L_m^{(+)}} \mu_h^{(+)}(\vec{r}_s) H_0^{(2)}(k_{\perp 1}^{(+)}r')ds - \int_{L_m^{(-)}} \mu_h^{(-)}(\vec{r}_s) H_0^{(2)}(k_{\perp 1}^{(-)}r')ds +$$

$$+b^{(+)} \int_{L_m^{(+)}} \mu_e^{(+)}(\vec{r}_s) H_0^{(2)}(k_{\perp 2}^{(+)}r')ds - b^{(-)} \int_{L_m^{(-)}} \mu_e^{(-)}(\vec{r}_s) H_0^{(2)}(k_{\perp 2}^{(-)}r')ds = 0 . \tag{9.50}$$

The boundary condition (9.42) is written in this form (Fig.9.8*b*):

$$\Bigg\{ -\frac{ih}{\Delta_{PF}^{(-)}}\left[k^2 \left(\varepsilon_{xx}^{(-)}\mu_{xx}^{(-)} + \varepsilon_{xy}^{(-)}\mu_{xy}^{(-)} \right) - h^2 \right] +$$

$$+\frac{ih}{\Delta_{PF}^{(+)}}\left[k^2 \left(\varepsilon_{xx}^{(+)}\mu_{xx}^{(+)} + \varepsilon_{xy}^{(+)}\mu_{xy}^{(+)} \right) - h^2 \right] \Bigg\} \times$$

$$\times \left\{ -a^{(-)} k_{\perp 1}^{(-)} \int_{L_m} \mu_h^{(-)}(\vec{r}_s) \, RH_1^{(2)}(k_{\perp 1}^{(-)} r') \frac{\partial r'}{\partial s} \, ds - \right.$$

$$-k_{\perp 2}^{(-)} \int_{L_m} \mu_e^{(-)}(\vec{r}_s) \, RH_1^{(2)}(k_{\perp 2}^{(-)} r') \frac{\partial r'}{\partial s} \, ds -$$

$$\left. -\frac{2i}{\pi} \operatorname{Re} e^{j\theta_0} \int_{L_m} \frac{a^{(-)} \mu_h^{(-)} + \mu_e^{(-)}(\vec{r}_s)}{t_0 - t} \, ds \right\} +$$

$$+\left\{ \frac{\omega \mu_0}{\Delta_{PF}^{(-)}} \left[k^2 \varepsilon_{xy}^{(-)} \left(\mu_{xx}^{(-)2} - \mu_{xy}^{(-)2} \right) + h^2 \mu_{xy}^{(-)} \right] - \right.$$

$$\left. -\frac{\omega \mu_0}{\Delta_{PF}^{(+)}} \left[k^2 \varepsilon_{xy}^{(+)} \left(\mu_{xx}^{(+)2} - \mu_{xy}^{(+)2} \right) + h^2 \mu_{xy}^{(+)} \right] \right\} \times$$

$$\times \left\{ -k_{\perp 1}^{(-)} \int_{L_m} \mu_h^{(-)}(\vec{r}_s) RH_1^{(2)}(k_{\perp 1}^{(-)} r') \frac{\partial r'}{\partial s} \, ds - \right.$$

$$-b^{(-)} k_{\perp 2}^{(-)} \int_{L_m} \mu_e^{(-)}(\vec{r}_s) \, RH_1^{(2)}(k_{\perp 2}^{(-)} r') \frac{\partial r'}{\partial s} \, ds -$$

$$-\frac{2i}{\pi} \operatorname{Re} e^{j\theta_0} \int_{L_m} \frac{\mu_h^{(-)}(\vec{r}_s) + b^{(-)} \mu_e^{(-)}(\vec{r}_s)}{t_0 - t} \, ds -$$

$$-\frac{hk^2}{\Delta_{PF}^{(-)}} \left(\varepsilon_{xx}^{(-)} \mu_{xy}^{(-)} + \varepsilon_{xy}^{(-)} \mu_{xx}^{(-)} \right) \left\{ -a^{(-)} k_{\perp 1}^{(-)} \int_{L_m} \mu_h^{(-)}(\vec{r}_s) \, RH_1^{(2)}(k_{\perp 1}^{(-)} r') \frac{\partial r'}{\partial n} \, ds - \right.$$

$$-k_{\perp 2}^{(-)} \int_{L_m} \mu_e^{(-)}(\vec{r}_s) \, RH_1^{(2)}(k_{\perp 2}^{(-)} r') \frac{\partial r'}{\partial n} \, ds + \frac{2i}{\pi} \left[-\pi (a^{(-)} \mu_h^{(-)}(\vec{r}_0) + \right.$$

$$\left. \left. +\mu_e^{(-)}(\vec{r}_0)) + \operatorname{Im} e^{j\theta_0} \int_{L_m} \frac{a^{(-)} \mu_h^{(-)}(\vec{r}_s) + \mu_e^{(-)}(\vec{r}_s)}{t_0 - t} \, ds \right] \right\} -$$

$$-\frac{i\omega \mu_0 \mu_{xx}^{(-)}}{\Delta_{PF}^{(-)}} \left(k^2 \varepsilon_{xx}^{(-)} \mu_{(-)}^{ef} - h^2 \right) \left\{ -k_{\perp 1}^{(-)} \int_{L_m} \mu_h^{(-)}(\vec{r}_s) \, RH_1^{(2)}(k_{\perp 1}^{(-)} r') \times \right.$$

$$\times \frac{\partial r'}{\partial n} \, ds - b^{(-)} k_{\perp 2}^{(-)} \int_{L_m} \mu_e^{(-)}(\vec{r}_s) RH_1^{(2)}(k_{\perp 2}^{(-)} r') \frac{\partial r'}{\partial n} \, ds +$$

$$+\frac{2i}{\pi} \left[-i\pi \left(\mu_h^{(-)}(\vec{r}_0) + b^{(-)} \mu_e^{(-)}(\vec{r}_0) \right) + \right.$$

$$+\operatorname{Im}e^{j\theta_0}\int_{L_m}\frac{\mu_h^{(-)}\left(\vec{r}_s\right)+b^{(-)}\mu_e^{(-)}\left(\vec{r}_s\right)}{t_0-t}ds\Bigg]\Bigg\}+\frac{hk^2}{\Delta_{PF}^{(+)}}(\varepsilon_{xx}^{(+)}\mu_{xy}^{(+)}+$$

$$+\varepsilon_{xy}^{(+)}\mu_{xx}^{(+)})\Bigg\{-a^{(+)}k_{\perp1}^{(+)}\int_{L_m^{(-)}}\mu_h^{(+)}\left(\vec{r}_s\right)RH_1^{(2)}(k_{\perp1}^{(+)}r')\frac{\partial r'}{\partial n}ds-$$

$$-k_{\perp2}^{(+)}\int_{L_m}\mu_e^{(+)}\left(\vec{r}_s\right)RH_1^{(2)}(k_{\perp2}^{(+)}r')\frac{\partial r'}{\partial n}ds+\frac{2i}{\pi}\Big[\pi(a^{(+)}\mu_h^{(+)}\left(\vec{r}_0\right)+$$

$$+\mu_e^{(+)}\left(\vec{r}_0\right))+\operatorname{Im}e^{j\theta}\int_{L_m}\frac{a^{(+)}\mu_h^{(+)}\left(\vec{r}_s\right)+\mu_e^{(+)}\left(\vec{r}_s\right)}{t_0-t}ds\Big]\Bigg\}+$$

$$+\frac{i\omega\mu_0\mu_{xx}^{(+)}}{\Delta_{PF}^{(+)}}\left(k^2\varepsilon_{xx}^{(+)}\mu_{(+)}^{ef}-h^2\right)\Bigg\{-k_{\perp1}^{(+)}\int_{L_m}\mu_h^{(+)}\left(\vec{r}_s\right)RH_1^{(2)}(k_{\perp1}^{(+)}r')\times$$

$$\times\frac{\partial r'}{\partial n}ds-b^{(+)}k_{\perp2}^{(+)}\int_{L_m}\mu_e^{(+)}\left(\vec{r}_s\right)RH_1^{(2)}(k_{\perp2}^{(+)}r')\frac{\partial r'}{\partial n}ds+$$

$$+\frac{2i}{\pi}\Bigg[\pi\left(\mu_h^{(+)}\left(\vec{r}_0\right)+b^{(+)}\mu_e^{(+)}\left(\vec{r}_0\right)\right)+$$

$$+\operatorname{Im}e^{j\theta_0}\int_{L_m}\frac{\mu_h^{(+)}\left(\vec{r}_s\right)+b^{(+)}\mu_e^{(+)}\left(\vec{r}_s\right)}{t_0-t}ds\Bigg]\Bigg\}=0. \tag{9.51}$$

And the boundary condition (9.43) is written in this form (Fig.9.8*b*):

$$\Bigg\{\frac{\omega\varepsilon_0}{\Delta_{PF}^{(-)}}\Big[k^2\mu_{xy}^{(-)}\left(\varepsilon_{xx}^{(-)2}-\varepsilon_{xy}^{(-)2}\right)+h^2\varepsilon_{xy}^{(-)}\Big]-$$

$$-\frac{\omega\varepsilon_0}{\Delta_{PF}^{(+)}}\Big[k^2\mu_{xy}^{(+)}\left(\varepsilon_{xx}^{(+)2}-\varepsilon_{xy}^{(+)2}\right)+h^2\varepsilon_{xy}^{(+)}\Big]\Bigg\}\times$$

$$\times\Bigg\{-a^{(-)}k_{\perp1}^{(-)}\int_{L_m}\mu_h^{(-)}\left(\vec{r}_s\right)RH_1^{(2)}(k_{\perp2}^{(-)}r')\frac{\partial r'}{\partial s}ds-$$

$$-k_{\perp2}^{(-)}\int_{L_m}\mu_e^{(-)}\left(\vec{r}_s\right)RH_1^{(2)}(k_{\perp2}^{(-)}r')\frac{\partial r'}{\partial s}ds-\frac{2i}{\pi}\operatorname{Re}e^{j\theta_0}\times$$

$$\times\int_{L_m}\frac{a^{(-)}\mu_h^{(-)}\left(\vec{r}_s\right)+\mu_e^{(-)}\left(\vec{r}_s\right)}{t_0-t}ds\Bigg\}+\Bigg\{-\frac{ih}{\Delta_{PF}^{(-)}}\Big[\left(k^2(\varepsilon_{xx}^{(-)}\mu_{xx}^{(-)}+\right.$$

$$+\varepsilon_{xy}^{(-)}\mu_{xy}^{(-)})-h^2\Big]+\frac{ih}{\Delta_{PF}^{(+)}}\Big[k^2\Big(\varepsilon_{xx}^{(+)}\mu_{xx}^{(+)}-\varepsilon_{xy}^{(+)}\mu_{xy}^{(+)}\Big)-h^2\Big]\Big\}\times$$

$$\times\Bigg\{-k_{\perp1}^{(-)}\int_{L_m^{(+)}}\mu_h^{(-)}(\vec{r}_s)RH_1^{(2)}(k_{\perp1}^{(-)}r')\frac{\partial r'}{\partial s}ds-b^{(-)}k_{\perp2}^{(-)}\int_{L_m^{(+)}}\mu_e^{(-)}(\vec{r}_s)\times$$

$$\times RH_1^{(2)}(k_{\perp2}^{(-)}r')\frac{\partial r'}{\partial s}ds-\frac{2i}{\pi}\mathrm{Re}\,e^{j\theta_0}\int_{L_m}\frac{\mu_h^{(-)}(\vec{r}_s)+b^{(-)}\mu_e^{(-)}(\vec{r}_s)}{t_0-t}ds\Bigg\}+$$

$$+\frac{i\omega\varepsilon_0\varepsilon_{xx}^{(-)}}{\Delta_{PF}^{(-)}}(k^2\varepsilon_{(-)}^{ef}\mu_{(-)}^{ef}-h^2)\Bigg\{-a^{(-)}k_{\perp1}^{(-)}\int_{L_m}\mu_h^{(-)}(\vec{r}_s)RH_1^{(2)}(k_{\perp1}^{(-)}r')\frac{\partial r'}{\partial n}ds-$$

$$-k_{\perp2}^{(-)}\int_{L_m^{(+)}}\mu_e^{(-)}(\vec{r}_s)RH_1^{(2)}(k_{\perp2}^{(-)}r')\frac{\partial r'}{\partial s}ds+\frac{2i}{\pi}\Big[-\pi(a^{(-)}\mu_h^{(-)}(\vec{r}_0)+$$

$$+\mu_e^{(-)}(\vec{r}_0))+\mathrm{Im}\,e^{j\theta_0}\int_{L_m}\frac{a^{(-)}\mu_h^{(-)}(\vec{r}_s)+\mu_e^{(-)}(\vec{r}_s)}{t_0-t}ds\Big]\Bigg\}-$$

$$-\frac{hk^2}{\Delta_{PF}^{(-)}}\Big(\varepsilon_{xx}^{(-)}\mu_{xy}^{(-)}+\varepsilon_{xy}^{(-)}\mu_{xx}^{(-)}\Big)\Bigg\{-k_{\perp1}^{(-)}\int_{L_m}\mu_h^{(-)}(\vec{r}_s)RH_1^{(2)}(k_{\perp1}^{(-)}r')\frac{\partial r'}{\partial n}ds-$$

$$-b^{(-)}k_{\perp2}^{(-)}\int_{L_m}\mu_e^{(-)}(\vec{r}_s)RH_1^{(2)}(k_{\perp2}^{(-)}r')\frac{\partial r'}{\partial n}ds+\frac{2i}{\pi}\Big[-\pi(\mu_h^{(-)}(\vec{r}_0)+$$

$$+b^{(-)}\mu_0^{(-)}(\vec{r}_0))+\mathrm{Im}\,e^{j\theta_0}\int_{L_m}\frac{\mu_h^{(-)}(\vec{r}_s)+b^{(-)}\mu_e^{(-)}(\vec{r}_s)}{t_0-t}ds\Big]\Bigg\}-$$

$$-\frac{i\omega\varepsilon_0\varepsilon_{xx}^{(+)}}{\Delta_{PF}^{(+)}}\Big(k^2\varepsilon_{(+)}^{ef}\mu_{(+)}^{ef}-h^2\Big)\Bigg\{-a^{(+)}k_{\perp1}^{(+)}\int_{L_m}\mu_h^{(+)}(\vec{r}_s)RH_1^{(2)}(k_{\perp1}^{(+)}r')\frac{\partial r'}{\partial n}ds-$$

$$-k_{\perp2}^{(+)}\int_{L_m}\mu_e^{(+)}(\vec{r}_s)RH_1^{(2)}(k_{\perp2}^{(+)}r')\frac{\partial r'}{\partial n}ds+\frac{2i}{\pi}\Big[\pi\Big(a^{(+)}\mu_h^{(+)}(\vec{r}_0)+\mu_e^{(+)}(\vec{r}_0)\Big)+$$

$$+\mathrm{Im}\,e^{j\theta_0}\int_{L_m}\frac{a^{(+)}\mu_h^{(+)}(\vec{r}_s)+\mu_e^{(+)}(\vec{r}_s)}{t_0-t}ds\Big]\Bigg\}+\frac{hk^2}{\Delta_{PF}^{(+)}}\Big(\varepsilon_{xx}^{(+)}\mu_{xy}^{(+)}+\varepsilon_{xy}^{(+)}\mu_{xx}^{(+)}\Big)\times$$

$$\times\Bigg\{k_{\perp1}^{(+)}\int_{L_m}\mu_h^{(+)}(\vec{r}_s)\,RH_1^{(2)}\Big(k_{\perp1}^{(+)}r'\Big)\frac{\partial r'}{\partial n}ds-b^{(+)}k_{\perp2}^{(+)}\times$$

$$\times \int_{L_m} \mu_e^{(+)}(\vec{r}_s) R H_1^{(2)}(k_{\perp 2}^{(+)} r') \frac{\partial r'}{\partial n} ds + \frac{2i}{\pi}\left[\pi\left(\mu_h^{(-)}(\vec{r}_0) + b^{(-)}\mu_e^{(-)}(\vec{r}_0)\right) + \right.$$

$$\left.\left. + \operatorname{Im} e^{j\theta_0} \int_{L_m} \frac{\mu_h^{(+)}(\vec{r}_s) + b^{(+)}\mu_e^{(+)}(\vec{r}_s)}{t_0 - t} ds \right]\right\} = 0. \tag{9.52}$$

In formulae (9.51), (9.52) and (9.38), we have used the condition that the equality $\partial E_z^{(+)}/\partial s = \partial E_z^{(-)}/\partial s$ follows from the condition $E_z^{(+)} = E_z^{(-)}$ on the contour dividing the two media.

Further, the Krylov – Bogoliubov method was applied to the system of equations (9.47)–(9.52), which transferred it into the system of homogeneous linear equations.

In this particular case, when integration contours are made of straight line stretches that are parallel to the axes of the Cartesian coordinates after applied the Krylov–Bogoliubov method one can write the longitudinal field components:

$$E_z = a\sum_{j=1}^{n_x} \mu_h(s_j) \int_{L_x} H_0^{(2)}(k_{\perp 1} r')dx + \sum_{j=1}^{n_x} \mu_e(s_j) \int_{L_x} H_0^{(2)}(k_{\perp 2} r')dx +$$

$$+a\sum_{j=1}^{n_y} \mu_h(s_j) \int_{L_y} H_0^{(2)}(k_{\perp 1} r')dy + \sum_{j=1}^{n_y} \mu_e(s_j) \int_{L_y} H_0^{(2)}(k_{\perp 2} r')dy , \tag{9.53}$$

$$H_z = \sum_{j=1}^{n_x} \mu_h(s_j) \int_{L_x} H_0^{(2)}(k_{\perp 1} r')dx + b\sum_{j=1}^{n_x} \mu_e(s_j) \int_{L_x} H_0^{(2)}(k_{\perp 2} r')dx +$$

$$+\sum_{j=1}^{n_y} \mu_h(s_j) \int_{L_y} H_0^{(2)}(k_{\perp 1} r')dy + b\sum_{j=1}^{n_y} \mu_e(s_j) \int_{L_y} H_0^{(2)}(k_{\perp 2} r')dy. \tag{9.54}$$

Here the notations for regions "+" and "–" were dropped because the writing is valid for both of them.

The derivatives of the longitudinal components of the EM field (which are in the expressions for the transversal components (9.23)–(9.26)) after applying the Krylov–Bogoliubov method take the form:

$$\frac{\partial E_z}{\partial x_0} = \sum_{j=1}^{n_x} \mu_h(s_j) H_0^{(2)}(k_{\perp 1} r')\Big|_{\Delta L_x} +$$

$$+k_{\perp 1}\sum_{j=1}^{n_y} \mu_h(s_j) \int_{L_y} H_1^{(2)}(k_{\perp 1} r')\frac{x-x_0}{r'} dy - a\sum_{j=1}^{n_x} \mu_e(s_j) H_0^{(2)}(k_{\perp 2} r')\Big|_{\Delta L_x} +$$

$$+ak_{\perp 2}\sum_{j=1}^{n_y} \mu_e(s_j) \int_{L_y} H_1^{(2)}(k_{\perp 2} r')\frac{x-x_0}{r'} dy , \tag{9.55}$$

$$\frac{\partial E_z}{\partial y_0} = k_{\perp 1} \sum_{j=1}^{n_x} \mu_h(s_j) \int_{L_x} H_1^{(2)}(k_{\perp 1} r') \frac{y-y_0}{r'} dx -$$

$$-\sum_{j=1}^{n_y} \mu_h(s_j) H_0^{(2)}(k_{\perp 1} r')\Big|_{\Delta L_y} + a k_{\perp 2} \sum_{j=1}^{n_x} \mu_e(s_j) \times$$

$$\times \int_{L_x} H_1^{(2)}(k_{\perp 2} r') \frac{y-y_0}{r'} dx - a \sum_{j=1}^{n_y} \mu_e(s_j) H_0^{(2)}(k_{\perp 2} r')\Big|_{\Delta L_y}, \tag{9.56}$$

$$\frac{\partial H_z}{\partial x_0} = -b \sum_{j=1}^{n_x} \mu_h(s_j) H_0^{(2)}(k_{\perp 1} r')\Big|_{\Delta L_x} + b k_{\perp 1} \sum_{j=1}^{n_y} \mu_h(s_j) \times$$

$$\times \int_{L_y} H_1^{(2)}(k_{\perp 1} r') \frac{x-x_0}{r'} dy - \sum_{j=1}^{n_x} \mu_e(s_j) H_0^{(2)}(k_{\perp 2} r')\Big|_{\Delta L_x} +$$

$$+ k_{\perp 2} \sum_{j=1}^{n_y} \mu_e(s_j) H_1^{(2)}(k_{\perp 2} r') \frac{x-x_0}{r'} dy, \tag{9.57}$$

$$\frac{\partial H_z}{\partial y_0} = b k_{\perp 1} \sum_{j=1}^{n_x} \mu_h(s_j) \int_{L_x} H_1^{(2)}(k_{\perp 1} r') \frac{y-y_0}{r'} dx -$$

$$-b \sum_{j=1}^{n_y} \mu_h(s_j) H_0^{(2)}(k_{\perp 1} r')\Big|_{\Delta L_y} + k_{\perp 2} \sum_{j=1}^{n_x} \mu_e(s_j) \int_{L_x} H_1^{(2)}(k_{\perp 2} r') \frac{y-y_0}{r'} dx -$$

$$-\sum_{j=1}^{n_y} \mu_e(s_j) H_0^{(2)}(k_{\perp 2} r')\Big|_{\Delta L y}. \tag{9.58}$$

Putting (9.55)–(9.58) into these formulae (9.23)–(9.26) we obtain the expressions for the transversal components of the EM field. The expressions are long and we choose not to provide them here. Here we would like you to notice that the leap value (according the Sokhotsky–Plemelj limit formulae) will have only the derivative $\partial E_z/\partial y$ in the expression of the transversal component E_x. And in a similar way the leap values will have only the derivatives $\partial H_z/\partial y$, $\partial E_z/\partial x$ and $\partial H_z/\partial x$ in the expressions of transversal components H_x, E_y and Hy correspondently.

Here we will write the leap values (according the Sokhotsky–Plemelj limit formulae) when the functions $\mu_h^+(s_j)$, $\mu_e^+(s_j)$ or $\mu_h^-(s_j)$, $\mu_e^-(s_j)$ are not at the corner points of the counter. Then the transversal components of the EM field with these leap values have the form:

$$E_x^{\mp} = \pm \frac{2ia_1^{ex}}{k_{\perp 2}^{(\mp)}} \cos\theta \sum_{j=1}^{n} \mu_e^{(\mp)}(s_j) + \frac{2ia_2^{ex}}{k_{\perp 1}^{(\mp)}} \cos\theta \sum_{j=1}^{n} \mu_h^{(\mp)}(s_j) + E_x, \tag{9.59}$$

$$E_y^{\mp} = \pm \frac{2ia_1^{ex}}{k_{\perp 2}^{(\mp)}} \sin\theta \sum_{j=1}^{n} \mu_e^{(\mp)}(s_j) + \frac{2ia_2^{ex}}{k_{\perp 1}^{(\mp)}} \sin\theta \sum_{j=1}^{n} \mu_h^{(\mp)}(s_j) + E_y, \tag{9.60}$$

$$H_x^{\mp} = \pm \frac{2ia_1^{hx}}{k_{\perp 2}^{(\mp)}} \cos\theta \sum_{j=1}^{n} \mu_e^{(\mp)}(s_j) + \frac{2ia_2^{hx}}{k_{\perp 1}^{(\mp)}} \cos\theta \sum_{j=1}^{n} \mu_h^{(\mp)}(s_j) + H_x, \tag{9.61}$$

$$H_y^{\mp} = \pm \frac{2ia_2^{hx}}{k_{\perp 2}^{(\mp)}} \sin\theta \sum_{j=1}^{n} \mu_e^{(\mp)}(s_j) + \frac{2ia_2^{hx}}{k_{\perp 1}^{(\mp)}} \sin\theta \sum_{j=1}^{n} \mu_h^{(\mp)}(s_j) + H_y. \tag{9.62}$$

The formulae (9.59)–(9.62) were obtained according to the formulae (9.23)–(9.26) taking the principle values of the integrals for the contour points on the smooth contour parts. E_x^-, E_y^-, H_x^-, H_y^- are the limited values of the field components approaching the contour point from the region "−". And E_x^+, E_y^+, H_x^+, H_y^+ are the limited values of the field components approaching the contour point from the region "+". The coefficients in the formulae (9.59)–(9.62) looks like these:

$$a_1^{ex} = (\psi_1 b - \psi_2)k_{\perp 2},$$

$$a_2^{ex} = (\psi_1 - a\psi_2)k_{\perp 1},$$

$$\psi_1 = i\omega\mu_0\mu_{xx}(h^2 - k^2\varepsilon_{xx}\mu^{ef})/\Delta_{PF},$$

$$\psi_2 = hk^2(\varepsilon_{xy}\mu_{xx} + \varepsilon_{xx}\mu_{xy})/\Delta_{PF},$$

$$a_1^{hx} = (\psi_3 - \psi_4 b)k_{\perp 2},$$

$$a_2^{hx} = (a\psi_3 - \psi_4)k_{\perp 1},$$

$$\psi_3 = -i\omega\varepsilon_0\varepsilon_{xx}(h^2 - k^2\mu_{xx}\varepsilon^{ef})/\Delta_{PF},$$

$$\psi_4 = ih\left[k^2(\varepsilon_{xx}\mu_{xx} + \varepsilon_{xy}\mu_{xy}) - h^2 \right]/\Delta_{PF}.$$

So we obtained the linear algebraic system to determine the unknown density functions $\mu_h^+(s_j)$, $\mu_h^-(s_j)$, $\mu_e^+(s_j)$, $\mu_e^-(s_j)$. Equalizing this algebraic system determinant to zero we obtain the dispersion equation in order to find the propagation constant h.

9.3. Some Examples of Testing our Algorithm

We now consider the stability of our algorithm that was based on the SIE method (described in Section 9.2) with respect to the small changes of the system's coefficients. We calculated the waveguide dispersion dependences when the physical parameters and geometry of the structure were changed minutely. And we saw that the calculated results would change minutely every time.

In Table 9.2 we present the dependence of the calculated results (h, $\xi_\perp$, $k_{\perp 1}$, $k_{\perp 2}$) on the external constant magnetic field H for a mirror ferrite waveguide (see Fig.8.29).

Table 9.2. The testing of our algorithm on the external constant magnetic field H for a mirror ferrite waveguide

Mode number	$H\,(kA/m)$	μ_{xx}	μ_{xy}	$h\,(m^{-1})$	$\xi_\perp\,(m^{-1})$	$k_{\perp 1}\,(m^{-1})$	$k_{\perp 2}\,(m^{-1})$	Coefficients Formulae (9.7 & 9.8)	
								a	b
1	2	3	4	5	6	7	8	9	10
The	119.05	0.99761	-0.04016	1938.4	1268	4111	4198	108.4	7.47
main	198.41	0.99597	-0.04042	1937.6	1267	4108	4106	104.3	7.19
mode	277.78	0.99429	-0.04081	1936.8	1266	4105	4195	100.3	6.92
	357.14	0.99256	-0.04135	1937.0	1266	4101	4193	96.57	6.66
The	39.68	0.99926	-0.04003	1840.4	1112	4161	4242	112.4	7.70
first	119.05	0.99761	-0.04016	1839.8	1112	4158	4240	107.7	7.42
higher	158.73	0.99679	-0.04027	1839.6	1111	4157	4239	105.4	7.27
mode	198.41	0.99597	-0.04042	1839.4	1111	4155	4238	103.3	7.12
	238.10	0.99514	-0.04060	1839.0	1110	4154	4237	101.1	6.97
	277.78	0.99429	-0.04081	1838.8	1110	4152	4237	99.07	6.83
	317.46	0.99344	-0.04107	1838.4	1109	4151	4236	97.05	6.69
The	119.05	0.99761	-0.04016	1887.4	1189	4136	4220	108	7.45
second	198.41	0.99597	-0.04042	1887.0	1488	4133	4218	103.2	7.15
higher	277.78	0.99429	-0.04081	1886.4	1187	4130	4217	99.69	6.87
mode	357.14	0.99256	-0.04135	1885.6	1186	4127	4215	95.81	6.61
The	39.68	0.99926	-0.04003	1749.6	954.8	4202	4278	111.9	7.72
third	119.05	0.99761	-0.04016	1749.4	954.5	4199	4276	106.9	7.37
higher	158.73	0.99679	-0.04027	1749.2	954.1	4198	4275	104.6	7.21
mode	198.41	0.99597	-0.04042	1749.0	953.7	4196	4274	102.3	7.05

Our calculations shown in Table 9.2 were completed for the mirror ferrite waveguide with sizes: $d = 0.3 \cdot 10^{-3}$ m, $l = 0.7 \cdot 10^{-3}$ m, $t = 10^{-6}$ m at frequency $f = 70$ GHz when the ferrite material parameters were: $4\pi M_s = 79.365$ kA/m, the half of the width of the resonance curve $\delta \approx 7.94$ kA/m and $\varepsilon_r = 9.8$.

The calculations shown in Table 9.3 were completed for the mirror ferrite waveguide with sizes $d = 0.35 \cdot 10^{-3}$ m, $t = 10^{-6}$ m at frequency $f = 70$ GHz, $4\pi M_s = 79.365$ kA/m when the components of the permeability tensor $\ddot{\mu}$ of ferrite material were: $\mu_{xx} = 0.99926$, $\mu_{xy} = 0.04003$ and $\mu_{zz} = 1$.

Analyzing the data given in the Tables 9.2 and 9.3, we came to conclusion that the electrodynamical problem solving for the gyrotropic structures is stable, with respect to small changes of both the medium parameters and structure geometry. The calculations that we will present also confirm this conclusion.

Table 9.3. *The testing of our algorithm with respect to small changes of the mirror ferrite waveguide width l .*

Mode number	l , m	h , (m^{-1})	$\xi_\perp$, (m^{-1})	k_{11} , (m^{-1})	k_{12} , (m^{-1})
The main mode	0.30	1936.4	1265	4115	4201
	0.31	1936.8	1266	4115	4200
	0.32	1937.4	1267	4114	4200
	0.34	1938.6	1268	4114	4200
	0.345	1938.8	1269	4114	4200
	0.35	1939.2	1269	4113	4199
	0.36	1939.8	1270	4113	4199
	0.37	1940.4	1271	4113	4199
	0.38	1941.0	1272	4112	4198
	0.39	1942.0	1274	4112	4198
	0.40	1942.8	1275	4112	4198
	0.41	1943.6	1276	4111	4197
	0.42	1944.4	1277	4111	4197
	0.44	1946.2	1280	4110	4196
	0.46	1948.0	1283	4109	4195
	0.48	1950.0	1286	4108	4195
The third higher mode	0.30	1750.6	956.7	4201	4277
	0.31	1750.2	955.9	4201	4278
	0.32	1750,0	955.6	4202	4278
	0.34	1749.8	955.2	4202	4278
	0.345	1749.8	955.2	4202	4278
	0.35	1749.8	955.2	4202	4278
	0.36	1749.8	955.2	4202	4278
	0.37	1750.0	955.6	4202	4278
	0.38	1750.2	955.9	4201	4278
	0.39	1750.6	956.7	4201	4277
	0.40	1750.8	957.0	4201	4277
	0.41	1751.2	957.8	4201	4277
	0.42	1751.8	958.9	4201	4277
	0.44	1752.4	960.0	4200	4277
	0.46	1754.2	963.2	4200	4276
	0.48	1756.0	966.5	4199	4275

9.4. Electrodynamical Characteristic for $\ddot{\mu}$ – Gyrotropic Slot Lines

We numerically investigated the dispersion characteristics of a slot line (SL) on a longitudinally magnetized ferrite substrate. The saturation magnetization of the substrate $4\pi M_s =377$ kA/m. The nondiagonal permeability tensor component μ_{xy} determined by formulae [9.1] and the diagonal components were assumed to be $\mu_{xx}=\mu_{zz}=1$ and $\varepsilon_f =13.5\left(1-i5\cdot10^{-4}\right)$. The relative permittivity of the demagnetized ferrite material was $\varepsilon_r = 13.5$. In Figs.9.9–9.11 the sizes of the SLs were the same: the substrate height $d =10^{-3}$ m, the slot width $s =0.2\cdot10^{-3}$ m and the thickness of metal strip $t =2.5\cdot10^{-6}$ m.

Fig.9.9 presents the dependences of the longitudinal propagation constant h and the transversal $ik_{\perp1}$, $k_{\perp2}$ propagation constants on the frequency in the geometrically symmetric SL with the metal strips not reaching the substrate edges. In Fig.9.9 we see that only one mode can propagate in the SL at the frequency band under our consideration. The solid line is $ik_{\perp1}$ and the dash line is $k_{\perp2}$.

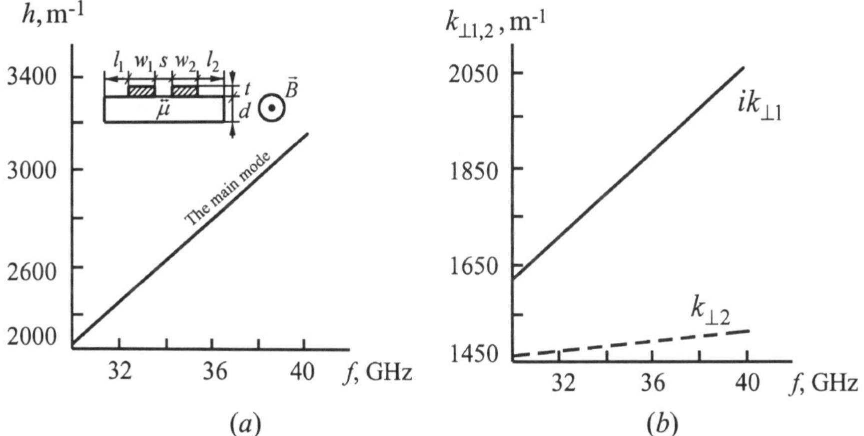

Fig. 9.9. *Dispersion dependences of the SL with the ferrite substrate at $4\pi M_s = 377$ kA/m. The sizes are: $w_1 = w_2 = 2\cdot10^{-3}$ m, $l_1 =l_2 = 2\cdot10^{-3}$ m, $d =10^{-3}$ m, $s =0.2\cdot10^{-3}$ m, $t =2.5\cdot10^{-6}$ m.*

Fig.9.10 presents the same dependences for an asymmetric SL and is one–mode as we saw in Fig.9.9. We see that this SL has a larger wavelength λ_{SL} ($h =2\pi/\lambda_{SL}$) than the previous one (Fig.9.9a).

Fig.9.11 presents the dispersion dependences of the main mode and the first higher mode for an asymmetric SL. The metal strips of the SL are placed at the substrate edges. One of the metal strips is three times wider. We see that the longitudinal propagation constants of the main and first higher modes are a little different in their values. Figs.9.9b–9.11b show the transversal propagation constants $ik_{\perp 1}$ and $k_{\perp 2}$ of the SLs. We know that the SL hybrid wave is formed on the surface quasi – H –wave and the volume quasi – E –wave with the transversal propagation constants $ik_{\perp 1}$ and $k_{\perp 2}$ correspondently.

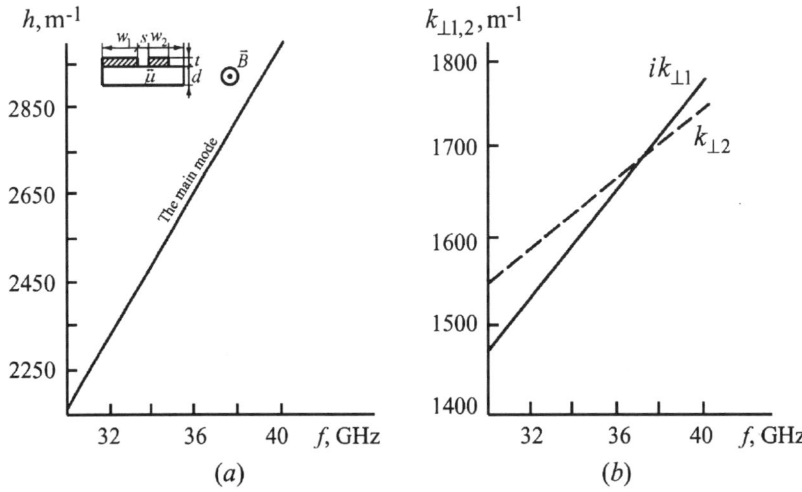

Fig. 9.10. Dispersion dependence of the asymmetric SL with the ferrite substrate at $4\pi M_s = 377$ *kA/m. The sizes are:* $w_1 = 3\cdot10^{-3}$ m, $w_2 = 10^{-3}$ m, $l_1 = 0$, $l_2 = 2\cdot10^{-3}$ m, $d = 10^{-3}$ m, $s = 0.2\cdot10^{-3}$ m, $t = 2.5\cdot10^{-6}$ m.

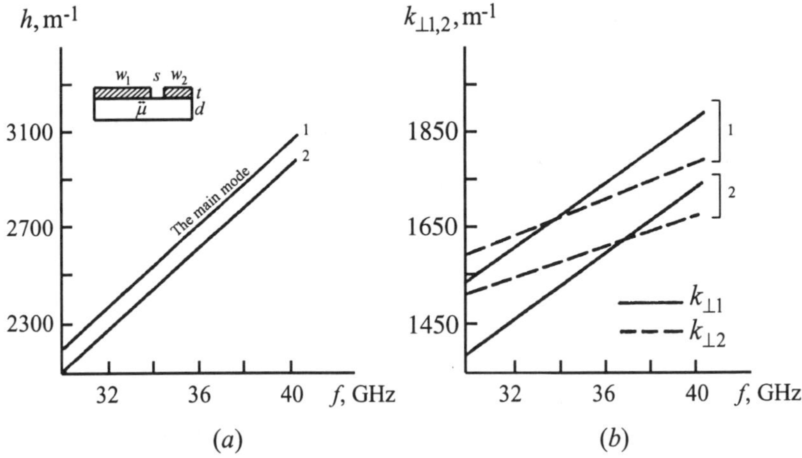

Fig. 9.11. Dispersion dependences of the asymmetric SL with the ferrite substrate for the main mode (curve 1) and the first higher mode (curve 2). The sizes are: $w_1 = 3\cdot10^{-3}$ m, $w_2 = 10^{-3}$m, $l_1 = l_2 = 0$, $d = 10^{-3}$ m, $s = 0.2\cdot10^{-3}$ m, $t = 2.5\cdot10^{-6}$ m.

Comparing the propagation constants (Figs.9.9–9.11) for the SLs with a different metal distribution on the substrate we see that they differ although the total width of metal strips $w_1 + w_2 = 4 \cdot 10^{-3}$ m is the same. Thus, it is possible to choose the SL topology to obtain the device characteristics needed.

9.5. Electrodynamical Characteristics for Square–Shaped $\ddot{\varepsilon}$ – Gyrotropic Waveguides

Fig.9.12a presents the dispersion dependences of the outside wave number for the first four modes of the square semiconductor *n–InSb* (Electron Indium Antimonide) waveguide placed into the longitudinal magnetic field. Below we used in our calculations electrophysical parameters of the corresponding semiconductor *n–InSb* with: an effective mass $m^* = 0.014 m_{el}$, the relative permittivity of lattice $\varepsilon_L = 17.8$ and mobility $\mu_{el} = 40$ m²/(V·s). The waveguide modes were investigated in the order in which they appeared and not according to the wave symmetry. Fig.9.12b presents the dispersion dependences of the transversal propagation constants $k_{\perp 1} = k_{\perp 2}$ for the main mode when the charge carrier concentration n was small. The main mode is a hybrid wave. Fig.9.13 presents the dispersion dependences of the longitudinal propagation constants of the first four modes for the square waveguide.

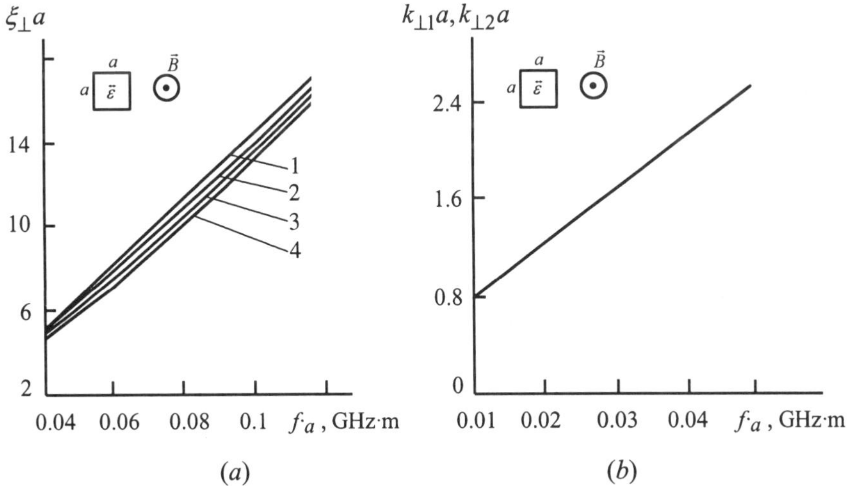

Fig. 9.12. *Dispersion dependences (a) the outside wave number $\xi_\perp$ of the main mode (curve 1) and three higher modes (curves 2, 3, 4). (b) transversal propagation constants $k_{\perp 1}$, $k_{\perp 2}$ of the main mode. The parameters are: the charge carrier concentration $n = 10^{18}$ m⁻³, the magnetic field induction $B = 1$ T and the waveguide width $a = 10^{-3}$ m.*

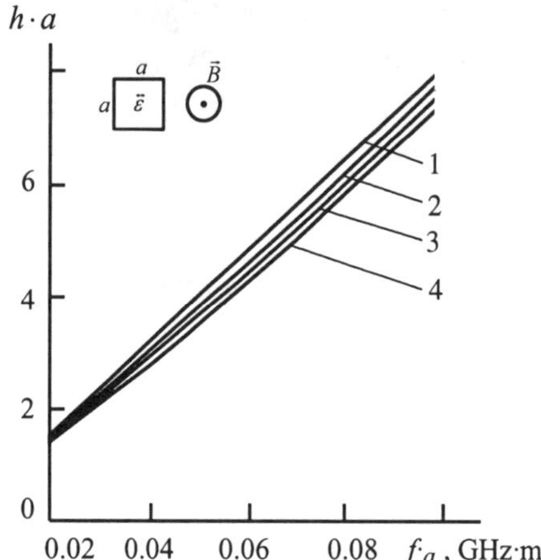

Fig. 9.13. *Dispersion dependences of the longitudinal propagation constant h of the main mode (curve 1) and three higher modes (2, 3, 4). The parameters are: the charge carrier concentration n = 10^{18} m^{-3}, the magnetic field induction B =1T and the waveguide width a = $2 \cdot 10^{-3}$ m.*

In Fig.9.14 we see the dependence of the longitudinal propagation constant h on the free charge carrier concentration n at the frequency $fa = 0.03\,\text{GHz·m}$. We see here that the dependence is strong and nonlinear and remind the reader that the charge concentration can be changed by light from different devices.

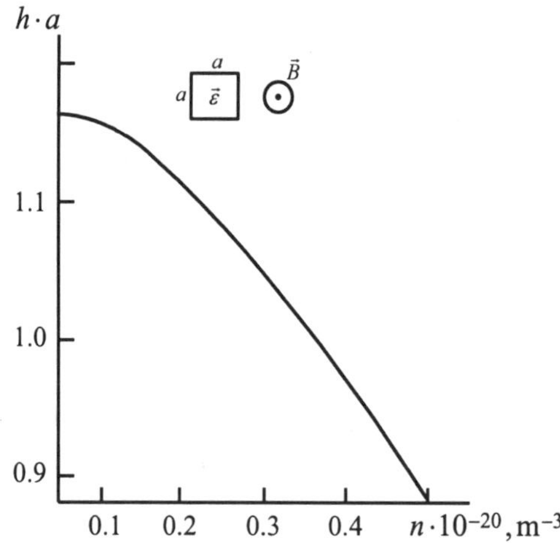

Fig. 9.14. *Dependence of the longitudinal propagation constant h of the main mode on the free charge carrier concentration n . The parameters are: the frequency fa = 0.03 GHz·m, the magnetic field induction B =1T and the waveguide width a = $2 \cdot 10^{-3}$ m.*

Fig.9.15 presents the dependence of the transversal wave numbers and outside wave number on the charge carrier concentration for the main mode of the same square plasma waveguide.

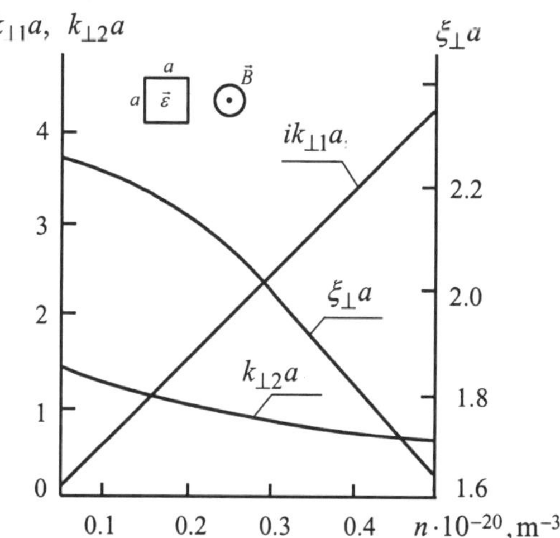

Fig. 9.15. Dependence of the transversal wave numbers $k_{\perp 1}$, $k_{\perp 2}$ and the outside wave number $\xi_{\perp}$ on the free charge carrier concentration for the square plasma waveguide. The parameters are: $fa = 0.03\,\mathrm{GHz\cdot m}$, the magnetic field induction $B = 1\,\mathrm{T}$ and the waveguide width $a = 2\cdot 10^{-3}$ m.

9.6. Electrodynamical Characteristics for the $\bar{\bar{\varepsilon}}-$ & $\bar{\bar{\mu}}-$ Gyrotropic One–Comb and Two–Comb Waveguides

The comb waveguides when compared to other dielectric waveguide types have certain advantages and these are:
1) they can only be a one–mode waveguide.
2) the comb waveguide shape is a better fastener to other solid state devices, which have a variety of electronic, optoelectronic, and magnetic applications.
3) using the waveguide with combs on the opposite sides of the common substrate, it is possible to form two independent channels coupled on certain stretches [9.6].

Here we numerically investigate the dispersion characteristics of two types of longitudinally magnetized ferrite and semiconductor comb waveguides (Fig.9.16*a*,*b*).

The ferrite saturation magnetization is $4\pi M_s$ = 377 kA/m. The nondiagonal permeability tensor component μ_{xy} is determined by formulae [9.1] and the diagonal components are assumed to be $\mu_{xx} = \mu_{zz} = 1$ and $\varepsilon_f = 13.5\left(1 - i5\cdot10^{-4}\right)$. The relative permittivity of the demagnetized ferrite material is $\varepsilon_r = 13.5$.

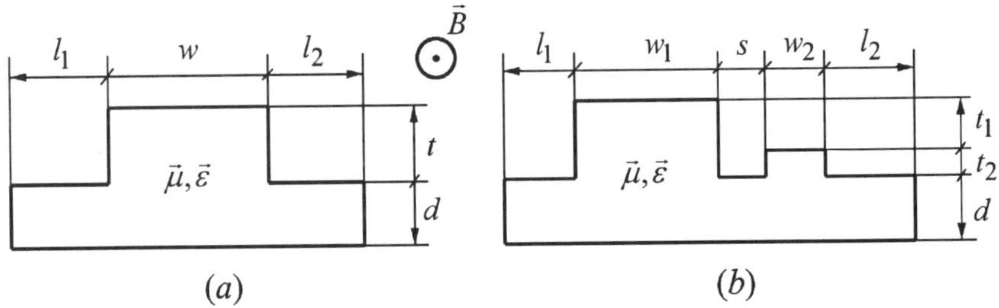

Fig. 9.16. The topology of (a) one - comb waveguide and (b) two – comb waveguide.

Fig.9.17a, b reveals the dependence of the longitudinal propagation constant h and the transversal propagation constants $k_{\perp1}$ and $k_{\perp2}$ for the ferrite one-comb waveguide (Fig.9.16a).

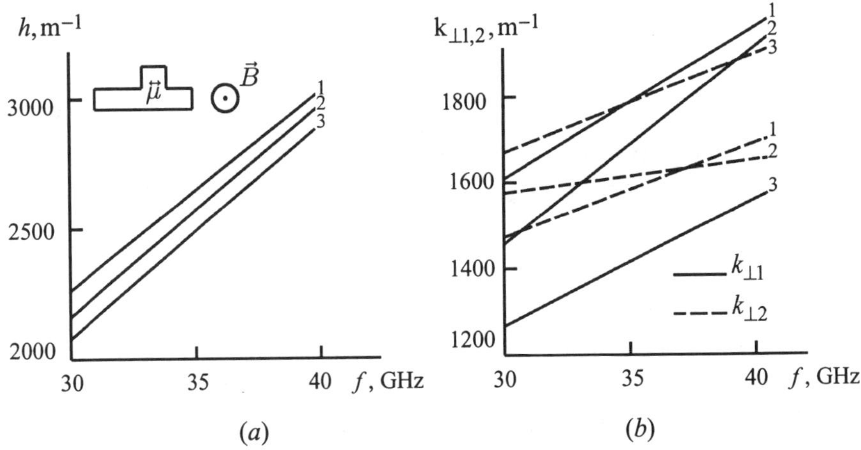

Fig. 9.17. Dispersion dependences of the main mode (curve 1) and two higher modes (2, 3) for the ferrite comb waveguide (Fig. 9.16a). The parameters are: $d = 10^{-3}$ m, $l_1 = w = t = 3\cdot10^{-3}$m, $l_2 = 2\cdot10^{-3}$ m, and the ferrite magnetization is $4\pi M_s = 377$ kA/m.

Fig.9.18a,b presents the similar dependence for a two–combs ferrite waveguide (Fig.9.16b).

One can see from the Figs.9.17a and 9.18a that three hybrid waves can propagate in the waveguides.

The Figs.9.19–9.22 present the electrodynamical characteristics of open longitudinally magnetized semiconductor (semiconductor plasma) one- and two-comb waveguides (Fig.9.16a,b). In our calculations, the parameters we used were: the relative permittivity $\varepsilon_L = 17.8$ of the semiconductor n–$InSb$, the effective mass of the charge carrier $m^* = 0.14 m_{el}$ and the mobility of the free charge carrier $\mu_{el} = 40$ m^2/(V·sec).

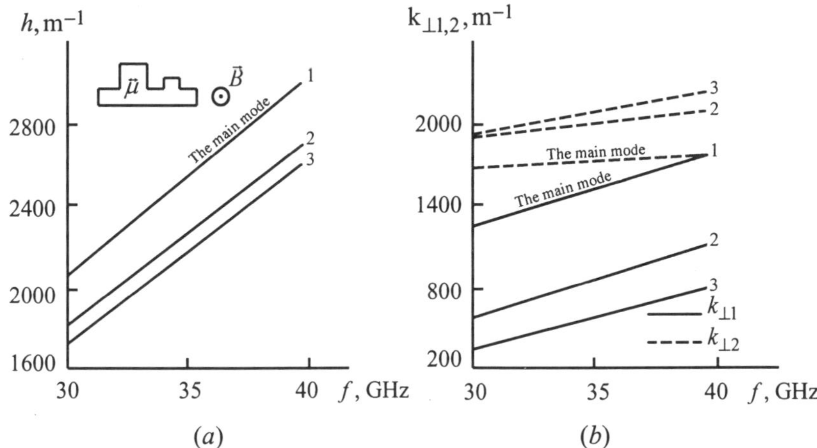

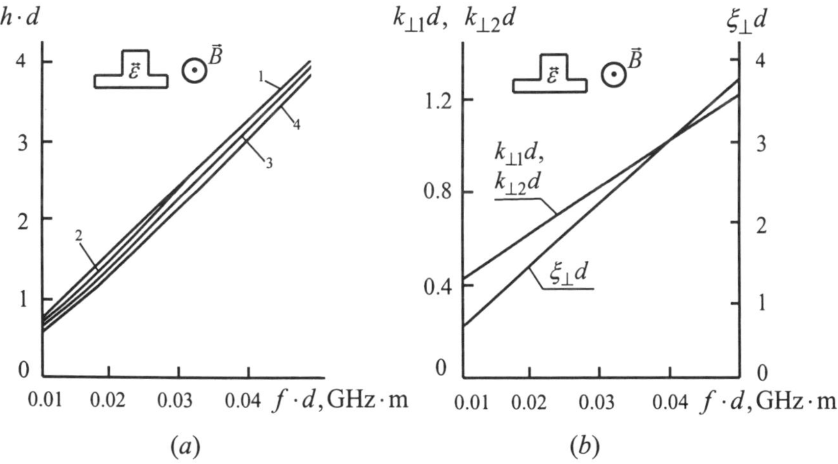

Fig. 9.18. *Dispersion dependences of the main mode (curve 1) and two higher modes (2, 3) for the two-comb ferrite waveguide (Fig.9.16b). The parameters are:* $l_1 = l_2 = t_1 = w_1 = 3\cdot10^{-3}$ m, $d = s = t_2 = w_2 = 10^{-3}$ m *and* $4\pi M_s = 377$ kA/m.

Fig. 9.19. *Dispersion dependences for a one–comb plasma waveguide (Fig.9.16a): (a) of the longitudinal propagation constant* h *of the main mode (curve 1) and three higher modes (2, 3, 4). (b) of the transversal propagation constants* $k_{\perp 1,2}$ *and of the outside wave number* $\xi_\perp$. *The parameters are:* $l_1 = w = t = 3\cdot10^{-3}$ m, $l_2 = 2\cdot10^{-3}$ m, $d = 10^{-3}$ m, *charge carrier concentration* $n = 10^{18}$ m^{-3} *and the magnetic field induction* $B = 1$T.

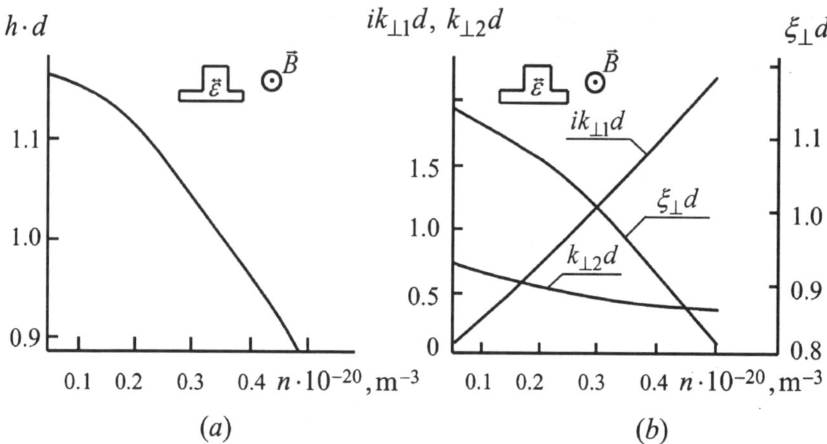

Fig. 9.20. *Dependence for a one–comb plasma waveguide: (a) the longitudinal propagation constant h, (b) the transversal wave numbers $k_{\perp 1,2}$ and the outside wave number $\xi_\perp$ of the main mode on the charge carrier concentration. The parameters are: $l_1 = w = t = 3\cdot10^{-3}$ m, $l_2 = 2\cdot10^{-3}$ m, $d = 10^{-3}$ m, frequency $f \cdot d = 0.015$GHz·m and the magnetic field induction $B = 1$T.*

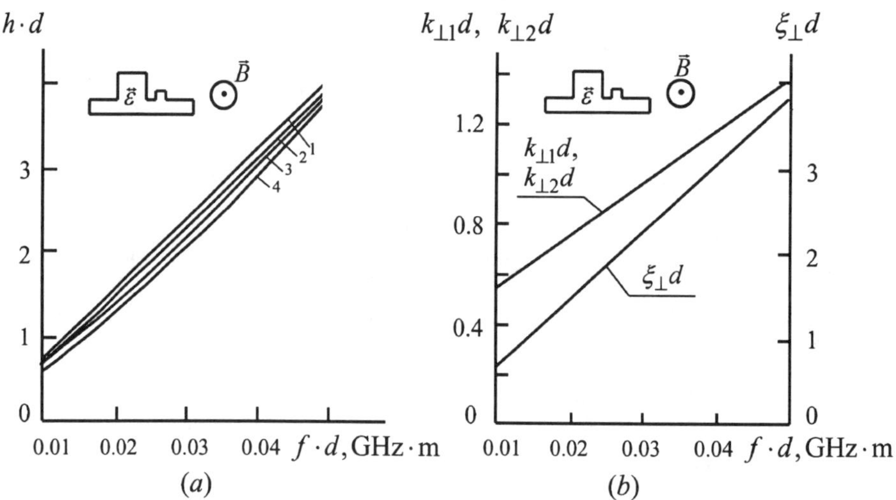

Fig. 9.21. *Dispersion dependences of the two–comb waveguide (Fig.9.16b): (a) the longitudinal propagation constant h of the main mode (curve 1) and three higher modes (2, 3, 4); (b) the transversal wave numbers $k_{\perp 1,2}$ and of the outside wave number $\xi_\perp$ of the main mode. The parameters are: $w_1 = l_1 = l_2 = t_1 = 3\cdot10^{-3}$ m, $w_2 = s = t_2 = d = 10^{-3}$ m, the charge carrier concentration $n = 10^{18}$m⁻³ and the magnetic field induction $B = 1$T.*

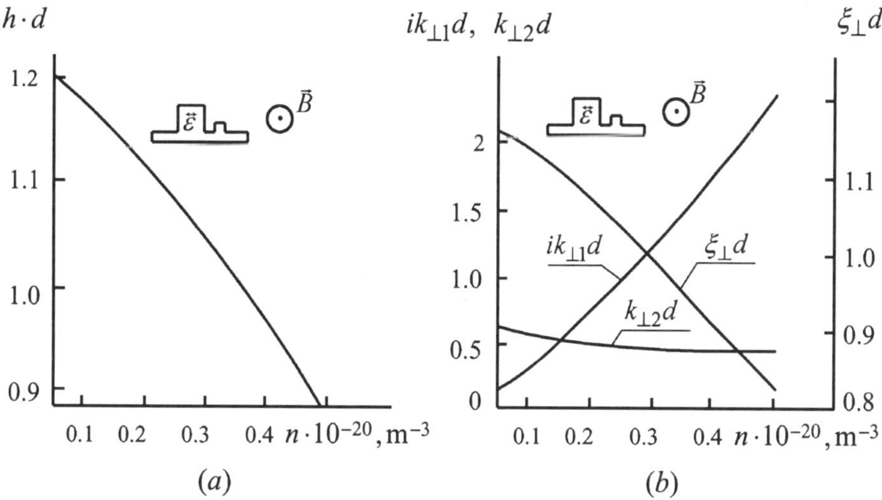

Fig. 9.22. *Dependence the two–comb plasma waveguide: (a) the longitudinal propagation constant h ; (b) the transversal wave numbers $k_{\perp 1,2}$ and the outside wave number $\xi_\perp$ on the carrier concentration for the main mode. The parameters are: $w_1 = l_1 = t_1 = 3\cdot10^{-3}\,m$, $w_2 = s = t_2 = d = 10^{-3}\,m$, $l_2 = 2\cdot10^{-3}\,m$ the magnetic field induction $B = 1T$ and the frequency $f\cdot d = 0.015$ GHz·m*

Fig.9.19*a* presents the dispersion dependences of the longitudinal propagation constants of the main mode and three higher modes for the plasma waveguide with one-comb (Fig.9.16*a*). In Fig.9.19*b*, we present the dispersion dependences of the transversal propagation constants and of the outside wave number for the main mode of the same one-comb plasma waveguide. Now, the first higher mode (curve 2) has in addition a high frequency cut off in the band, which is shown in Fig.9.19*a*.

Fig.9.20 *a,b* reveals the wave numbers of the main mode as a function of the charge carrier concentration n for a one-comb waveguide with the same sizes as in Fig.9.19. When we analyzed the dependence of the transversal wave numbers $k_{\perp 1}$ and $k_{\perp 2}$ on the charge carrier concentration ($n = 5\cdot10^{18} \div 5\cdot10^{19}$ m^{-3}) we saw that the wave number $k_{\perp 1}$ of the quasi–E–wave was changed extensively as compared to the wave number $k_{\perp 2}$. So we saw that the quasi–E–wave, which is characterized by the wave number $k_{\perp 1}$ is the cause of the main changes to the waveguide electrodynamical characteristics. The wave number $k_{\perp 1}$ is an imaginary number and the wave number $k_{\perp 2}$ is a real number in which both belong to an argument of the Hankel function. We understand this to mean that the quasi–E–wave is a surface wave. This also means the quasi–H–wave is a volume wave and that the EM wave energy is distributed into the volume of.

semiconductor material. The dependence of the longitudinal constant on the concentration (Fig.9.20*a*) is nonlinear, which means that the phase shift of the waveguide is a nonlinear function of the concentration as well.

Figs.9.21*a,b* and 9.22*a,b* present similar dependences (as Fig.9.20) for the open two–combs *n–InSb* waveguide that was placed into the external longitudinal constant magnetic field. The two-comb waveguide does not have a high frequency cut off at the frequency interval as was observed for the one–comb waveguide. And we see that nonlinearity of the main mode phase shift is weaker when we change the concentration as compared to the one–comb waveguide. Note that changing the charge carrier concentration one can change the waveguide electrodynamical characteristics and particularly the differential shift of the waveguide.

9.7. Electrodynamical Characteristics of Mirror Magnetized Ferrite Waveguides and Comparison of Calculated Results with Experimental Data

Here we present our calculations for a mirror ferrite waveguide (a longitudinally magnetized ferrite rod placed on a metal plate of finite sizes) [9.7]. The view of a mirror ferrite waveguide is shown in Fig.9.23.

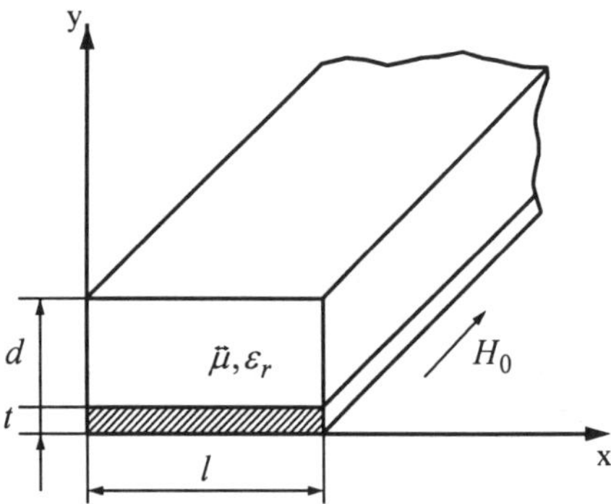

Fig. 9.23. *The topology of the mirror ferrite waveguide.*

Fig.9.24 presents the dispersion dependences of the first four modes (the main mode and three higher modes) for a waveguide with sizes: the width $l = 0.7 \cdot 10^{-3}$ m and height $d = 0.35 \cdot 10^{-3}$ m of the ferrite rod, which was placed on the metal plate of the width $l = 0.7 \cdot 10^{-3}$ m and the thickness $t = 10^{-6}$ m. We see the dispersion dependences on the frequency are linear.

Fig.9.25. reveals the propagation constants h of the main mode and three higher modes as a function of the saturation magnetization $4\pi M_s$ of the ferrite material for (as Fig.9.24) the same waveguide.

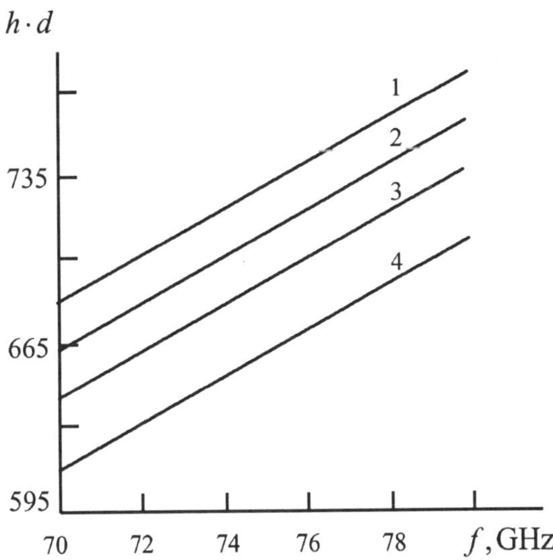

Fig. 9.24. *Dispersion dependences of the mirror ferrite waveguide. The parameters are: $d = 0.35 \cdot 10^{-3}$ m, $l = 0.7 \cdot 10^{-3}$ m, $t = 10^{-6}$ m, the saturation magnetization $4\pi M_s = 79.37$ kA/m, $\varepsilon_r = 9.8$ and the external constant magnetic field, $H_0 = 39.68$ kA/m.*

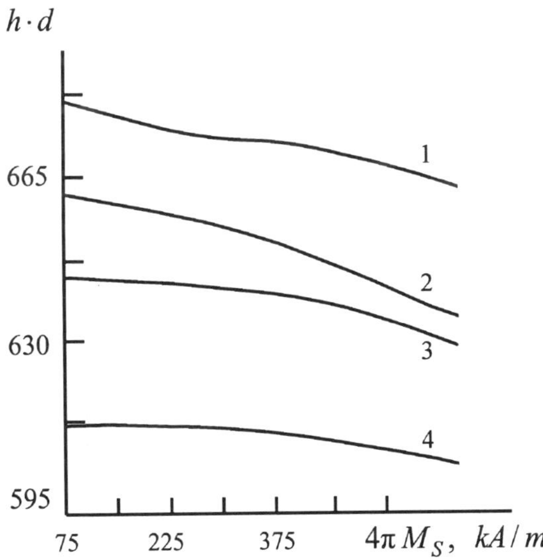

Fig. 9.25. *Dependence of the longitudinal propagation constant h of the main mode (curve 1) and three higher modes (2, 3, 4) on the saturation magnetization. The parameters of the magnetized ferrite mirror waveguide are: $d = 0.35 \cdot 10^{-3}$ m, $l = 0.7 \cdot 10^{-3}$ m, $t = 10^{-6}$ m, $\varepsilon_r = 9.8$, constant magnetic field $H_0 = 39.68$ kA/m and the frequency $f = 70$ GHz.*

Here the parameters are: the external constant magnetic field $H_0 = 39.68$ kA/m and the frequency $f = 70$ GHz. We see from the Fig.9.25 when the saturation magnetization (or just magnetization [9.4]) increases the longitudinal propagation constants h of the main mode and three higher modes decrease and the wavelengths $\lambda = 2\pi/h$ in the waveguide increases.

Fig.9.26 presents the longitudinal propagation constant as a function of the waveguide width for the magnetized ferrite waveguide (Fig.9.23) at the ferrite saturation magnetization $4\pi M_s = 79.37$ kA/m, the external constant magnetic field $H_0 = 39.68$ kA/m and the frequency $f = 70$ GHz.

Figs.9.24–9.26 also confirm the stability of our algorithm (Section 9.3) as we see our calculated results change minutely. Figs.9.27 and 9.28 present the calculated results for the mirror ferrite waveguide (Fig.9.23), which is created of ferrite material *1SCh4*.

Fig.9.27*a* presents the dependence of the propagation constants h on the frequency for the main mode and two higher modes of the mirror ferrite waveguide at $M = 0$. Fig.9.27*b* shows the dispersion dependences of the mirror ferrite waveguide main mode at the several ferrite magnetization values.

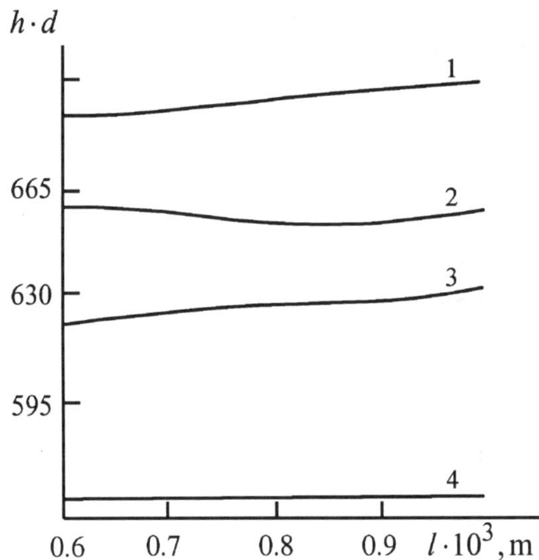

Fig. 9.26. *Dependence of the longitudinal propagation constant of the main mode (curve 1) and three higher modes (2, 3, 4) on the width of the magnetized ferrite mirror waveguide. The parameters are:* $d = 0.35 \cdot 10^{-3}$ m, $t = 10^{-6}$ m, $\varepsilon_r = 9.8$, *the saturation magnetization* $4\pi M_s = 79.37$ kA/m, *the external constant magnetic field* $H_0 = 39.68$ kA/m *and the frequency* $f = 70$ GHz.

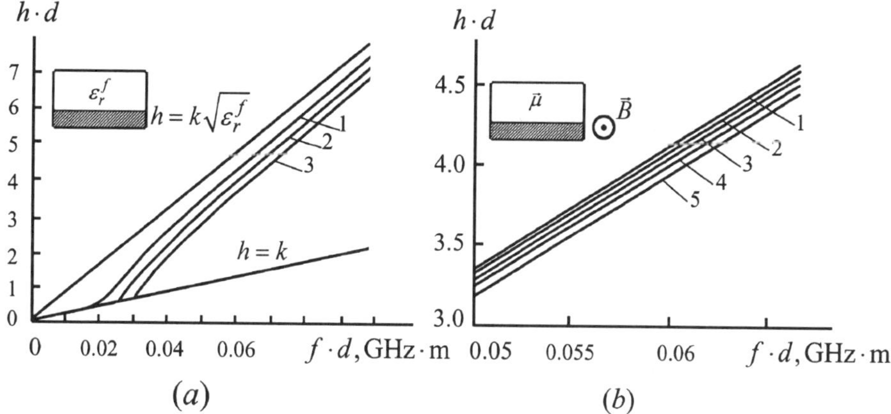

Fig. 9.27. *(a) Dispersion dependences of the demagnetized ferrite waveguide: curve 1 for the main mode; curves 2 and 3 for the first and second higher modes; (b) dispersion dependences of the main mode propagating in the ferrite waveguide at several ferrite magnetization values. The parameters are: $d = 10^{-3}$ m, $l = 2 \cdot 10^{-3}$ m, $t = 20 \cdot 10^{-6}$ m, $\varepsilon_r = 13.5$.*

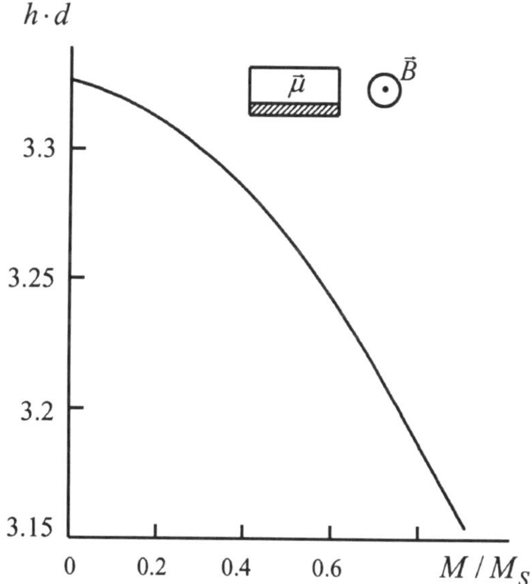

Fig. 9.28. *Dependence of the longitudinal propagation constant h of the main mode on the ferrite magnetization. The parameters of the mirror ferrite waveguide are: $d = 10^{-3}$ m, $l = 2 \cdot 10^{-3}$ m, $t = 20 \cdot 10^{-6}$ m, $\varepsilon_r = 13.5$ and $f \cdot d = 0.05$ GHz·m .*

Our calculations were fulfilled at M/M_s from 0.1 through 0.9 with the interval-step 0.1. In Fig.9.27b we show only the results for some M/M_s. The curve 1 corresponds to $M/M_s = 0$, curves 2, 3, 4, 5 correspond to $M/M_s = 0.3$,

M/M_s =0.5, M/M_s =0.7, M/M_s =0.9 accordingly. We found that by changing the direction of the external constant magnetic field (which is equal to changing the tensor nondiagonal component sign) the sign and the absolute value of the differential phase shift remain the same. Then Fig.9.28 presents the longitudinal propagation constant as a function of the normalized ferrite magnetization for the main mode of the mirror ferrite waveguide.

The calculated and measured differential phase shift (in degrees) of the waveguide section length L/d =20, as a function of the relative ferrite magnetization are shown in Fig.9.29. The waveguide phase shift is expressed by $\Delta h = \left[(h)_M - (h)_{M=0}\right] \cdot L$. Where the first term in the brackets corresponds with the magnetized ferrite state and the second term corresponds with the demagnetized ferrite state. Here we see in Fig.9.29 that the experimental data confirms the theoretically discovered nonlinearity of the differential phase shift dependence on the ferrite magnetization.

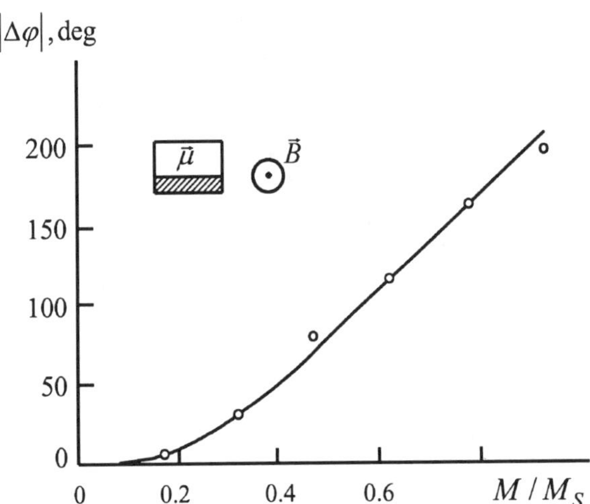

Fig. 9.29. Differential phase shift of the waveguide stretch. The solid line shows the calculated results and the circles show the measured results. The waveguide parameters are: $d = 10^{-3}$ m, $l = 2 \cdot 10^{-3}$ m, $t = 2 \cdot 10^{-5}$ m, $\varepsilon_r = 13.5$ and $f \cdot d$ = 0.05GHz·m.

Microwave devices controlled by a magnetic field were usually created on the basis of gyrotropic waveguides. For the creation of these devices we saw it was possible to use longitudinal magnetized gyrotropic waveguides. Devices controlled by a magnetic field also were created on the basis of transversally magnetized gyrotropic waveguides. So transversally magnetized waveguides are also interesting topics of scientific investigations. In the next chapter we will describe a new SIE method for calculating waveguides having transversally magnetized ferrite or semiconductor material.

10. SOLUTION OF MAXWELL'S EQUATIONS BY THE SIE METHOD FOR OPEN TRANSVERSALLY MAGNETIZED GYROTROPIC WAVEGUIDES

10.1. Introduction

The numerical investigation of the open dielectric waveguides having transversally magnetized gyrotropic layers is of practical interest in this day and age. They make up the basis for the creation of many reciprocal and nonreciprocal devices controlled by magnetic fields at the EHF range. Some examples of these created devices are valves, phase–shifters and modulators [10.1]–[10.3].

In this chapter, we describe a new method based on the SIE theory [3.3]. By this, one is enabled to theoretically investigate open waveguides of complex (arbitrary) cross–section geometry having transversally magnetized ferrite or semiconductor material (Fig.10.1). In this chapter the SIE method yields the solution to the problem in the rigorous electrodynamical formulation. This method enables one to analyze the main and higher modes propagating in the investigated waveguide [10.4].

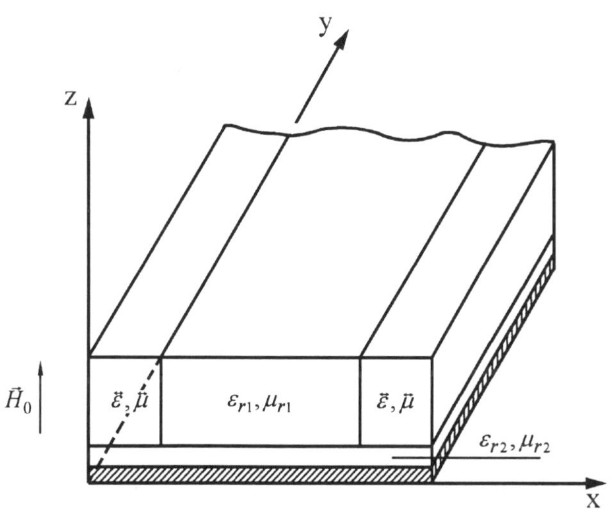

Fig. 10.1. The waveguide topology.

The dependence on time t and coordinate y is presented in the form $e^{i(\omega t - hy)}$ where $\omega = 2\pi f$ is the cyclic frequency and h is the propagation constant.

10.2. The Integral Representation of Maxwell's Equations Solution

From Maxwell's equations written in Cartesian coordinates the transversal electromagnetic field components are expressed through the longitudinal components as follows:

$$E_x = \frac{1}{\Delta(\varepsilon)}\left[i\omega\mu_0\mu_{zz}\left(\frac{\partial H_y}{\partial z} - \omega\varepsilon_0\varepsilon_{xy}E_y \right) - ih\frac{\partial E_y}{\partial x} \right], \tag{10.1}$$

$$E_z = \frac{1}{\Delta(\mu)}\left[-i\omega\mu_0\mu_{xx}\frac{\partial H_y}{\partial x} - ih\left(\frac{\partial E_y}{\partial z} + \omega\mu_0\mu_{xy}H_y \right) \right], \tag{10.2}$$

$$H_x = \frac{1}{\Delta(\mu)}\left[-i\omega\varepsilon_0\varepsilon_{zz}\left(\frac{\partial E_y}{\partial z} + \omega\mu_0\mu_{xy}H_y \right) - ih\frac{\partial H_y}{\partial x} \right], \tag{10.3}$$

$$H_z = \frac{1}{\Delta(\varepsilon)}\left[i\omega\varepsilon_0\varepsilon_{xx}\frac{\partial E_y}{\partial x} - ih\left(\frac{\partial H_y}{\partial z} - \omega\varepsilon_0\varepsilon_{xy}E_y \right) \right]. \tag{10.4}$$

Here the values are:

$$\Delta(\varepsilon) = k^2\varepsilon_{xx}\mu_{zz} - h^2,$$

$$\Delta(\mu) = k^2\mu_{xx}\varepsilon_{zz} - h^2.$$

Using (10.1)–(10.4) we obtained the system of two coupled equations for the longitudinal components E_y and H_y. The solution is presented in the form of hybrid waves, which after diminishing the gyrotropy are transformed into the pure $E-$ and $H-$waves.

In the SIE method we use the fundamental solutions of the differential equations for the functions $E_y(x,z)$ and $H_y(x,y)$. In the right sides of these equations we added the Dirac delta functions:

$$\frac{\varepsilon_{xx}}{\Delta(\varepsilon)}\frac{\partial^2 E_y}{\partial x^2} - \frac{\varepsilon_{zz}}{\Delta(\mu)}\frac{\partial^2 E_y}{\partial z^2} + \left(\varepsilon_{xx} - \frac{k^2\varepsilon_{xy}^2\mu_{zz}}{\Delta(\varepsilon)} \right)E_y =$$

$$= \frac{1}{\omega\varepsilon_0}\frac{\partial}{\partial z}\left[\left(\frac{1}{\Delta(\varepsilon)} - \frac{1}{\Delta(\mu)} \right)h\frac{\partial H_y}{\partial x} - \left(\frac{\varepsilon_{xy}\mu_{zz}}{\Delta(\varepsilon)} + \frac{\mu_{xy}\varepsilon_{zz}}{\Delta(\mu)} \right)k^2 H_y \right] + A\delta(x)\delta(z),$$

$$\frac{\mu_{xx}}{\Delta(\mu)}\frac{\partial^2 H_y}{\partial x^2} + \frac{\mu_{zz}}{\Delta(\varepsilon)}\frac{\partial^2 H_y}{\partial z^2} + \left(\mu_{xx} - \frac{k^2\mu_{xy}^2\varepsilon_{zz}}{\Delta(\mu)} \right)H_y =$$

$$= \frac{1}{\omega\mu_0}\frac{\partial}{\partial z}\left[\left(\frac{\varepsilon_{xy}\mu_{zz}}{\Delta(\varepsilon)}+\frac{\mu_{xy}\varepsilon_{zz}}{\Delta(\mu)}\right)k^2 E_y +\left(\frac{1}{\Delta(\varepsilon)}-\frac{1}{\Delta(\mu)}\right)h\frac{\partial E}{\partial x}\right]+B\delta(x)\delta(z). \quad (10.5)$$

Here the values A and B are constants. The solution of this system of differential equations (10.5) at $B = 0$ describes the quasi–E wave and at $A =0$ describes the quasi–H -wave.

The Fourier transformation:

$$E_y(x,z) = \frac{1}{2\pi}\int\limits_{-\infty}^{+\infty}\int\limits_{-\infty}^{+\infty}d\chi_x d\chi_z e^{-i\chi_x x - i\chi_z z}E_y(\chi_x,\chi_z), \quad (10.6)$$

$$H_y(x,z) = \frac{1}{2\pi}\int\limits_{-\infty}^{+\infty}\int\limits_{-\infty}^{+\infty}d\chi_x d\chi_z e^{-i\chi_x x - i\chi_z z}H_y(\chi_x,\chi_z), \quad (10.7)$$

$$\delta(x) = \frac{1}{\sqrt{2\pi}}\int\limits_{-\infty}^{+\infty}d\chi_x e^{-i\chi_x x}, \quad (10.8)$$

$$\delta(z) = \frac{1}{\sqrt{2\pi}}\int\limits_{-\infty}^{+\infty}d\chi_z e^{-i\chi_z z}, \quad (10.9)$$

turns the differential equations into a system of uniform algebraic equations. In the formulae (10.6) and (10.7) the integration path goes in the right half $\mathrm{Re}(\chi_z)\succ 0$ of the complex plane χ_z above the cut line and in the left half $\mathrm{Re}(\chi_z)\prec 0$ the complex plane χ_z below the cut line. We write the solution of this system in the form:

$$E_y(\chi_x,\chi_z) = \frac{A}{4\pi^2}\frac{f_1}{f}+\frac{B}{4\pi^2\omega\varepsilon_0}\frac{\chi_z f_2}{f},$$

$$H_y(\chi_x,\chi_z) = \frac{B}{4\pi^2}\frac{f_3}{f}+\frac{A}{4\pi^2\omega\mu_0}\frac{\chi_z f_4}{f}, \quad (10.10)$$

where

$$f = \varepsilon_{zz}\mu_{zz}\chi_z^4 + \chi_z^2\left[\left(\varepsilon_{xx}\mu_{zz}+\mu_{xx}\varepsilon_{zz}\right)\chi_x^2 - \mu_{xx}\varepsilon_{zz}\Delta(\varepsilon)-\right.$$

$$\left.-\varepsilon_{xx}\mu_{zz}\Delta(\mu)-2k^2\varepsilon_{xy}\mu_{xy}\varepsilon_{zz}\mu_{zz}\right]+\varepsilon_{xx}\mu_{xx}(\chi_x^2-\Delta(\varepsilon_{ef}))\ (\chi_x^2-\Delta(\mu_{ef})),$$

$$f_1 = -\mu_{xx}\Delta(\varepsilon)\chi_x^2 - \Delta(\mu)\mu_{zz}\chi_z^2 + \mu_{xx}\Delta(\varepsilon)\Delta(\mu_{ef}),$$

$$f_2 = \left(\Delta(\varepsilon)-\Delta(\mu)\right)h\chi_x + ik^2(\varepsilon_{xy}\mu_{zz}\Delta(\mu)+\mu_{xy}\varepsilon_{zz}\Delta(\varepsilon)),$$

$$f_3 = -\varepsilon_{xx}\Delta(\mu)\chi_x^2 - \Delta(\varepsilon)\varepsilon_{zz}\chi_z^2 + \varepsilon_{xx}\Delta(\mu)\Delta(\varepsilon_{ef}),$$

$$f_4 = \left(\Delta(\varepsilon)-\Delta(\mu)\right)h\chi_x - ik^2\left(\varepsilon_{xy}\mu_{zz}\Delta(\mu)+\mu_{xy}\varepsilon_{zz}\Delta(\varepsilon)\right),$$

$$\varepsilon_{ef} = \varepsilon_{xx} - \varepsilon_{xy}^2/\varepsilon_{xx}, \qquad \mu_{ef} = \mu_{xx} - \mu_{xy}^2/\mu_{xx}. \quad (10.11)$$

The function f has four roots: two roots in the right half of the plane χ_z and two roots in left half of the plane χ_z. One can write these expressions like this:

$$f(\chi_z) = \varepsilon_{zz}\mu_{zz}\left(\chi_z - \chi_{z_1}\right)\left(\chi_z - \chi_{z_2}\right)\left(\chi_z - \chi_{z_3}\right)\left(\chi_z - \chi_{z_4}\right),$$

$$\chi_{z_1}(\chi_x) = \sqrt{0.5\left[\left(\varepsilon_{xx}\mu_{zz} + \mu_{xx}\varepsilon_{zz}\right)\xi(\chi_x) + \left(\varepsilon_{xx}\mu_{zz} - \varepsilon_{zz}\mu_{xx}\right)\eta(\chi_x)\right]},$$

$$\chi_{z_2}(\chi_x) = \sqrt{0.5\left[\left(\varepsilon_{xx}\mu_{zz} + \mu_{xx}\varepsilon_{zz}\right)\xi(\chi_x) - \left(\varepsilon_{xx}\mu_{zz} - \varepsilon_{zz}\mu_{xx}\right)\eta(\chi_x)\right]},$$

$$\chi_{z_3} = -\chi_{z_1}, \quad \chi_{z_4} = -\chi_{z_2}, \quad \xi(\chi_x) = a_1 - \chi_x^2, \quad \eta(\chi_x) = \sqrt{\chi_x^4 - 2a\chi_x^2 + b},$$

$$a_1 = \frac{1}{\varepsilon_{xx}\mu_{zz} - \mu_{xx}\varepsilon_{zz}}\left[\mu_{xx}\varepsilon_{zz}\Delta(\varepsilon) + \varepsilon_{xx}\mu_{zz}\Delta(\mu) + 2k^2\varepsilon_{xx}\mu_{xy}\mu_{zz}\varepsilon_{zz}\right],$$

$$a = \frac{1}{\left(\varepsilon_{xx}\mu_{zz} + \mu_{xx}\varepsilon_{zz}\right)^2}\left[\left(\varepsilon_{xx}\mu_{zz} + \mu_{xx}\varepsilon_{zz}\right)\left(\mu_{xx}\varepsilon_{zz}\Delta(\varepsilon) + \right.\right.$$

$$\left.\left. + 2k^2\varepsilon_{xy}\mu_{xy}\varepsilon_{zz}\mu_{zz} + \varepsilon_{xx}\mu_{zz}\Delta(\mu)\right) - 2\varepsilon_{xx}\mu_{xx}\left(\Delta\left(\varepsilon_{ef}\right) + \Delta\left(\mu_{ef}\right)\right)\right],$$

$$b = \frac{1}{\left(\varepsilon_{xx}\mu_{zz} + \mu_{xx}\varepsilon_{zz}\right)^2}\left[\left(\mu_{xx}\varepsilon_{zz}\Delta(\varepsilon) - \varepsilon_{xx}\mu_{zz}\Delta(\mu)\right)^2 + \right.$$

$$\left. + 4k^2\left(\varepsilon_{xx}\mu_{zz}\mu_{xy} + \mu_{xx}\varepsilon_{zz}\varepsilon_{xy}\right)\left(\varepsilon_{xy}\Delta(\mu) + \mu_{xy}\Delta(\varepsilon)\right)\right]. \tag{10.12}$$

When $z \succ 0$ the integration contour is closed by the half–circle below. When $z \prec 0$ the integration contour is closed by the half–circle from above. After this we obtained:

$$E_y(x,y) = \frac{iA}{2\pi\left(\varepsilon_{xx}\mu_{zz} - \mu_{xx}\varepsilon_{zz}\right)}\int\limits_{-\infty}^{\infty} d\chi_x e^{ix\chi_x}\left[-\frac{f_1\left(\chi_{z_1}\right)e^{-i|z|\chi_{z_1}}}{\chi_{z_1}\eta(\chi_x)} + \frac{f_1\left(\chi_{z_2}\right)e^{-i|z|\chi_{z_2}}}{\chi_{z_2}\eta(\chi_x)}\right] -$$

$$- \frac{z}{|z|}\frac{iB}{2\pi\varepsilon_0\omega\left(\varepsilon_{xx}\mu_{zz} - \mu_{xx}\varepsilon_{zz}\right)}\int\limits_{-\infty}^{\infty} d\chi_x e^{-ix\chi_x}\left[\frac{f_2\left(\chi_{z_1}\right)e^{-i|z|\chi_{z_1}}}{\eta(\chi_x)} - \frac{f_2\left(\chi_{z_2}\right)e^{-i|z|\chi_{z_2}}}{\eta(\chi_x)}\right],$$

$$H_y(x,y) = \frac{iB}{2\pi\left(\varepsilon_{xx}\mu_{zz} - \mu_{xx}\varepsilon_{zz}\right)}\int\limits_{-\infty}^{\infty} d\chi_x e^{-ix\chi_x} \times$$

$$\times\left[-\frac{f_3\left(\chi_{z_1}\right)e^{-i|z|\chi_{z_1}}}{\chi_{z_1}\eta(\chi_x)} + \frac{f_3\left(\chi_{z_2}\right)e^{-i|z|\chi_{z_2}}}{\chi_{z_2}\eta(\chi_x)}\right] - \frac{z}{|z|}\frac{iA}{2\pi\mu_0\omega\left(\varepsilon_{xx}\mu_{zz} - \mu_{xx}\varepsilon_{zz}\right)} \times$$

$$\times\int\limits_{-\infty}^{\infty} d\chi_x e^{-ix\chi_x}\left[\frac{f_4\left(\chi_{z_1}\right)e^{-i|z|\chi_{z_1}}}{\eta(\chi_x)} - \frac{f_4\left(\chi_{z_2}\right)e^{-i|z|\chi_{z_2}}}{\eta(\chi_x)}\right]. \tag{10.13}$$

As follows from the definition (10.12) of the expression $f(\chi_z)$ the functions $\chi_{z_{1,2}}(\chi_x)$ turns into zeros at $\chi_x^2 = \Delta(\varepsilon_{ef})$ and $\chi_x^2 = \Delta(\mu_{ef})$. Here the four points $\chi_x = \pm\sqrt{\Delta(\varepsilon_{ef})}$ and $\chi_x = \pm\sqrt{\Delta(\mu_{ef})}$ are the branch points for the functions $\chi_{z_1}(\chi_x)$ and $\chi_{z_2}(\chi_x)$.

In our electrodynamical calculations it is convenient to draw the cut from the branching points along the lines on which the imaginary part of the function is zero. Then the imaginary part of the function has on one sheet the positive values and on the other sheet the negative values [10.5]. The functions $\chi_{z_{1,2}}(\chi_x)$ according to the definition (10.12) were being expressed through magnitudes ξ and η. The first magnitude is a one–valued function and the second magnitude is a two–valued function.

10.3. The Choice of a One–Valued Branch for the Function $\eta(\chi_x)$

We now discuss in more detail the choice of a one–valued branch for the function $\eta(\chi_x)$ when the media is lossless material. In all four branching points the function $\eta(\chi_x)$ is equal to zero when the argument of this function is:

$$\chi_{x_{1,2}}^2 = a \pm \sqrt{a^2 - b}. \tag{10.14}$$

We draw the cut along the line where $\text{Im}\,\eta = 0$. On the line $\text{Re}\,\eta \leq 0$ therefore:

$$\text{Im}\,\eta^2 = xy\left(x^2 - y^2\right) = 0, \tag{10.15}$$

$$\text{Re}\,\eta^2 = \left(x^2 - y^2\right) - 4x^2 y^2 - 2a\left(x^2 - y^2\right) + b \geq 0, \tag{10.16}$$

here we use the notation $\chi_x = x + iy$ and the constants a and b are real. There are five cases in all. Here we will consider every one of these five cases.

Case 1. When the constant $b \prec 0$. In this case one root of (10.14) is positive and the other root is negative. And we see that there are two branching points on the real axis and two branching points on the imaginary axis (Fig.10.2). Also that on the straight line $x = 0$ the inequality of (10.16) is satisfied from the points above the branch–point B for $y \succ 0$ and below the branch–point C for $y \prec 0$. On the straight line $y = 0$ the cuts should be directed to the right from the point D and to the left from the point E. On the hyperbola:

$$x^2 - y^2 = a. \tag{10.17}$$

When the function $\mathrm{Re}\,\eta^2 \prec 0$ then the cut cannot be here. For the point A the angles are: $\alpha_1 = \pi$, $\alpha_2 = \alpha$, $\beta_1 = \pi$, $\beta_2 = \pi - \alpha$ and   so   we   write: $e^{i(\alpha_1 + \alpha_2 + \beta_1 + \beta_2)/2} = e^{j3\pi/2}$.

Thus $\mathrm{Re}\,\eta(A) = 0$ and $\mathrm{Im}\,\eta(A) \prec 0$ in all the points that are shown in the sheet of the Riemann surface $\mathrm{Im}\,\eta(\chi_x) \leq 0$ and equality takes place only at the points of the cuts (Fig.10.2). At the points of the lower cut bank, which start at point D the value is $\mathrm{Re}\,\eta \succ 0$. For the points of the upper cut bank the value is $\mathrm{Re}\,\eta \prec 0$.

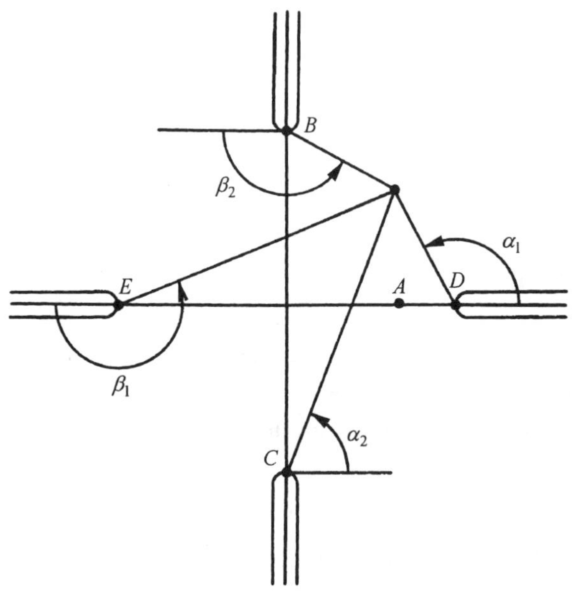

Fig. 10.2. *The sheet* $\mathrm{Im}\,\eta \prec 0$ *for* $b \prec 0$.

Case 2. When constants $a \succ 0$ and $0 \prec b \prec a^2$. In this case all four branch–points (10.14) are on the real axis. $\mathrm{Re}\,\eta^2 \succ 0$ is on the whole straight line $x = 0$. Between the points B and C the value is $\mathrm{Re}\,\eta^2 \prec 0$ here there is not a cut. One must keep in mind that the cuts should be oriented so that the integration path in the formulae (10.12) and (10.13) are drawn from the third quarter into the first quarter of the plane. Now we obtained the cuts, which are shown in Fig.10.3. For the point A angles are: $\alpha_2 = 0$, $\beta_1 = \pi$, $\beta_2 = \pi$ and therefore $\mathrm{Re}\,\eta(A) = 0$ and $\mathrm{Im}\,\eta(A) \prec 0$. In the same way we see that $\mathrm{Re}\,\eta(D) \prec 0$ and $\mathrm{Re}\,\eta(E) \succ 0$.

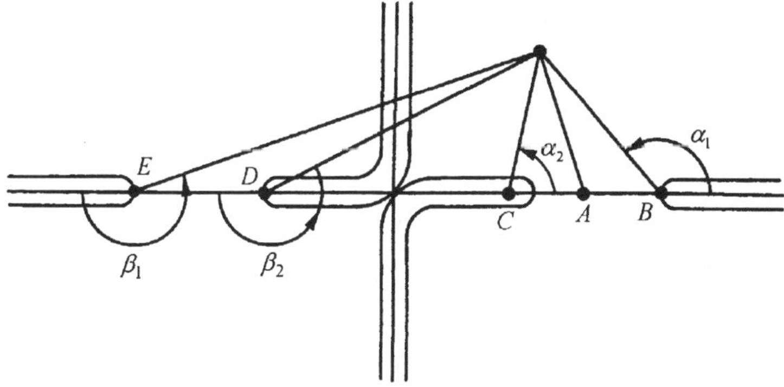

Fig. 10.3. *The sheet* $\mathrm{Im}\,\eta \prec 0$ *for* $a \succ 0$, $0 \prec b \prec a^2$.

Case 3. When constants $a \prec 0$ and $0 \prec b \prec a^2$. In this case the four branch–points are on the imaginary axis. On the straight line $y = 0$ the value is $\mathrm{Re}\,\eta^2 \succ 0$. The value is $\mathrm{Re}\,\eta^2 \prec 0$ on the imaginary axis at the points between the roots that have the same sign of y–coordinate on the hyperbola (10.17). Taking into account the integration contour form in the formulae (10.12) and (10.13) we receive the cuts shown in Fig.10.4. For the point A the angles are: $\alpha_1 = -3\pi/2$, $\alpha_2 = \pi/2$, $\beta_1 = -\pi/2$, $\beta_2 = \pi/2$ and the magnitudes are $\mathrm{Re}\,\eta(A) = 0$ and $\mathrm{Im}\,\eta(A) \prec 0$. In the same way we determined the point C where the magnitudes are: $\mathrm{Re}\,\eta(C) \prec 0$ and $\mathrm{Im}\,\eta(C) = 0$.

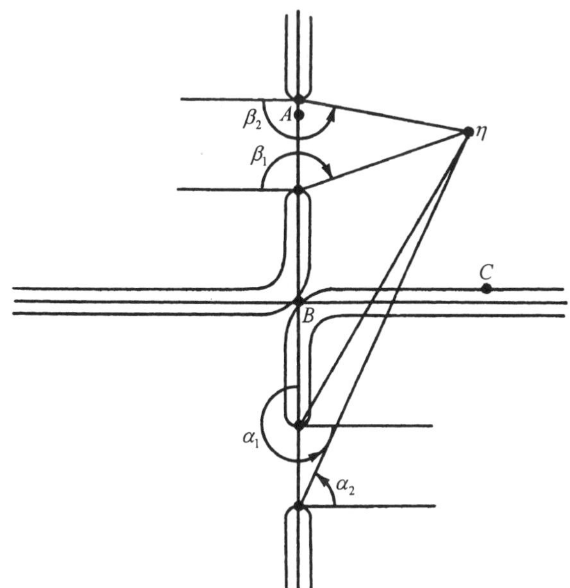

Fig. 10.4. *The sheet* $\mathrm{Im}\,\eta \prec 0$ *for* $a \prec 0$ *and* $0 \prec b \prec a^2$.

Case 4. When constants $a \succ 0$ and $b \succ a^2$. The roots determined by formula (10.14). In this case the roots are complex and accordingly to the four branch points do not belong to the axes $\operatorname{Re}\chi_x$ and $\operatorname{Im}\chi_x$. Although with respect to the original of the coordinate system they are located symmetrically. For the points $x = 0$ the value $\operatorname{Re}\eta^2(0,y) = \left(y^2 + a^2\right) + b - a^2 \succ 0$ for any values y and so the cut can be drawn along the imaginary axis. At points $y = 0$ the value $\operatorname{Re}\eta^2(x,0) = \left(x^2 - a^2\right) - b - a^2 \succ 0$ at any values x and so the cut can be drawn along the real axis. Finally along the hyperbola (10.17):

$$\operatorname{Re}\eta^2 = -4x^2\left(x^2 - a\right) + b - a^2 \tag{10.18}$$

is positive when $|x|$ is small enough and so we have the cut here. When $|x|$ is large the value (10.18) is negative and so the cut cannot be drawn along the hyperbola into infinity. In the branch points the function $\operatorname{Re}\eta^2$ turns into zero. Bearing this in mind the integration contour form in (10.13) we obtain the cuts, which are seen in the Fig.10.5. The angles for the point A are: $\alpha_1 = \alpha$, $\alpha_2 = -2\pi + \alpha$, $\beta_1 = -2\pi - \beta$ and $\beta_2 = \pi + \beta$. Therefore the values are: $\operatorname{Re}\eta(A) \prec 0$, $\operatorname{Im}\eta(A) = 0$ and for the point B the value is $\operatorname{Re}\eta(B) \succ 0$.

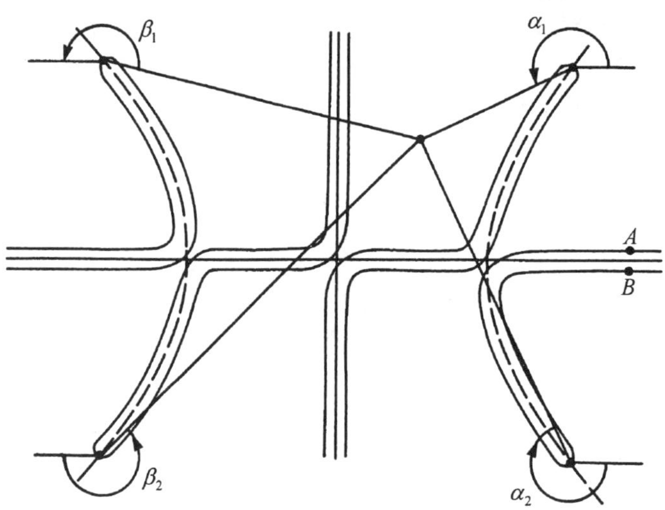

Fig. 10.5. The sheet $\operatorname{Im}\eta \prec 0$ *for* $a \succ 0$ *and* $b \succ a^2$.

Case 5. When constants $a \prec 0$ and $b \prec a^2$. This case differs from the previous cases in the hyperbola position. The cuts are shown in Fig.10.6. The sign of the real part in the point A are determined by the angles: $\alpha_1 = -\alpha$, $\alpha_2 = \alpha$, $\beta_1 = 2\pi - \beta$ and $\beta_2 = \beta$, therefore the value is $\operatorname{Re}\eta(A) \succ 0$.

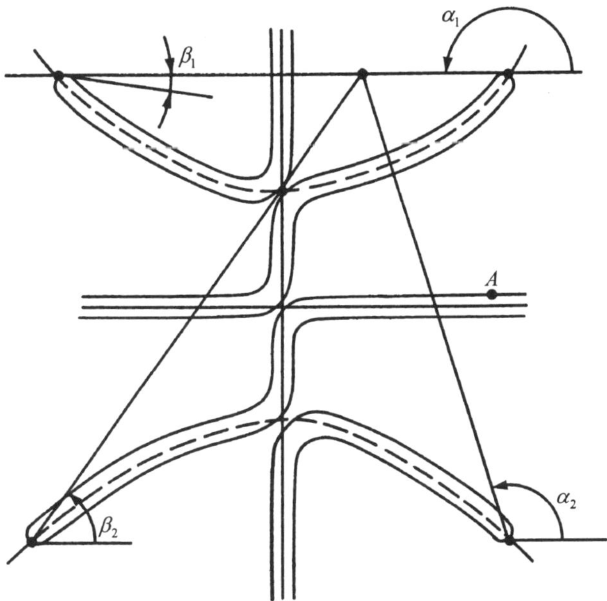

Fig.10.6. *The sheet* $\mathrm{Im}\,\eta \prec 0$ *for* $a \prec 0$ *and* $b \succ a^2$.

10.4. Integration in the Complex Plane of the Variable χ_x and Examples

When we integrate in the formulae (10.13) then the integration contour in the left half–plane $\mathrm{Im}\,\chi_x \prec 0$ lies below and in the right half–plane above the cuts. The functions $\chi_{z_{1,2}}(\chi_x)$ are not one–valued. The cut (along the Riemann surface sheets join) starts in the branch points $\chi_x = \pm\sqrt{\Delta(\varepsilon_{ef})}$ and $\chi_x = \pm\sqrt{\Delta(\mu_{ef})}$.

When there are no losses the expressions under the square root values are real. In this situation there can be only three cases:

The first case is when the values $\Delta(\varepsilon_{ef}) \succ 0$ and $\Delta(\mu_{ef}) \succ 0$. The branch points are located on the real axis.

The second case is when the values $\Delta(\varepsilon_{ef}) \prec 0$ and $\Delta(\mu_{ef}) \prec 0$. Therefore the cuts start at the points on the imaginary axis.

Finally, the third case is when the values $\Delta(\varepsilon_{ef}) \succ 0$ and $\Delta(\mu_{ef}) \prec 0$ or $\Delta(\varepsilon_{ef}) \prec 0$ and $\Delta(\mu_{ef}) \succ 0$. Now the branch points are located in pairs on the real and the imaginary axes. The cut direction from each branch point is determined by the condition $\operatorname{Im}\chi_{z_{1,2}}(\chi_x) \prec 0$ on the first sheet of the Riemann surface and by the shape of the integration contour.

Here we have an example of our calculations. The parameters that we used in our calculations are: $\mu_{xx}=0.5$, $\mu_{xy}=0.2$, $\varepsilon_{xx}=1$, $\varepsilon_{xy}=0$, $h=0$, $k=1$. Thus $\Delta(\varepsilon)=\Delta(\varepsilon_{ef})=1$, $\Delta(\mu)=0.5$ and $\Delta(\mu_{ef})=0.42$.

The function $\eta(\chi_x)$ is determined by the formula (10.12) with coefficients $a=0.32$, $b=0.64$, which correspond to the fourth case (in section 10.3). The shapes of the cuts are shown in Fig.10.5. The cuts start at the points $\chi_x = \pm 0.75 \pm i0.49$.

Now we determine the form of the cut for the function $\chi_{z1}(\chi_x)$. Its branch points are $\chi_x = \pm 1$ and $\chi_x = \pm 0.648$ on the real axis that are on the cut line of the function $\eta(\chi_x)$ where χ_x is real. A simple calculation yields $0.5\left(\varepsilon_{xx}-\mu_{xx}\right)\eta(0.648)=0.185$ and $0.5\left(\varepsilon_{xx}+\mu_{xx}\right)\xi(0.648)=0.5025$.

Therefore, the branch point $\chi_x=0.648$ belonging to the upper bank of the function $\eta(\chi_x)$ cut line where the value $0.5\left(\varepsilon_{xx}-\mu_{xx}\right)\eta(0.648)=-0.185$ is negative. In a similar way, we calculate $0.5\left(\varepsilon_{xx}+\mu_{xx}\right)\xi(1)=-0.25$ and $0.5(\varepsilon_{xx}-\mu_{xxx})\eta(1)=0.25$. We then find that the branch point $\chi_x=1$ of the function $\chi_{z_1}(\chi_x)$ is on the lower bank of the function $\eta(\chi_x)$ cut line where the last function has a positive value.

The cut line of the function $\chi_{z_1}(\chi_x)$ is determined by the equation:

$$\operatorname{Im}\chi_{z_1}^2 = -(\varepsilon_{xx}+\mu_{xx})xy+0.5(\varepsilon_{xx}-\mu_{xy})\operatorname{Im}\eta(\chi_x)=0, \tag{10.19}$$

on the condition that:

$$\operatorname{Re}\chi_{z_1}^2 = 0.5(\varepsilon_{xx}+\mu_{xx})(a_1-x^2+y^2)+0.5(\varepsilon_{xx}-\mu_{xx})\operatorname{Re}\eta(\chi_x)>0,$$

where we begin again to write $\chi_x=x+iy$. The equation (10.19) is satisfied on the coordinate axes because both terms (first term ($xy=0$) and the second term) turn into zero as follows from Fig.10.5. To the right from the point $\chi_x=1$ the function $\xi(\chi_x)<0$ is negative and its modulus increases as a square. The function $\eta(\chi_z)$ also increases as a square at a large real magnitude of χ_x, but with a lesser coefficient $0.5(\varepsilon_{xx}-\mu_{xx})\prec 0.5(\varepsilon_{xx}+\mu_{xx})$. Thus, the additional condition is not satisfied and there is no cut line at $\chi_x \succ 1$ so the cut line is drawn from $\chi_x=1$ along the real axis toward the origin of the coordinate system. The Fig.10.7 shows this cut line on that cut side of function $\eta(\chi_x)$ where the last function is positive. A similar consideration shows that from the point $\chi_x=0.648$ the cut line goes toward the origin of the coordinate system as well.

When calculating the value $\chi_{z_1}(\chi_x)$ one has to take the value $\eta(\chi_x)$ on the cut side where the magnitude of $\mathrm{Re}\,\eta \prec 0$. And the cut line from the point $\chi_x = 0.648$ first goes along the upper bank of the value $\eta(\chi_x)$ cut line (which starts at $\chi_x = 0.75 - i0.49$) and then along the lower bank of the value $\eta(\chi_x)$ cut line (which starts at $\chi_x = 0.75 + i0.49$). After that the cut line goes near the origin of the coordinate system and turns in the direction of the negative imaginary axis downward (Fig. 10.7).

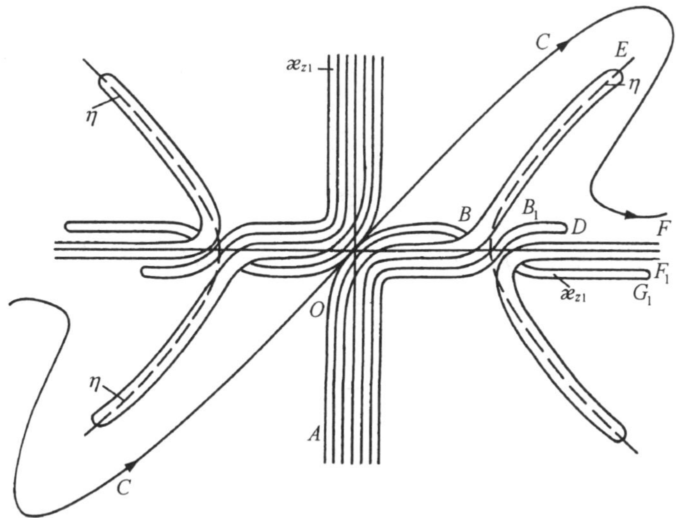

Fig. 10.7. *The function* $\chi_{z_1}(\chi_x)$ *cuts shown on the sheet* $\eta(\chi_x)$, $\mathrm{Im}\,\chi_{z_1} \le 0$, *where C is the integration contour in the formula (10.19). This function* $\eta(\chi_x)$ *cuts start at the points* $\chi_x = \pm 0.75 \pm i0.49$, *the function* $\chi_{z_1}(\chi_x)$ *cuts start at the points* $\chi_x = \pm 1$ *and* $\chi_x = \pm 0.648$.

The cuts of the function χ_{z2} are shown in Fig. 10.8. The difference from Fig. 10.7 is that the branch points are on the other bank of the function $\eta(\chi_x)$ cut line. Because the value $\eta(\chi_x)$ appears in the formulae (10.12) with the other sign.

For $x \succ 0$ the integration contour in the formula (10.13) is pressing the negative imaginary axis from the left half–plane and close to the cuts from above in the right half–plane. For $x \prec 0$ the integration contour is pressing the cuts from below in the left half–plane and close to the positive imaginary axis from the right in the right half–plane and then goes to the infinity. The result obtained is the same in both cases so that E_y and H_y depend on $|x|$.

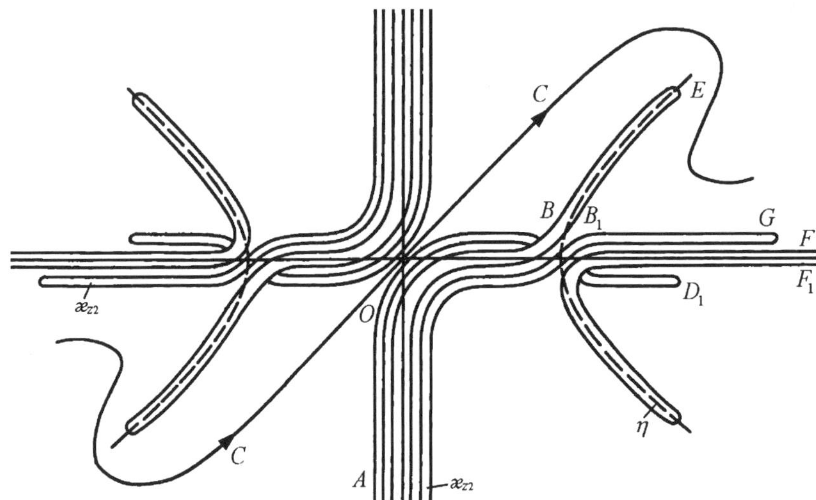

Fig. 10.8. *The function $\chi_{z_2}(\chi_x)$ cuts shown on the sheet $\eta(\chi_x)$ and $\mathrm{Im}\,\chi_{z_2} \prec 0$, where C is the integration contour in the formula (10.19). The function $\eta(\chi_x)$ cuts start at the points $\chi_x = \pm 0.75 \pm i0.49$, the function $\chi_{z_2}(\chi_x)$ cuts start at the points $\chi_x = \pm 1$ and $\chi_x = \pm 0.648$.*

It is convenient to write the final results in this form when one uses only the positive values of functions $\eta(\chi_x)$ and $\chi_{z_1,z_2}(\chi_x)$. One has to take into account that the function $\chi_{z_1}(\chi_x)$ is calculated on one bank of the function $\eta(\chi_x)$ cut, which coincides with the function $\chi_{z_2}(\chi_x)$. When the function $\chi_{z_2}(\chi_x)$ is calculated on the opposite bank of the function $\eta(\chi_x)$ cut and as follows (10.12). An example of an integral:

$$I = \int\limits_{AGEB_1DF} e^{-i|x|\chi_x} d\chi_x \frac{e^{-i|z|\chi_x} f_1(\chi_{z_1})}{\chi_{z_1}\eta(\chi_x)} \tag{10.20}$$

along the initial contour $ABEB_1DF$ Here the point A is in the infinity of the negative imaginary axis and F is in the infinity of the positive real axis (Fig. 10.7). On the contour segment AB the values $\chi_{z_1}(\chi_x)$ and $\eta(\chi_x)$ are positive. On the contour segment BE the function $\chi_{z_1}(\chi_x)$ has complex values and $\eta(\chi_x) \succ 0$. On the opposite bank of the function $\eta(\chi_x)$ cut (in the points of the *arc* EB_1) the values $\eta(\chi_x)$ are negative and the values $\chi_{z_1}(B_1) = \chi_{z_2}(B)$. Here B and B_1 can be opposite points of the cut BEB_1. Then on the contour segment B_1D the function χ_{z_1} is positive and $\eta(\chi_x)$ is negative. Therefore the values χ_{z_1} coincide with the values χ_{z_2} on the contour segment $B_1 D_1$ in Fig.10.8 where $\eta(\chi_x)$ is positive. Here the integrals (10.20) can be written as:

$$I = \int\limits_{ABE} d\chi_x e^{-i|x|\chi_r} \frac{e^{-i|z|\chi_{z_1}} f_1\left(\chi_{z_1}\right)}{\chi_{z_1}\eta\left(\chi_x\right)} - \int\limits_{EB_1 D_1 F_1} d\chi_x e^{-i|x|\chi_x} \frac{e^{-i|z|\chi_{z_2}} f_1\left(\chi_{z_2}\right)}{\chi_{z_2}\eta\left(\chi_x\right)}, \qquad (10.21)$$

where one has to put in the positive values of $\eta(\chi_x)$ in the points of cuts.

The convergence on the infinite segment AO is ensured by the first exponent of the integrand because χ_x on this contour segment is purely imaginary and has a negative imaginary part. On the contour segment $D_1 F_1$ is ensured by the second exponent because here the function $\chi_{z_2}\left(\chi_x\right)$ is purely imaginary and has a negative imaginary part. As a result we obtained the formulae for the integral representation of the longitudinal components:

$$E_y = \frac{iA}{2\pi\left(\varepsilon_{xx}\mu_{zz} - \mu_{xx}\varepsilon_{zz}\right)} \left\{ - \int\limits_{ABG_1 F_1} d\chi_x e^{-i|x|\chi_x} \frac{e^{-i|z|\chi_{z_1}} f_1\left(\chi_{z_1}\right)}{\chi_{z_1}\eta\left(\chi_x\right)} + \right.$$

$$\left. + \int\limits_{ABD_1 F_1} d\chi_x e^{-i|x|\chi_x} \frac{e^{-i|z|\chi_{z_2}} f_1\left(\chi_{z_2}\right)}{\chi_{z_2}\eta\left(\chi_x\right)} \right\} +$$

$$+ \frac{z}{|z|} \frac{iB}{2\pi\left(\varepsilon_{xx}\mu_{zz} - \mu_{xx}\varepsilon_{zz}\right)\omega\varepsilon_0} \left\{ - \int\limits_{ABG_1 F_1} d\chi_x e^{-i|x|\chi_x} \frac{e^{-i|z|\chi_{z_1}} f_2\left(\chi_{z_1}\right)}{\chi_{z_1}\eta\left(\chi_x\right)} + \right.$$

$$\left. + \int\limits_{ABD_1 F_1} d\chi_x e^{-i|x|\chi_x} \frac{e^{-i|z|\chi_{z_2}} f_2\left(\chi_{z_2}\right)}{\chi_{z_2}\eta\left(\chi_x\right)} \right\},$$

$$H_y = \frac{iB}{2\pi\left(\varepsilon_{xx}\mu_{zz} - \mu_{xx}\varepsilon_{zz}\right)} \left\{ - \int\limits_{ABG_1 F_1} d\chi_x e^{-i|x|\chi_x} \frac{e^{-i|z|\chi_{z_1}} f_3\left(\chi_{z_1}\right)}{\chi_{z_1}\eta\left(\chi_x\right)} + \right.$$

$$\left. + \int\limits_{ABD_1 F_1} d\chi_x e^{-i|x|\chi_x} \frac{e^{-i|z|\chi_{z_2}} f_3\left(\chi_{z_2}\right)}{\chi_{z_2}\eta\left(\chi_x\right)} \right\} +$$

$$+ \frac{z}{|z|} \frac{iA}{2\pi\left(\varepsilon_{xx}\mu_{zz} - \mu_{xx}\varepsilon_{zz}\right)\omega\mu_0} \left\{ - \int\limits_{ABG_1 F_1} d\chi_x e^{-i|x|\chi_x} \frac{e^{-i|z|\chi_{z_1}} f_4\left(\chi_{z_1}\right)}{\chi_{z_1}\eta\left(\chi_x\right)} + \right.$$

$$+ \int\limits_{ABD_1F_1} d\chi_x e^{-i|x|\chi_x} \frac{e^{-i|z|\chi_{z_2}} f_4\left(\chi_{z_2}\right)}{\chi_{z_2} \eta\left(\chi_x\right)} \Bigg\}. \qquad (10.22)$$

In the formula (10.22), we have used the integrals along the arc BEB_1 and they mutually cancelled out each other.

10.5. Singling Out of the Singularity

The solutions $E_y(x,z)$ and $H_y(x,z)$ have the logarithmic singularity at $x \to 0$ and $z \to 0$ because there is a δ – function in the equations (10.1)–(10.4). The transversal components will also have the singular terms, which appear after the differentiating of the logarithm. When applying our SIE method it is convenient to single out the singularities.

It is obvious that the singularity of $E_y(x,z)$ rises from the term that consists of the factor A. And the singularity of $H_y(x,z)$ rises from the term that consists of the factor B. The singularity itself is described by the Hankel function of the zeroth order of the second kind. Adding and subtracting the corresponding terms we find from the formula (10.10):

$$E_y\left(\chi_x,\chi_z\right) = \frac{A}{4\pi^2}\left[\frac{\Delta(\mu)\Delta(\varepsilon)}{f_3} + \frac{\chi_z^2 f_2 f_4}{f_3 f}\right] + \frac{B}{4\pi^2}\frac{\chi_z f_2}{f},$$

$$H_y\left(\chi_x,\chi_z\right) = \frac{B}{4\pi^2}\left[\frac{\Delta(\mu)\Delta(\varepsilon)}{f_4} + \frac{\chi_z^2 f_2 f_4}{f_1 f}\right] + \frac{A}{4\pi^2}\frac{\chi_z f_4}{f}. \qquad (10.23)$$

After having integrated the last expressions with respect to χ_x and χ_z the logarithmic singularity rises only from the first terms in the formula (10.23). Thus, we arrive at:

$$E_{y1} = -\frac{A}{4\pi^2}\Delta(\mu)\Delta(\varepsilon)\iint d\chi_x d\chi_z \frac{e^{-i\chi_x x - i\chi_z z}}{\varepsilon_{xx}\Delta(\mu)\chi_x^2 + \Delta(\varepsilon)\chi_z^2 - \varepsilon_{xx}\Delta(\mu)\Delta\left(\varepsilon_{ef}\right)}. \qquad (10.24)$$

Introducing the new variables:

$$\varepsilon_{xx}\Delta(\mu)\chi_x^2 = k_x^2, \qquad \Delta(\varepsilon)\chi_z^2 = k_z^2$$

and the notation

$$\varepsilon_{xx}\Delta(\mu)\Delta\left(\varepsilon_{ef}\right) = k_{e\perp}^2$$

we obtained the integral in this form:

$$E_y = -\frac{A}{4\pi^2}\sqrt{\frac{\Delta(\mu)\Delta(\varepsilon)}{\varepsilon_{xx}}}\iint dk_x dk_z \frac{e^{-ik_x\frac{x}{\sqrt{\varepsilon_{xx}\Delta(\mu)}}\,-ik_z\frac{z}{\sqrt{\mu_{xx}\Delta(\varepsilon)}}}}{k_x^2 + k_z^2 + k_{e\perp}^2}\,.$$

So we find:

$$E_{y_1}(x,z) = \frac{iA}{4}\sqrt{\frac{\Delta(\mu)\Delta(\varepsilon)}{\varepsilon_{xx}}}\,H_0^{(2)}\left(\sqrt{\Delta\!\left(\varepsilon_{ef}\right)x^2 + \varepsilon_{xx}\Delta(\mu)\frac{\Delta\!\left(\varepsilon_{ef}\right)}{\Delta(\varepsilon)}z^2}\right),$$

$$H_{y_1}(x,z) = \frac{iB}{4}\sqrt{\frac{\Delta(\mu)\Delta(\varepsilon)}{\mu_{xx}}}\,H_0^{(2)}\left(\sqrt{\Delta\!\left(\mu_{ef}\right)x^2 + \mu_{xx}\Delta(\varepsilon)\frac{\Delta\!\left(\mu_{ef}\right)}{\Delta(\mu)}z^2}\right), \qquad (10.25)$$

where $H_0^{(2)}$ is the Hankel function with accuracy to a factor.

Integrating the second terms in the formulae (10.23) one deals with singularities of the function $1/f$ as described in the previous sections. The singularities of the function $1/f_3$ (or $1/f_1$) lead to the function of the type $F(\chi_x) = \sqrt{a^2 - \chi_x^2}$, which is described in detail in [10.5].

For the parameters of certain problem we have to determine the coefficients a and b of the function $\eta(\chi_x)$. According to a and b it is possible to choose one of the five cases for $\eta(\chi_x)$, described in section 10.2. It is possible to execute the integration with respect to χ_x as in the section 10.3.

Even if the waveguide consists of lossy material the approach to the solution remains the same. But the branch points in the right half–plane descend and in the left half–plane they ascend and they do not coincide with the coordinate lines. The strategy of this solution will be the same as the solution described in the ninth chapter.

The fundamental solution of the Maxwell's equations has been found for the bigyrotropic medium in the case of transversal magnetization. The developed method enables one to (in a rigorously electrodynamical problem formulation) investigate the dispersion dependences and the electromagnetic fields of the main and higher modes in the regular waveguide of arbitrary cross–section geometry.

In chapters 6–10 we used methods of SIE in order to solve 2D (two dimension) electrodynamical problems for regular waveguides. In the final chapter we will describe a method of SIE in which we will solve 3D (three dimension) electrodynamical problems. We will have to determine the reflected and transmitted electromagnetic waves for 3D magneto–dielectric structures placed in a known electromagnetic field.

11. SOLUTION OF MAXWELL'S EQUATIONS BY THE SIE METHOD FOR THREE–DIMENSIONAL SCATTERING PROBLEMS

In this chapter we will describe the rigorous method to solving Maxwell's equations for three–dimensional (3D) structures. We will determine the reflected and transmitted electromagnetic waves (EM) for 3D magneto–dielectric structures placed in a known EM field. This structure may be of any complicated (arbitrary) shape. One or more sources of EM waves can be used inside or outside of the structure that is under radiation. The approach to solving Maxwell's equations by our SIE methods in this chapter is the same as our approach in the previous chapters 5–10.

The modern development of telecommunications and different satellite systems has lead to the creation of a multitude of microwave devices [11.1]. Some of these devices are: receiving antennas and microstrip reflectors of microwaves. The detection of EM fields near antenna feeders, waveguides, microwave home devices and anechoic chambers are current topics now days [11.2]–[11.3].

Microwaves are used in therapeutic treatments [11.4]–[11.6] as well as in receiving images of organs in the body [11.7]–[11.10]. But the development of modern microwave biomedical devices has lead to an increase in electrodynamical problems. In biomedical engineering there are certain diffraction problems that can be solved using the SIE method presented in this chapter.

11.1. The Fundamental Solution of Maxwell's Equations for Constracting the SIE Method

In order to calculate an EM wave and 3D structure (a scattering body) interactions we need to solve a diffraction problem. A known EM wave (which is radiated by an antenna) influences the 3D structure. A solution to this electrodynamical problem allows us to find the EM fields inside and outside of the 3D structure. Here we will describe the solution to this problem by use of our SIE method. To present EM fields in integral form we will use the solution of Maxwell's equations:

$$rot\vec{H} = i\omega\varepsilon_0\varepsilon\vec{E} + \vec{j}_e,$$

$$rot\vec{E} = -i\omega\mu_0\mu\vec{H} - \vec{j}_m \qquad (11.1)$$

with the electric and magnetic point sources $\vec{j}_e = \vec{n}\delta_e(\vec{r})$ and $\vec{j}_m = \vec{n}\delta_m(\vec{r})$.

Applying Fourier transformation we obtained an electric type wave:

$$e(\vec{r}) = \left[\frac{1}{k^2 \varepsilon\mu} \nabla(\vec{n}(\vec{r}_0),\nabla) + \vec{n}(\vec{r}_0) \right] h_0(k\vec{r}\sqrt{\varepsilon\mu}) ,$$

$$h(\vec{r}) = \frac{i}{Z_0}(\sqrt{\frac{\varepsilon}{\mu}} \left[\vec{n}(\vec{r}_0), \frac{\vec{r}-\vec{r}_0}{|\vec{r}-\vec{r}_0|} \right] h_1(k\vec{r}\sqrt{\varepsilon\mu}) , \tag{11.2}$$

when $\vec{j}_m = 0$ and $\vec{j}_e \neq 0$. Also similar expressions will be for a magnetic type field when $\vec{j}_m \neq 0$ and $\vec{j}_e = 0$. Because of the linearity the general solution is the sum of the solutions for $\vec{j}_e \neq 0$, $\vec{j}_m = 0$ and $\vec{j}_e = 0$, $\vec{j}_m \neq 0$ [11.11].

The first expression of the solution (11.2) describes the electrical type wave and the second expression describes the magnetic type wave. The unknown scattered wave and inner field in the scattering body originate from the bordering surfaces. Therefore we write them as surface integrals:

$$\vec{E}_e(\vec{r}) = \int_{S_i} \mu_e(\vec{r}_0) \left\{ \frac{1}{k^2 \varepsilon_i\mu_i} \nabla(\vec{n}(\vec{r}_0),\nabla) + \vec{n}(\vec{r}_0) \right\} h_0\left(k\sqrt{\varepsilon_i\mu_i}|\vec{r}-\vec{r}_0|\right) dS , \tag{11.3}$$

$$\vec{H}_e(\vec{r}) = \frac{i}{Z_0}\sqrt{\frac{\varepsilon_i}{\mu_i}} \int_{S_i} \mu_m(\vec{r}_0) \left[\vec{n}(\vec{r}_0), \frac{\vec{r}-\vec{r}_0}{|\vec{r}-\vec{r}_0|} \right] h_1\left(k\sqrt{\varepsilon_i\mu_i}|\vec{r}-\vec{r}_0|\right) dS . \tag{11.4}$$

Here $\mu_e(\vec{r}_0)$ and $\mu_m(\vec{r}_0)$ respectively are electric and magnetic source densities in the point $\vec{r}_0$ on the surface for the electric wave. The value $\vec{n}(\vec{r}_0)$ is a unit normal vector to the surface at the same point. Quantity $|\vec{r}-\vec{r}_0|$ is the distance from the point $\vec{r}$ where we look for the fields $\vec{E}(\vec{r})$, $\vec{H}(\vec{r})$ to the electric and magnetic source densities ($\mu_e(\vec{r}_0)$, $\mu_m(\vec{r}_0)$) in the point $\vec{r}_0$. The function $h_0\left(k\sqrt{\varepsilon_i\mu_i}|\vec{r}-\vec{r}_0|\right) \sim e^{-ik\sqrt{\varepsilon_i\mu_i}|\vec{r}-\vec{r}_0|}/k\sqrt{\varepsilon_i\mu_i}|\vec{r}-\vec{r}_0|$ is the spherical Hankel function of the zeroth order and the second kind. The function $h_1\left(k\sqrt{\varepsilon_i\mu_i}|\vec{r}-\vec{r}_0|\right) = -\frac{d}{dz}h_0\left(k\sqrt{\varepsilon_i\mu_i}|\vec{r}-\vec{r}_0|\right)$ is the spherical Hankel function of the first order and the second kind [11.12]. The value $k = 2\pi/\lambda$ is the wave number and λ is wavelength of the EM wave propagating in a vacuum. The value dS is the infinitesimal patch of area. The sign ∇ is the gradient operator. In the formulae (11.3) and (11.4) the notations $(.,.)$ and $[.,.]$ are the scalar and vector products of two vectors, respectively. The value $Z_0 = \sqrt{\mu_0/\varepsilon_0}$ is the characteristic impedance of a vacuum (free space). By analogy we write the magnetic wave:

$$\vec{H}_m(\vec{r}) = \int_{S_i} \mu_m(\vec{r}_0) \left\{ \frac{1}{k^2 \varepsilon_i \mu_i} \nabla\big(\vec{n}(\vec{r}_0), \nabla\big) + \vec{n}(\vec{r}_0) \right\} h_0\left(k\sqrt{\varepsilon_i \mu_i}\,|\vec{r}-\vec{r}_0|\right) dS, \qquad (11.5)$$

$$\vec{E}_m(\vec{r}) = -iZ_0\sqrt{\frac{\mu_i}{\varepsilon_i}} \int_{S_i} \mu_e(\vec{r}_0) \left[\vec{n}(\vec{r}_0), \frac{\vec{r}-\vec{r}_0}{|\vec{r}-\vec{r}_0|} \right] h_1\left(k\sqrt{\varepsilon_i \mu_i}\,|\vec{r}-\vec{r}_0|\right) dS. \qquad (11.6)$$

The general solution of Maxwell's equations for an EM field is constructed as the superposition of hybrid quasi–E waves and hybrid quasi–H waves. Hybrid quasi–E waves are described by the expressions (11.3) and (11.5). Hybrid quasi–H waves are described by the expressions (11.4) and (11.6).The components of an EM field are:

$$\vec{E}(\vec{r}_1) = \int_{S_i} \mu_e(\vec{r}_0) \left\{ \frac{1}{k^2 \varepsilon_i \mu_i} \nabla\big(\vec{n}(\vec{r}_0), \nabla\big) + \vec{n}(\vec{r}_0) \right\} h_0\left(k\sqrt{\varepsilon_i \mu_i}\,|\vec{r}_1-\vec{r}_0|\right) dS -$$

$$-iZ_0\sqrt{\frac{\mu_i}{\varepsilon_i}} \int_{S_i} \mu_m(\vec{r}_0) \left[\vec{n}(\vec{r}_0), \frac{\vec{r}_1-\vec{r}_0}{|\vec{r}_1-\vec{r}_0|} \right] h_1\left(k\sqrt{\varepsilon_i \mu_i}\,|\vec{r}_1-\vec{r}_0|\right) dS,$$

$$\vec{H}(\vec{r}_1) = \frac{i}{Z_0}\sqrt{\frac{\varepsilon_i}{\mu_i}} \int_{S_i} \mu_m(\vec{r}_0) \left[\vec{n}(\vec{r}_0), \frac{\vec{r}_1-\vec{r}_0}{|\vec{r}_1-\vec{r}_0|} \right] h_1\left(k\sqrt{\varepsilon_i \mu_i}\,|\vec{r}_1-\vec{r}_0|\right) dS +$$

$$+ \int_{S_i} \mu_e(\vec{r}_0) \left\{ \frac{1}{k^2 \varepsilon_i \mu_i} \nabla\big(\vec{n}(\vec{r}_0), \nabla\big) + \vec{n}(\vec{r}_0) \right\} h_0\left(k\sqrt{\varepsilon_i \mu_i}\,|\vec{r}_1-\vec{r}_0|\right) dS. \qquad (11.7)$$

Here we will write the boundary conditions only on one of the structure surfaces, for example, for surface S_3 (Fig.11.1a). The surface S_3 divides the area into two parts described by the values ε_3, μ_3 (for air) and ε_1, μ_1 (for myocardium).

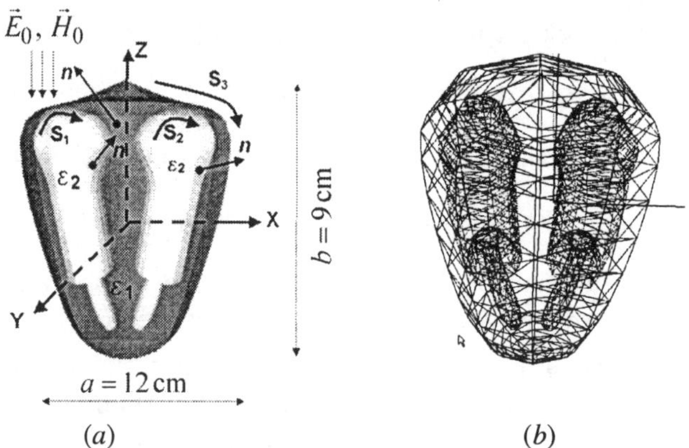

Fig. 11.1. *(a) 3D structure (scattered body) used in our calculations, which was limited by the surfaces S_1, S_2, S_3; (b) the 3D structure surface was constructed of triangles.*

Equating the vector products $\left[\vec{n}(\vec{r}_1),\vec{E}(\vec{r}_1)\right]^{+} = \left[\vec{n}(\vec{r}_1),\vec{E}(\vec{r}_1)\right]^{-}$ we arrive at the integral equation for the surface densities $\mu_e^{+}(\vec{r}_0)$, $\mu_e^{-}(\vec{r}_0)$ and $\mu_m^{+}(\vec{r}_0)$, $\mu_m^{-}(\vec{r}_0)$:

$$\left[\vec{n}(\vec{r}_1),\vec{E}_{(source\_in)}(\vec{r}_1)\right]^{+} + \int\limits_{S_i}\mu_e^{+}(\vec{r}_0)\left\{\left(\vec{n}(\vec{r}_0),\frac{\vec{r}_1-\vec{r}_0}{|\vec{r}_1-\vec{r}_0|}\right)\left[\vec{n}(\vec{r}_1),\frac{\vec{r}_1-\vec{r}_0}{|\vec{r}_1-\vec{r}_0|}\right]\times\right.$$

$$\times\left(\frac{3}{k\sqrt{\varepsilon_1\mu_1}\,|\vec{r}_1-\vec{r}_0|}\cdot h_1\left(k\sqrt{\varepsilon_1\mu_1}\,|\vec{r}_1-\vec{r}_0|\right)-h_0\left(k\sqrt{\varepsilon_1\mu_1}\,|\vec{r}_1-\vec{r}_0|\right)\right)+$$

$$\left.+\left[\vec{n}(\vec{r}_1),\vec{n}(\vec{r}_0)\right]\left(h_0\left(k\sqrt{\varepsilon_1\mu_1}\,|\vec{r}_1-\vec{r}_0|\right)-\frac{h_1\left(k\sqrt{\varepsilon_1\mu_1}\,|\vec{r}_1-\vec{r}_0|\right)}{k\sqrt{\varepsilon_1\mu_1}\,|\vec{r}_1-\vec{r}_0|}\right)\right\}dS-$$

$$-iZ_0\sqrt{\frac{\mu_1}{\varepsilon_1}}\int\limits_{S_i}\mu_m^{+}(\vec{r}_0)\left\{\vec{n}(\vec{r}_0)\left(\vec{n}(\vec{r}_1),\frac{\vec{r}_1-\vec{r}_0}{|\vec{r}_1-\vec{r}_0|}\right)-\right.$$

$$\left.-\frac{\vec{r}_1-\vec{r}_0}{|\vec{r}_1-\vec{r}_0|}(\vec{n}(\vec{r}_1),\vec{n}(\vec{r}_0))\,h_1\left(k\sqrt{\varepsilon_1\mu_1}\,|\vec{r}_1-\vec{r}_0|\right)\right\}dS=$$

$$=\left[\vec{n}(\vec{r}_1),\vec{E}_{(source\_out)}(\vec{r}_1)\right]^{-} + \int\limits_{S_i}\mu_e^{-}(\vec{r}_0)\left\{\left(\vec{n}(\vec{r}_0),\frac{\vec{r}_1-\vec{r}_0}{|\vec{r}_1-\vec{r}_0|}\right)\left[\vec{n}(\vec{r}_1),\frac{\vec{r}_1-\vec{r}_0}{|\vec{r}_1-\vec{r}_0|}\right]\times\right.$$

$$\times\left(\frac{3}{k\sqrt{\varepsilon_3\mu_3}\,|\vec{r}_1-\vec{r}_0|}\cdot h_1\left(k\sqrt{\varepsilon_3\mu_3}\,|\vec{r}_1-\vec{r}_0|\right)-h_0\left(k\sqrt{\varepsilon_3\mu_3}\,|\vec{r}_1-\vec{r}_0|\right)\right)+$$

$$\left.+\left[\vec{n}(\vec{r}_1),\vec{n}(\vec{r}_0)\right]\left(h_0\left(k\sqrt{\varepsilon_3\mu_3}\,|\vec{r}_1-\vec{r}_0|\right)-\frac{h_1\left(k\sqrt{\varepsilon_3\mu_3}\,|\vec{r}_1-\vec{r}_0|\right)}{k\sqrt{\varepsilon_3\mu_3}\,|\vec{r}_1-\vec{r}_0|}\right)\right\}dS-$$

$$-iZ_0\sqrt{\frac{\mu_3}{\varepsilon_3}}\int\limits_{S_i}\mu_m^{-}(\vec{r}_0)\left\{\vec{n}(\vec{r}_0)\left(\vec{n}(\vec{r}_1),\frac{\vec{r}_1-\vec{r}_0}{|\vec{r}_1-\vec{r}_0|}\right)-\right.$$

$$\left.-\frac{\vec{r}_1-\vec{r}_0}{|\vec{r}_1-\vec{r}_0|}(\vec{n}(\vec{r}_1),\vec{n}(\vec{r}_0))\,h_1\left(k\sqrt{\varepsilon_3\mu_3}\,|\vec{r}_1-\vec{r}_0|\right)\right\}dS. \tag{11.8}$$

In the formula (11.8) the value $\vec{E}_{(source\_in)}(\vec{r}_1)$ is the electric field of a radiated EM wave when an antenna (microwave catheter) is located inside of the 3D structure (Fig.11.1a). And the value $\vec{E}_{(source\_out)}(\vec{r}_1)$ is the electric field of an incident EM wave when an antenna (source of EM wave) is located outside of the 3D structure.

Equating the vector products $\left[\vec{n}(\vec{r}_1), \vec{H}(\vec{r}_1)\right]^+ = \left[\vec{n}(\vec{r}_1), \vec{H}(\vec{r}_1)\right]^-$ the integral equation takes the form:

$$\left[\vec{n}(\vec{r}_1), \vec{H}_{(source\_in)}(\vec{r}_1)\right]^+ + \int_{S_i} \mu_m^+(\vec{r}_0)\left\{\left(\vec{n}(\vec{r}_0), \frac{\vec{r}_1 - \vec{r}_0}{|\vec{r}_1 - \vec{r}_0|}\right)\left[\vec{n}(\vec{r}_1), \frac{\vec{r}_1 - \vec{r}_0}{|\vec{r}_1 - \vec{r}_0|}\right]\times\right.$$

$$\times\left(\frac{3}{k\sqrt{\varepsilon_1\mu_1}\,|\vec{r}_1 - \vec{r}_0|}\cdot h_1\left(k\sqrt{\varepsilon_1\mu_1}\,|\vec{r}_1 - \vec{r}_0|\right) - h_0\left(k\sqrt{\varepsilon_1\mu_1}\,|\vec{r}_1 - \vec{r}_0|\right)\right) +$$

$$+ [\vec{n}(\vec{r}_1), \vec{n}(\vec{r}_0)]\left(h_0\left(k\sqrt{\varepsilon_1\mu_1}\,|\vec{r}_1 - \vec{r}_0|\right) - \frac{h_1\left(k\sqrt{\varepsilon_1\mu_1}\,|\vec{r}_1 - \vec{r}_0|\right)}{k\sqrt{\varepsilon_1\mu_1}\,|\vec{r}_1 - \vec{r}_0|}\right)\right\}dS +$$

$$+ \frac{i}{Z_0}\sqrt{\frac{\varepsilon_1}{\mu_1}}\int_{S_i} \mu_e^+(\vec{r}_0)\left\{\vec{n}(\vec{r}_0)\left(\vec{n}(\vec{r}_1), \frac{\vec{r}_1 - \vec{r}_0}{|\vec{r}_1 - \vec{r}_0|}\right) -\right.$$

$$\left. - \frac{\vec{r}_1 - \vec{r}_0}{|\vec{r}_1 - \vec{r}_0|}(\vec{n}(\vec{r}_1), \vec{n}(\vec{r}_0))\, h_1\left(k\sqrt{\varepsilon_1\mu_1}\,|\vec{r}_1 - \vec{r}_0|\right)\right\}dS =$$

$$= \left[\vec{n}(\vec{r}_1), \vec{H}_{(source\_out)}(\vec{r}_1)\right]^- + \int_{S_i} \mu_m^-(\vec{r}_0)\left\{\left(\vec{n}(\vec{r}_0), \frac{\vec{r}_1 - \vec{r}_0}{|\vec{r}_1 - \vec{r}_0|}\right)\left[\vec{n}(\vec{r}_1), \frac{\vec{r}_1 - \vec{r}_0}{|\vec{r}_1 - \vec{r}_0|}\right]\times\right.$$

$$\times\left(\frac{3}{k\sqrt{\varepsilon_3\mu_3}\,|\vec{r}_1 - \vec{r}_0|}\cdot h_1\left(k\sqrt{\varepsilon_3\mu_3}\,|\vec{r}_1 - \vec{r}_0|\right) - h_0\left(k\sqrt{\varepsilon_3\mu_3}\,|\vec{r}_1 - \vec{r}_0|\right)\right) +$$

$$+ [\vec{n}(\vec{r}_1), \vec{n}(\vec{r}_0)]\left(h_0\left(k\sqrt{\varepsilon_3\mu_3}\,|\vec{r}_1 - \vec{r}_0|\right) - \frac{h_1\left(k\sqrt{\varepsilon_3\mu_3}\,|\vec{r}_1 - \vec{r}_0|\right)}{k\sqrt{\varepsilon_3\mu_3}\,|\vec{r}_1 - \vec{r}_0|}\right)\right\}dS +$$

$$+ \frac{i}{Z_0}\sqrt{\frac{\varepsilon_3}{\mu_3}}\int_{S_i} \mu_m^-(\vec{r}_0)\left\{\vec{n}(\vec{r}_0)\left(\vec{n}(\vec{r}_1), \frac{\vec{r}_1 - \vec{r}_0}{|\vec{r}_1 - \vec{r}_0|}\right) -\right.$$

$$\left. - \frac{\vec{r}_1 - \vec{r}_0}{|\vec{r}_1 - \vec{r}_0|}(\vec{n}(\vec{r}_1), \vec{n}(\vec{r}_0))\, h_1\left(k\sqrt{\varepsilon_3\mu_3}\,|\vec{r}_1 - \vec{r}_0|\right)\right\}dS , \tag{11.9}$$

In this formula (11.9) the value $\vec{H}_{(source\_in)}(\vec{r}_1)$ is the magnetic field of a radiated EM wave when an antenna is located inside of the 3D structure (Fig.11.1*a*).

Then the value $\vec{H}_{(source\_out)}(\vec{r}_1)$ is the magnetic field of an incident EM wave when an antenna is located outside of the 3D structure.

Integrals with the spherical Hankel function h_1 are understood in the principal value sense with the removal of an infinitely small area around the point $\vec{r}_1$. We used the Sokhotsky–Plemelj formulas and the integral over the indented hemisphere turns to zero in this case [11.12]. Thus the free member (out of integral one) is in the equations. The equations like (11.8) and (11.9) are written for all surfaces that divide the areas into parts described by the different permittivities ε_i and where $i = 1,2,3....$. These integral equations were obtained by satisfying all boundary conditions, which allows us to determine the densities $\mu_e^{+}(\vec{r}_0)$, $\mu_e^{-}(\vec{r}_0)$, $\mu_m^{+}(\vec{r}_0)$ and $\mu_m^{-}(\vec{r}_0)$. The integral equations are reduced to an algebraic system of the linear equations and this system was decided numerically.

11.2. A Comparison of Calculated Results with Experimental Data of the Back Scattering Cross–Section Dependence for a Circular Metal Disc

The back scattering cross–section dependence for a circular metal disk was calculated in order to verify our computer algorithm bases on our SIE method. We began with a simple electrodynamical problem, when a polarized plane EM wave was falling normally on a circular metal disc. Our calculated results [11.13] were compared with the experimental data from Ref. [11.14] and [11.15].

Fig.11.2b gives the dependence of the back scattering cross–section $\sigma = 4\pi R^2 \left|\vec{E}\right|^2 \Big/ \left|\vec{E}_0\right|^2$ as a function of kr. The value kr is the product of the wave number k and the radius r of the circular metal disc. Here the value $k = 2\pi/\lambda$ and λ is the wavelength of the EM wave of the incident wave. $\vec{E}$ is a vector of the reflected electric field in the direction of the incident wave source. A vector $\vec{E}_0$ is the unit electric field of the incident plane wave at the circular metal disc (Fig.11.2a). The values: R is the distance from the EM field source to the circular metal disc, r is the circular metal disc radius and t_d is its thickness. In Fig.11.2b the experimental data is shown by circles and our computation is shown by a solid line. The material of the circular metal disc is a perfect conductor in our calculations.

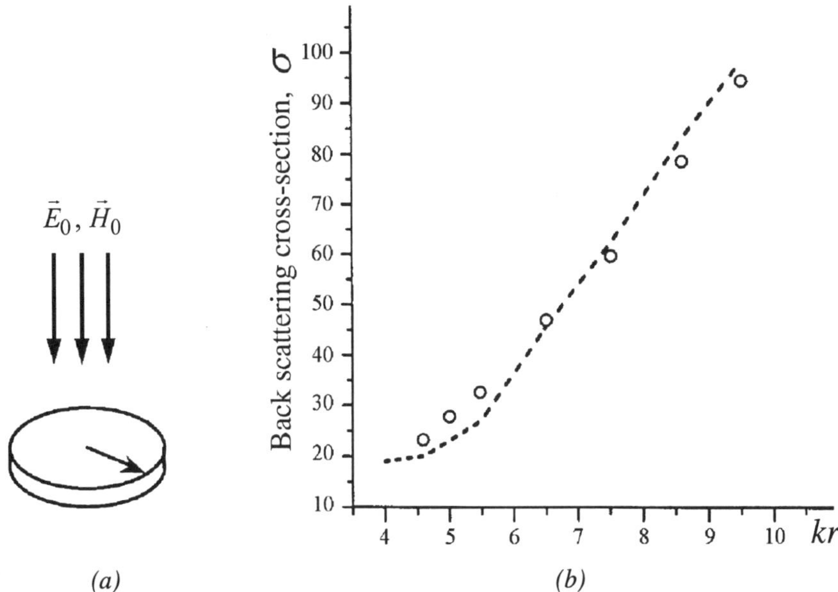

Fig. 11.2. *(a) Circular metal disc; (b) back scattering cross–section dependence on kr for a circular metal disc. Solid line shows our calculations and the experimental data is shown by circles.*

The geometrical parameters used in this section are $r = 10^{-2}\,\text{m}$, $t_d = 10^{-5}\,\text{m}$ and the distance $R = 2\,\text{m}$. A 3D surface model of the circular metal disk was created in the 3D StudioMax program. The circular metal disc surface was created using 48 triangles. We see that our calculations and experimental results are in agreement.

11.3. Diffraction Parameters of a Microstrip Reflecting Structure

Frequency selective structures are used for space detection of the EM wave sources. The main parameter of the structure is the frequency bandwidth. This and other characteristics depend on geometry and substrate parameters. The structure consists of microstrip metallic frames. There is agreement between our computations [11.16] and the results given in [11.17].

Our calculations were carried out for the structure (Fig.11.3) with and without the lower substrate layer. The incident microwave was a plane perpendicular–polarized one.

The ratio of the back reflected signal (toward the microwave source) and the incident microwave field was computed as a function of a frequency. The central frequency f_R, is the frequency in which the reflection coefficient is the greatest.

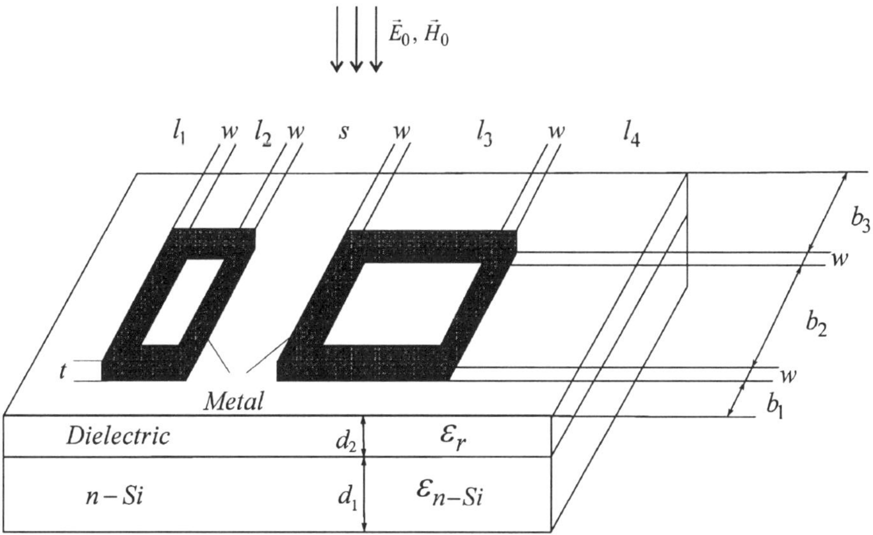

Fig. 11.3. *The frequency selective structure with rectangular metal strip frames on its surface.*

In Fig.11.4a we compared our calculations with data from [11.17]. Our calculations are presented as solid lines and the data from references is shown as points. The dependencies in Fig.11.4 characterize the reflective ability of the structure on the upper and lower layer substrate thicknesses d, d_1 as well as the relative permittivity ε_r of the upper layer dielectric material. Fig.11.4a also shows the dependences of the parameter $f_R \cdot p/c$ on the upper layer substrate thickness d, where $p = 2w + l_2 + s$. The values: w is the width of the metal strip, l_2 is the frames width, s is the distance between the frames and c is the velocity of light in a vacuum.

We can see in Fig.11.4a,b,d that the larger the upper layer thickness d and the relative permittivity of the upper dielectric layer ε_r the less the central frequency f_R. We note also that the central frequency f_R decreases when the metal strips width w becomes wider.

Fig.11.4c,d shows the dependences of the central frequency on substrate layer thicknesses d_1. Here the relative permittivity of the upper dielectric layer is $\varepsilon_r = 3$ and the lower semiconductor layer is $\varepsilon_{n-Si} = 12.7 - i10^{-3}$.
The dependencies in Fig.11.4c were calculated for these structure sizes:

Curve (1)–$l_1 = l_4 = b_3 = 2 \cdot 10^{-3}$ m, $l_2 = l_3 = b_1 = b_2 = 3 \cdot 10^{-3}$ m, $w = 2.5 \cdot 10^{-4}$ m,
$t = 10^{-5}$ m, $s = 1.5 \cdot 10^{-3}$ m, $d = 5 \cdot 10^{-6}$ m.
Curve (2)–the sizes are the same as for curve (1) except $d = 0$ (without the upper dielectric layer of the substrate).

Curve (3)–the sizes are the same as for curve (1) except $w = 5 \cdot 10^{-4}$ m.

Curve (4)–the sizes are the same as for curve (3) except $l_3 = 3.5 \cdot 10^{-3}$ m.

The dependencies in Fig.11.4d were calculated for structure sizes:

Curve (5)–the sizes are the same as for curve (1) except $d_1 = 0$ (without the lower semiconductor n–Si layer of the substrate).

Curve (6)–the same as for curve (5) except $l_2 = 10^{-3}$ m, $l_3 = 3 \cdot 10^{-3}$ m.

Curve (7)–the same as for curve (6) except $d_1 = 5 \cdot 10^{-6}$ m (with the lower semiconductor n–Si layer of the substrate).

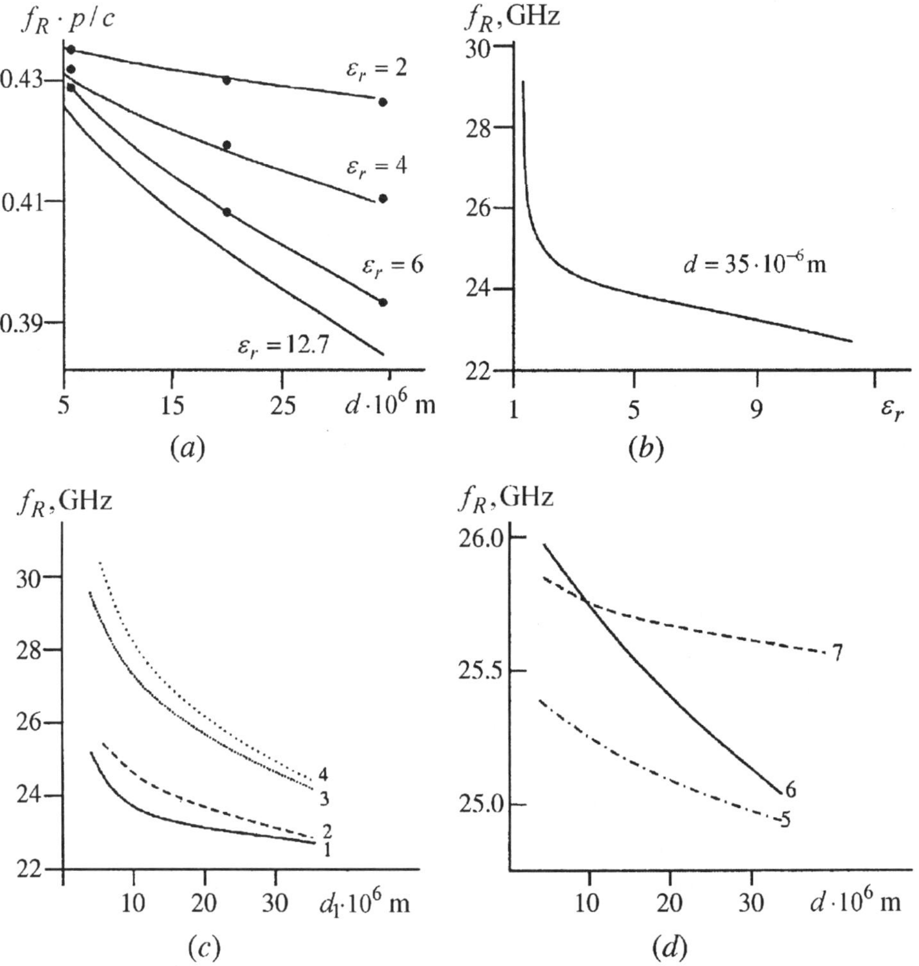

Fig. 11.4. *The central frequency dependences (a, c, d) on substrate layer thicknesses d, d_1 and (b) on the relative permittivity ε_r of the upper dielectric layer.*

11.4. Numerical Investigations of Three-Dimensional Semiconductor Structures Such as a Microwave Sensor

Our calculations were carried out for the structure (Fig.11.5). This structure is a semiconductor sensor. This one is a kind of dipole probe's sensitive element to detect microwave electric fields. The sensor is constructed as a symmetric vibrator made of two microstrip conductors with a semiconductor sample of cuboid form between them.

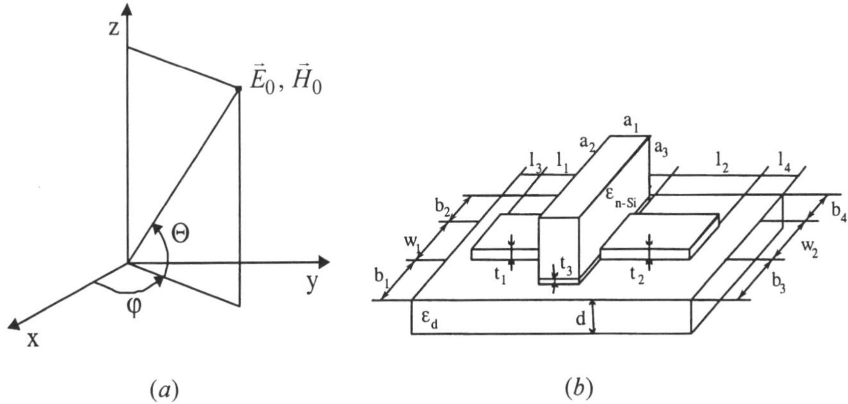

(a)   (b)

Fig. 11.5. *(a) polar coordinates; (b) the 3D structure of a sensor illuminated by microwaves* $E_0(\vec{r})e^{-ik\vec{r}}$ *and* $H_0(\vec{r})e^{-ik\vec{r}}$.

The sizes of the structure in Fig.11.5b are: a semiconductor cuboid sample $a_1 \times a_2 \times a_3 = (0.5 \times 1.25 \times 1.25) \cdot 10^{-3} \text{m}^3$, metal strip thicknesses are $t_1 = 5 \cdot 10^{-5}\text{m}$, $t_2 = 5 \cdot 10^{-6}\text{m}$, the width is $w_1 = w_2 = 10^{-3}\text{m}$ and the length is $l_1 = l_2 = 4 \cdot 10^{-3}\text{m}$. The dielectric substrate thickness is $d = 3 \cdot 10^{-4}$ m, with $b_1 = b_2 = b_3 = b_4 = 1.25 \cdot 10^{-3}\text{m}$, the whole length of the sensor is $l = l_1 + l_2 + l_3 + l_4 + a_1 = 8.502 \cdot 10^{-3}$ m and where $l_3 = l_4$. The semiconductor cuboid sample is placed in the hollow, which the depth is $t_3 = 5 \cdot 10^{-5}\text{m}$.

Fig.11.6 shows the dependence of the transmission coefficient module $|E^t|/|E_0|$ of the semiconductor sensor on the incidence angle θ for the linearly polarized plane wave coming at the angle of $\varphi = 60°$ and $90°$ $\varphi =$. Here φ and θ are the azimuthal and the polar angles respectively. In the transmission coefficient module the value $\vec{E}^t$ is the transmitted electric field into the semiconductor cuboid sample and $\vec{E}_0$ is the electric field of the incident wave.

The relative permittivity of the semiconductor cuboid sample is $\varepsilon_{n-Si} = 12.7 - i0.2$ and the relative permittivity of dielectric substrate is $\varepsilon_d = 5 - i0.01$. The sizes of the structure for all curves in Fig.11.6 are identical and all the curves correspond to the parallel polarization of the incident EM wave. Here data from [11.18] for a symmetric vibrator are presented as circles. The calculated results for the symmetric vibrator (curve 1) with the relative permittivities $\varepsilon_{n-Si} = \varepsilon_d = 1.05$ are presented by a dash line, which connects the circles. When relative permittivities of semiconductor and dielectric materials of the 3D structure (Fig.11.5b) became $\varepsilon_{n-Si} = \varepsilon_d = 1.05$, the metal strips was almost placed in free space. This converting of the relative permittivities allows us to transform our 3D structure into a symmetric microstrip vibrator, was placed in free space. Curves 2–4 show our calculations for the 3D structure (Fig.11.5b) when the relative permittivities are $\varepsilon_{n-Si} = 12.7 - i0.2$ and $\varepsilon_d = 5 - i0.01$.

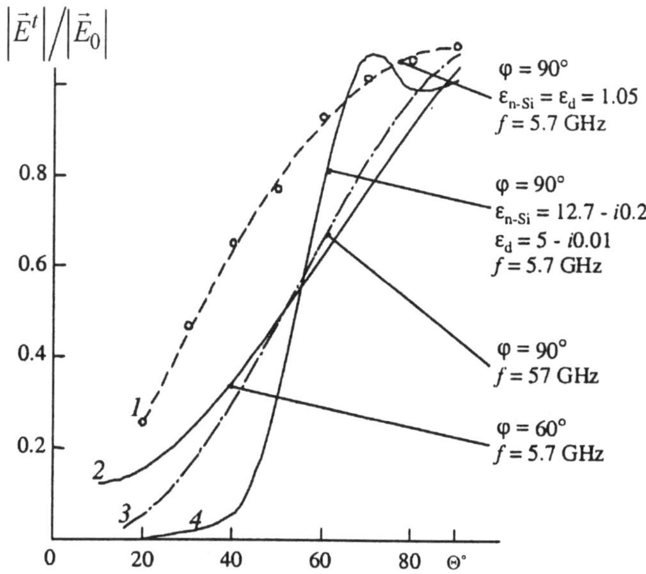

Fig. 11.6. *The transmission coefficient module* $\left|\vec{E}^t\right|/\left|\vec{E}_0\right|$ *dependence on polar angles* Θ *of the parallel– polarized incident EM wave.*

Curve 2 presents our calculated results for the structure under investigation when the frequency equals 5.7 GHz and the azimuthal angle is $\varphi = 60°$. The dot–and–dash line in Fig.11.6 (curve 3) shows our calculations at a frequency of $57GHz$, which is ten times greater then the frequency we used to calculate the other three curves. Curve 4 shows our calculation when all parameters were the same as curve 2, except with the azimuthal angle that was $\varphi = 90°$. We see that parameters for curve 4 were the same as curve 3 but the frequency $f = 5.7GHz$ was ten times less. One can see from curves 2–4 that there is a strong dependence on the frequency f and the azimuthal angle φ. Fig.11.6 shows the essential dependence of the semiconductor sensor transmission coefficient on the polar angle Θ of the incident microwave.

11.5. Electrodynamical Characteristics of a Three–Dimensional Human Heart Model

Certain microwave devices are now used in medical noninvasive diagnosis as well as in therapeutic treatment of heart patients. The EM wave source could be placed inside or outside of the heart. An example would be for the treatment of cardiac arrhythmias. The medical staff would use a microwave catheter device that is introduced inside the heart [11.14], [11.19] and [11.20]. The heart patient would then be under the influence of the EM waves and the treatment or diagnosis would begin. Our calculations were done when the source of the EM wave was placed either inside or outside of the heart model.

We numerically investigated electrodynamical characteristics of a heart model that was under microwave radiation. The diffraction and absorption of microwaves are governed by the complex permittivity properties of a heart model, shape and sizes. These electrodynamical characteristics depend on the polarization, amplitude and phase of the interacting microwaves.

Now we will present important expressions that will be useful in solving surface integrals in our numerical algorithm of the SIE method.

11.5.1. The surface integrals over triangular areas. The commercial computer program 3D Studio Max creates the heart model surface of triangles (Fig.11.1b). The position of a point in the plane can be given through two scalars α and β, and two nonparallel vectors a and b. In other words:

$\vec{r} = \alpha \, \vec{a} + \beta \, \vec{b}$, for $0 \le \alpha \le 1$ and $0 \le \beta \le 1$ when the point is inside the triangle OAB (Fig.11.7) as it follows from the similarity of the triangles OAB and Oab.

The triangle area built on vectors $d\vec{r} = \vec{a}d\alpha$ and $d\vec{r}' = \vec{b}d\beta$ is $dS = \left|\left[\vec{a},\vec{b}\right]\right|d\alpha d\beta$. The triangle area built on the vectors $\vec{a}$ and $\vec{b}$ is given by integral:

$$S = \int_{\Delta} dS = \left|\vec{a},\vec{b}\right| \int_{0}^{1} d\alpha \int_{0}^{\alpha} d\beta = \left[\left[\vec{a},\vec{b}\right]\right] \int_{0}^{1} \alpha d\alpha = \frac{1}{2}\left[\left[\vec{a},\vec{b}\right]\right]\alpha^{2}\Big|_{0}^{1} = \frac{1}{2}\left[\left[\vec{a},\vec{b}\right]\right].$$

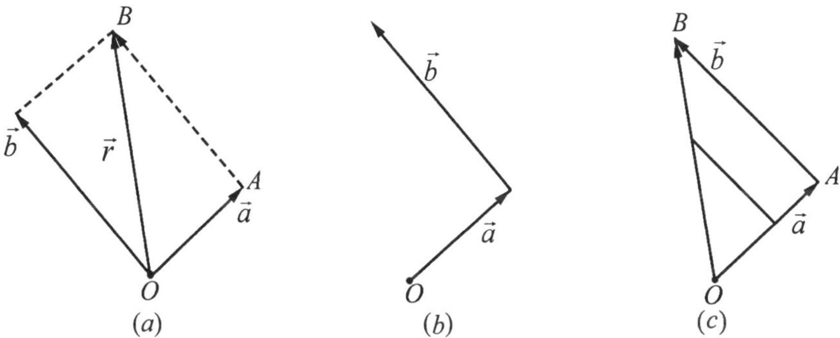

Fig. 11.7. *Description of the point position in a plane AOB.*

Integrating the function $f(x, y)$ over the triangle we have to write:

$$\int_\Delta f(x, y)ds = \int_0^1 \int_0^\alpha f(x_0 + \alpha a_x + \beta b_x, y_0 + \alpha a_y + \beta b_y)|\vec{a}, \vec{b}| d\alpha d\beta.$$

Similarly, in any coordinate system we have:

$$\int_\Delta f(x, y, z)dS = \left\|[\vec{r}_{21}, \vec{r}_{32}]\right\| \int_0^1 d\alpha \int_0^\alpha d\beta \ f(x_1 + \alpha(x_2 - x_1) +$$

$$+ \beta(x_3 - x_2), y_1 + \alpha(y_2 - y_1) + \beta(y_3 - y_2), \ z_1 + \alpha(z_2 - z_1) + \beta(z_3 - z_2)).$$

11.5.2. Electromagnetic wave scattering by a heart model. Here we have to solve a diffraction problem. We assumed that the permittivities ε_i and permeabilities μ_i of the relevant heart tissue and the environment are scalars, where $i = 1, 2, 3$. The heart model is surrounded by the surface S_3 (Fig.11.8). Inside of the heart model there are two atria with ventricles surrounded by the surfaces S_1 and S_2. The muscle region between the surfaces S_1, S_2 and S_3 is defined by the relative permittivity ε_1 and the relative permeability μ_1. We determined that two atria with ventricles were filled with blood characterized by the relative permittivities ε_2 and the relative permeability μ_2. We will assume that the heart is placed in an environment with the relative permittivity ε_3 and the relative permeability μ_3. Here we let $\varepsilon_3 = 1$ and $\mu_1 = \mu_2 = \mu_3 = 1$.

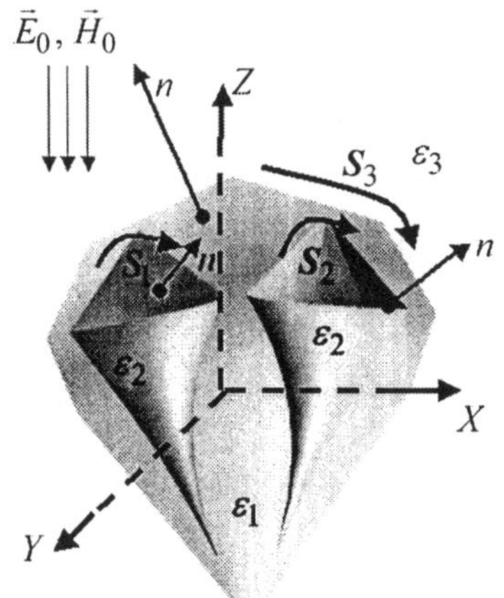

Fig. 11.8. The model of the heart with the myocardium, the left and right atria with ventricles that are limited by the surfaces S_1, S_2, S_3 and designations.

Now we will begin with a simple electrodynamical problem where an EM plane wave perpendicularly falls on a 3D model of the human heart [11.21]. The incident wave $\vec{E}_0(\vec{r})$, $\vec{H}_0(\vec{r})$ depends on time as $\exp(i\omega t)$ and here ω is the angular frequency. All surfaces of the heart are defined in the input file as a list of x, y, z coordinates of a triangular mesh on the heart's surface and normal unit vectors $\vec{n}$ at every triangle. Here the 3D heart model surface was created by use of the 3D StudioMax program. The external myocardium surface S_3 consisted of 42 triangles and the internal surfaces S_1 and S_2 (Fig.11.8) was constructed from 192 triangles.

The values of the real and imaginary parts of the relative permittivities for the myocardium $\dot{\varepsilon}_1 = \varepsilon_1' - i\varepsilon_1''$, as well as for the blood in the atria and ventricles $\dot{\varepsilon}_2 = \varepsilon_2' - i\varepsilon_2''$, are listed in Fig.11.9. The values $\dot{\varepsilon}_1$ and $\dot{\varepsilon}_2$ were taken from [11.22]–[11.24].

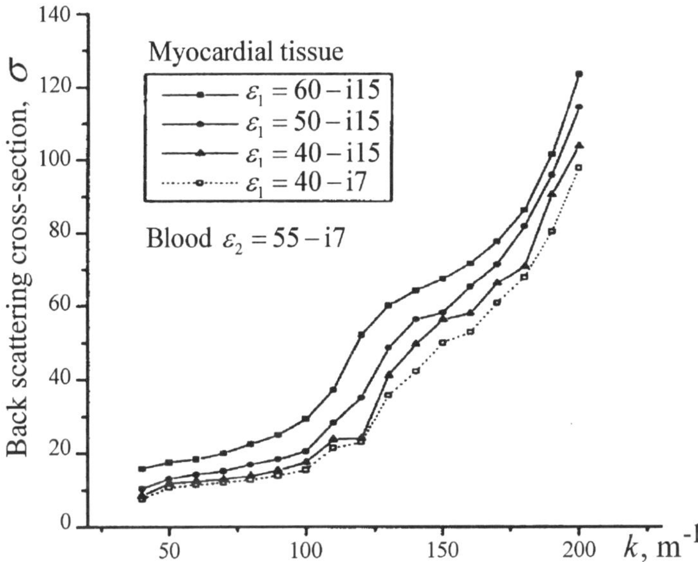

***Fig. 11.9.** Dependences of the back scattering cross–section for the heart model.*

Fig.11.9 shows the calculated dependences of the back scattering cross–section $\sigma = 4\pi R^2 |\vec{E}|^2 / |\vec{E}_0|^2$ (section 11.2) for a normal unit parallel–polarized plane wave $\vec{E}_0(\vec{r}), \vec{H}_0(\vec{r})$ with the wave number k (where $k = 2\pi/\lambda$ and λ is wavelength of the EM wave propagating into a vacuum). When microwaves fall on the top of the heart model surface some parts of the microwave reflect and some parts of it transmit into the heart. In our calculations the relative permittivity of the blood in the atria and ventricles was constant and equaled $\varepsilon_2 = 55 - i7$. We can see in Fig.11.9 that when the relative permittivity of the myocardium is changed from *40-i7* to *60-i15* (please look at the legend in Fig.11.9) the back scattering cross–section for the heart model increased remarkably.

Fig.11.10 shows the dependence of the calculated modulus of the normalized scattered electric field $|E_x|/|E_0|$ on the distance towards the axis *ox* (Fig.10.8).

The incident microwave $\vec{E}_0(\vec{r}), \vec{H}_0(\vec{r})$ is a normal unit parallel–polarized plane wave. The microwave falls on the top part of the heart model. As we see in Fig.11.10 there is a maximum of the scattered electric field value at the distance $x = 2.5 \cdot 10^{-2}$ m. The dependence of the scattered electric field on the distance towards the axis ox shows that the microwave energy is concentrate at certain points on the heart. The concentration of this microwave energy could be dangerous to the heart tissue. To avoid this situation it is necessary to change the microwave polarization or its azimuthal and polar angles.

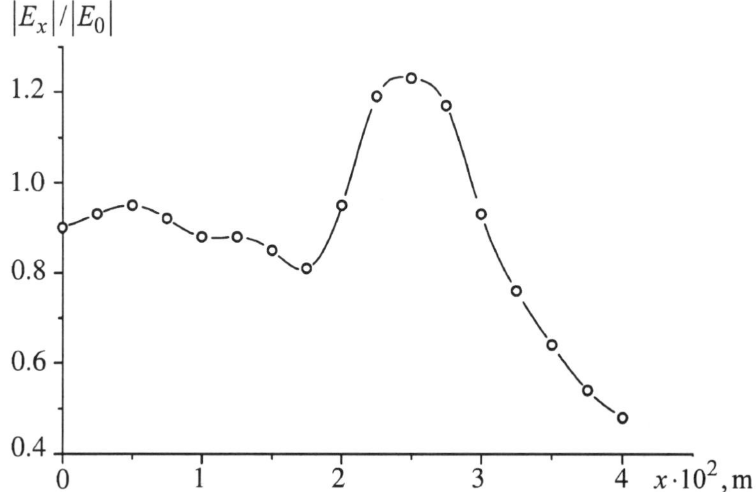

Fig. 11.10. *The normalized modulus of the scattered electric field as a function of the distance* x *for a parallel–polarized incident microwave at f=10 GHz, φ=30°, θ=0°, ε_1=40–i15, ε_2=55–i7.*

Fig.11.11 shows the dependence of the calculated normalized modulus of the scattered electric field $|E_y|/|E_0|$ on the distance towards the axis Ox. A normal unit perpendicular–polarized plane wave $\vec{E}_0(\vec{r}), \vec{H}_0(\vec{r})$ falls on the top of the heart. Comparisons of the last two figures show that a distribution of the electric field on the top of the heart along the axis ox is remarkably different depending on the polarization of the plane wave.

Fig.11.12 gives the dependence of the modulus of the normalized scattered electric field $|E_x|/|E_0|$ on the polar angles θ for a parallel–polarized incident microwave. We calculated the modulus of the scattered electric field at the top of the heart model where the coordinate y=0 (Fig.11.8). We see that changing of the polar angle θ allows us to change the value of the scattered electric field within a multitude of ranges.

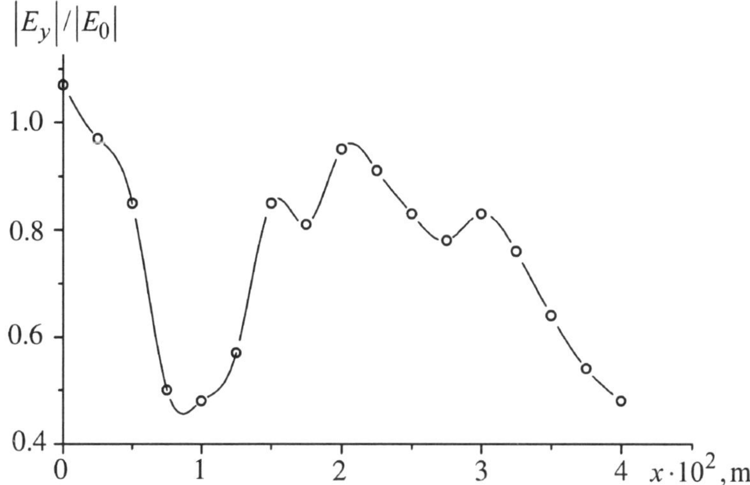

Fig. 11.11. *The normalized modulus of the scattered electric field as a function of the distance x for a perpendicular–polarized incident microwave at f =10 GHz, φ =30°, θ=0°, ε_1=40–i15, ε_2=55–i7.*

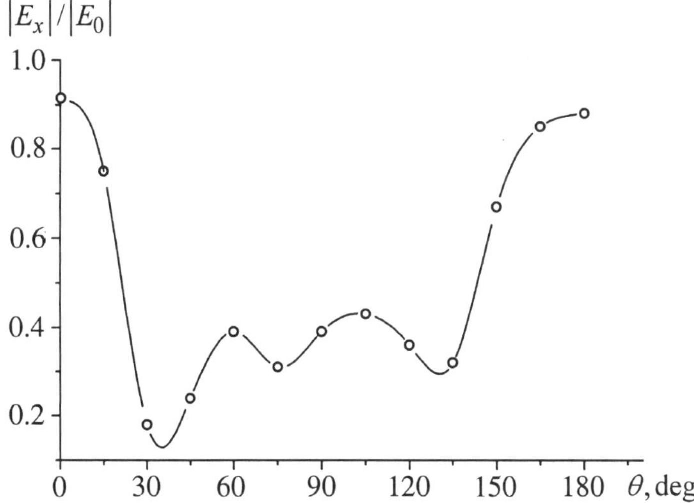

Fig. 11.12. *The normalized modulus of the scattered electric field as a function of polar angle θ for a parallel–polarized incident microwave at f=10GHz, φ =30°, ε_1=40–i15, ε_2=55–i7.*

11.5.3. The computation of electric fields on a heart model with a microwave catheter placed inside. Normally, electrical impulses flow throughout the heart in regularly measured rhythms. This electrical rhythm pattern is the basis for all heart muscle contractions. But at times these electrical flows get blocked or will travel the same pathways repeatedly creating something of a "short circuit", which creates a serious medical problem. These "short circuits" disturb normal heart rhythms in most all cases. At this present time the most effective treatment for this medical problem is to destroy the tissue causing the short circuit or abnormal heart rhythms. The procedure most commonly used is called microwave ablation. Microwave ablation is a non–surgical procedure and like most other cardiac procedures, no longer requires a full frontal chest opening and is a relatively non–invasive procedure. Microwave ablation is generally performed by making an incision in the chest, inserting a catheter and guiding it under the heart. Microwave energy is focused on the bottom side of the heart and heats from the bottom up. The most popular use of this procedure is to treat rapid heartbeats that begin in the upper chambers (atria) of the heart. Once a problem of this sort is discovered and confirmed, microwave energy is used to destroy small amounts of tissue causing the abnormal heart rhythms. Thus, ending the disturbance of electrical flow or short circuit throughout the heart and restoring a healthy and normal heart rhythm. Microwave energy works by use of dielectric heating and has a favorable thermal profile that permits such ablation. For many types of such arrhythmias, microwave ablation is successful and eliminates the need for open–heart surgery or long–term drug therapies [11.4], [11.5] and [11.19].

Section 11.5.2 we presented the calculated scattered electric field though use of a 3D human heart model when the point source of a microwave antenna was outside the boundary surfaces of the model. We did not take into account the shape and the placement of the antenna that radiated the microwaves on the heart model. Our computational algorithm was verified through the comparison of data given for the simplest diffraction problems Ref. [11.21]. When performing a microwave ablation, the microwave catheter is placed inside the heart in the medical procedure as well as in our theoretical algorithm. The heart model was composed of the myocardium as well as the right and left atria with ventricles that were filled with blood. The values of the real and imaginary parts of the relative permittivity for the myocardium and the blood were taken from [11.22]–[11.24]. The 3D surfaces of the heart model as well the catheter was created in the 3D StudioMAX. This tool exports the surfaces as a set of triangles with a normal vector on certain surface points. Here we have investigated the distribution of a modulus of the electric field at several parallel cross–sections of the heart model with the catheter inside of its right atrium (Fig.11.13)

All surfaces of the model heart are defined in input files as a list of x, y, z coordinates of a triangular mesh on the heart surfaces and normal unit vectors at every triangle. The external myocardium surface consisted of 42 triangles. The internal surface of the right and left atria with ventricles were constructed each having 48 triangles and the surface of the catheter consisted of 12 triangles. The values of the complex relative permittivities are: the myocardial $75 - i5$ and the blood in the cavities of the atria with ventricles $50 - i15$.

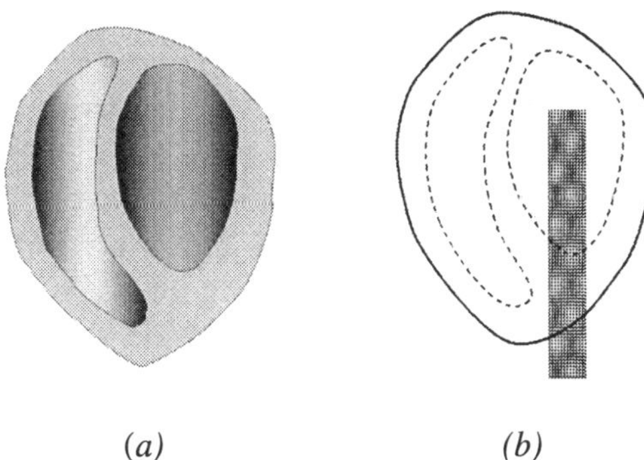

(a)   (b)

Fig. 11.13. (a) A schematic longitudinal section of the heart model consisting of the myocardium, the atria and ventricles; (b) the same heart model with a catheter (the shaded rod) inside of its right atria.

In our calculations the center of the Cartesian coordinate system is in the center of the heart. The maximum sizes of the heart model along the axis z is $12 \cdot 10^{-2}$ m and along axes x and y is $9 \cdot 10^{-2}$ m. In this coordinate system: the z-axis is upwards, x–axis is to the right and y–axis is away of the figure's plane (see Fig.11.8). In our theoretical approach we admit that the point source of the microwaves was localized at the top of the microwave catheter. The frequency of the microwave signal was f =10GHz. The distances between the source of the microwave and the surfaces of the heart regions (the atria with ventricles and an external heart surface) are small (only several wavelengths λ or less). The EM field nearer the zone of the catheter is more complicated. We have determined the hypothetical components of this radiated EM field by $E_x = E_y = 1$, $E_z = 0.5$ E_x, $H_x = H_y = 1$, $H_z = 0.5\ H_x$. We calculated the electric field modulus at the points inside the heart model for three parallel planes of its cross–sections. Fig.11.14 shows the dependences of the electric field modulus on the axis x in three cross–sections *xoy* of the heart model at $z = 2 \cdot 10^{-2}$ m, $z = 3 \cdot 10^{-2}$ m, $z = 5 \cdot 10^{-2}$ m and $y=0$ when the catheter is made of dielectric material. Every cross–section is at different distances (from the beginning of the coordinate system) in the direction toward the top of the heart. We see that the greatest value of the electric field modulus for the cross–section $z = 5 \cdot 10^{-2}$ m is when it is nearest to the tip of the dielectric catheter.

The heart model cross–section is an inhomogeneous structure. The structures regions consist of myocardium or blood substances with different relative permittivities. And microwaves reflected from the boundaries of these regions. We see that after the interference of reflected microwaves from the cross–section regions of the heart that the distribution of the electric field had resonance peaks. We also see (Fig.11.14) that resonance peaks at different cross–sections of the heart model have different locations and values. The two main reasons for this are: the microwave source distances are different on every cross–section and the contours that divide these different media at every cross-section are not the same. The value of the electric field become less when the distance is further from the tip of the microwave catheter because the heart regions consist of lossy media.

The curves in Fig.11.15 show the distribution of the electric field modulus on the x-axis at three cross–section $x0y$ of the heart model with a metal catheter inside. In our calculations we allowed the catheter to be made of perfect metal. This means that the electric field inside of the perfect metal was zero. We see that the distribution of the electric field modulus on the cross–section of the heart model with a metal catheter placed inside is different then that of a dielectric catheter. The value of the EM field modulus is concentrated along the side of the metal catheter tip. In Fig.11.15 the resonances were narrower then the same resonances for the heart model with a dielectric catheter (Fig.11.14). Distribution of the electric field modulus was quite different from the heart model with the dielectric and metal catheters nearer the catheters tips (at $z = 5 \cdot 10^{-2}$ m). We see that the electric field can concentrate in different places on the heart as well the width of the resonance peaks depend on the material of the catheter.

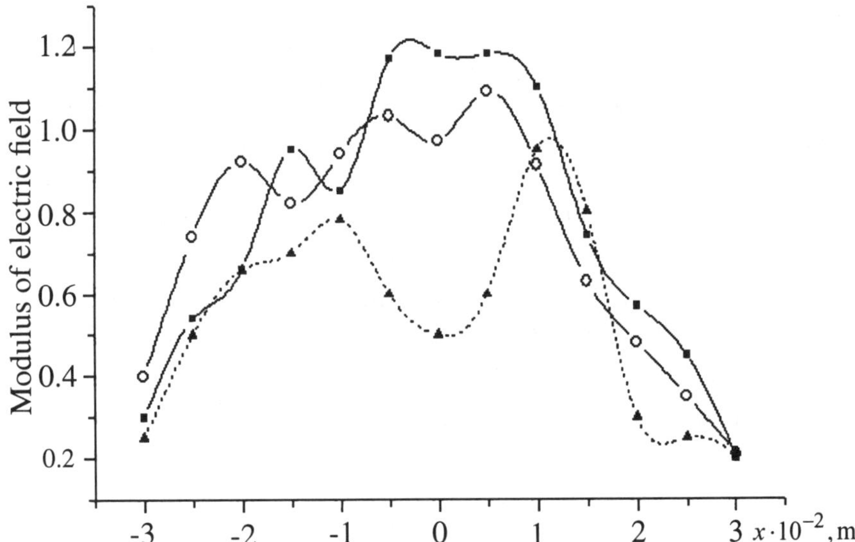

Fig. 11.14. *Distribution of the normalized electric field modulus on the x direction in three cross–sections of the heart model at the coordinates* $z = 2 \cdot 10^{-2}$ m *(triangles),* $z = 3 \cdot 10^{-2}$ m *(circles),* $z = 5 \cdot 10^{-2}$ m *(squares) and coordinate* $y=0$. *The microwave catheter is made of dielectric material with the relative permittivity of 10.*

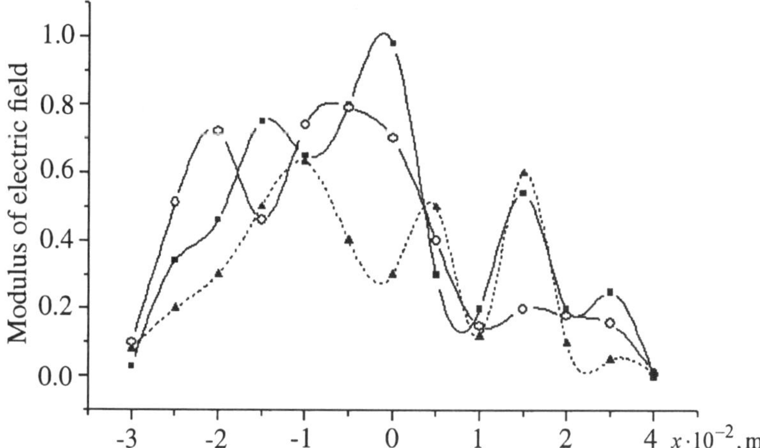

Fig. 11.15. *Distribution of the electric field modulus on the x direction in three cross-sections of the heart model at* $z = 2 \cdot 10^{-2}$ m *(triangles),* $z = 3 \cdot 10^{-2}$ m *(circles),* $z = 5 \cdot 10^{-2}$ m *(squares) and y=0 when the microwave catheter is made of the metal.*

We see that it is possible to concentrate the microwave electric field at different places inside the heart model depending on the material of the catheter. Also it is possible to regulate the width of the resonance peaks of the electric fields of microwaves. Our SIE method can be used to analyze the electrodynamical characteristics of hearts model as well as other organ models. Also our method could be used in obtaining characteristics when different medical devices are placed inside the heart.

References[*]

Chapter 2:

2.1. I. S. Gradshtein and I. M. Ryzhik, *The tables of integrals, sums, series and products*. GIFML, Moscow (1963) (*in Russian*).

2.2. E. Janke, F. Emde and F. Loesch, *Taffeln hoeherer funktionen*, B. G. Teubner Verlagsgelschaft, Stuttgard (1960).

2.3. H. Meinke and F. Gundah, *The reference book on radio engineering*, **1**, GEI, Moscow–Leningrad (1960) (*in Russian*).

2.4. I. I. Privalov, *Introduction into the complex variable function theory*, GITTL, Moscow–Leningrad (1945) (*in Russian*).

2.5. V. P. Ivashka and V. K. Shugurov, "The symmetric strip line with arbitrary finite number of central strips", *Lith. Phys. J.*, **13**, No. 5, 709–722 (1973).

Chapter 3:

3.1 V. P. Shestopalov, *The method of the Riemann–Hilbert problem in the diffraction and electromagnetic wave propagation theory*, Kharkov Univerity, Kharkov (1971) (*in Russian*).

3.2 L. Lewin, *Theory of waveguides,* Newnes–Butterworths, London (1975).

3.3 N. I. Muskhelishvili, *The singular integral equations*, Science, Moscow (1968) (*in Russian*)

3.4 F. D. Gakhov, *The boundary problems*, Science, Moscow (1977) (*in Russian*).

3.5 L. V. Kantorovich and V. I. Krylov, *The approximate methods of the higher analysis*, GIFML, Moscow (1962) (*in Russian*).

3.6 D. Mac–Cracken and U. Dorn, *The numerical methods and FORTRAN programming*, **1**, Peace, Moscow (1977) (*in Russian*).

3.7 V. I. Krylov, V. V. Bobkov and P. I. Monastyrny, *The computation methods,* **1**, Science, Moscow (1976) (*in Russian*).

3.8 R. Tuarson, *The sparse matrices*, Peace, Moscow (1977) (*in Russian*)

Chapter 4:

4.1 A. L. Mikaelyan, *Theory and application of the super high frequency ferrite*, GEI, Moscow (1963) (*in Russian*).

4.2 L. D. Landau and E. M. Lifschitz, "On the theory of the dispersion of magnetic permeability in ferromagnetic bodies", *Phys. Z. Sowjetunion*, **B. 8**, H. 2, S. 153–169 (1935).

---

[*] (L. Nickelson *formerly spelled as:* L.Kniševskaja / L. Knisevskaja / L. Knishevskaja / L. Knishevskaya)

4.3   R. K. Wangsness, "Magnetic resonance for arbitrary field strengths", *Phys. Rev.*, **98**, No. 4, 927–933 (1955).

4.4   F. Bloch, "Nuclear induction", *Phys. Rev.*, **70**, No. 7, 460–474 (1946).

4.5   R. K. Wangsness, "Magnetic resonance and minimum entropy production. Macroscopic equation of motion", *Phys. Rev.*, **104**, No. 4, 857–866 (1956).

4.6   N. S. Akulov, *Ferromagnetism*, GITTL, Moscow–Leningrad (1939) (*in Russian*).

4.7   E. A. Turov and A. I. Micek, "On theory of ferromagnetic anisotropy temperature dependence", *JETPh*, **37**, No. 4 (10), 1127–1132 (1959).

4.8   S. V. Vonsovsky, *Magnetism.* Science, Moscow (1971) (*in Russian*).

4.9   V. D. Krivchenkov and A. I. Pilshchikov, "Magnetostatic precession types in anisotropic sphere", *JETPh,* **43**, No. 2 (8), 573–580 (1962).

4.10 V. K. Shugurov, "The phase surfaces in the anisotropic ferrite", *Lith. Phys. J.*, **16**, No. 1, 89 – 95 (1976).

4.11 E. D. Palik and J. K. Furdina, "Infrared and microwave magnetoplasma effects in semiconductors", *Rep. Progr. Phys.*, **33**, 1193–1322 (1970).

Chapter 5:

5.1   I. M. Gelfand and G. E. Shilov, *The generalized functions and operations on them,* GIFML, Moscow (1958) (*in Russian*).

5.2   A. M. Lerer and S. M. Tsvetkovskaya, "The dispersion characteristics of the microstrip lines coupled through a slot", *J. of High Schools (USSR), Radioelectrinics*, **23**, No. 9, 10–16 (1980).

5.3   L. A. Vainshtein, *Electromagnetic waves,* Sovradio, Moscow (1957) (*in Russian*).

5.4   V. P. Ivashka, "Phase and energetic characteristics of coupled microstrip lines", *Lith. Phys. J.*, **20**, No. 2, p. 85–94 (1980).

5.5   V. P. Ivashka and V. K. Shugurov, "The strip line with the symmetry center", *Lith. Phys. J.*, **14**, No. 2, 285–295 (1974).

Chapter 6:

6.1   V. P. Ivashka and V. K. Shugurov, "The strip line with the nonuniform dielectric filling in the approximation of the transversal electromagnetic waves", *Lith. Phys. J.*, **14**, No. 3, 517–523 (1974).

6.2   V. P. Ivashka and V. K. Shugurov, "A line with conductors of arbitrary cross-section", *Lith. Phys. J.*, **14**, No. 3, 507–515 (1974).

6.3   V. P. Ivashka, L. V. Knishevskaya, Yu. P. Kolmakov, L. D. Selyanskaya and V. K. Shugurov, "The solution of the inverse problem for the strip lines", *Lith. Phys. J.*, **16**, No. 5, 695–702 (1976).

6.4   D. S. Denisov, B. V. Kondratyev, N. N. Lesnik, N. V. Lyapunov, I. I. Saprykin and V. M. Sedykh, *The strip lines and the super high frequency devices,* Kharkov University, Kharkov (1974) (*in Russian*).

6.5   E. Yamashita, "Variational method for the analysis of microstrip–like transmission lines", *IEEE Trans.*, **MTT–16**, No. 8, 529–535 (1968).

6.6   R. A. Pucel and D. J. Masse, "Microstrip propagation on magnetic substrates. Part I: Design theory", *IEEE Trans.*, **MTT–20**, No. 5, 304–308 (1972).

6.7   L. G. Maloratsky, *Microminiaturization of the SHF elements and device,* Sovradio, Moscow (1976) (*in Russian*).

6.8   N. Dreiper and G. Smith, *Applied regression analysis*, Statistics, Moscow (1975) (*in Russian*).

6.9   G. Jakimavichius, L. Knishevskaya, S. Gechauskas and J. Kalinauskas, "Analysis of electrodynamical characteristics of MSLs on semiconductor substrate", *Proceedings of Saratov University*, **13**, 29–34 (1988).

6.10   L. Knisevskaja, M. Tamosiuniene and V. Shugurov, "Numerical study of a MSL with a two–layer dielectric substrate of finite width", *Intern. J. Infrared and Millimeter Waves*, **23**, No. 2, 325–336 (2002).

6.11   G. E. Ponchak, A. Margomenos and L. P. B. Katehi, "Low–loss CPW on low – resistivity Si substrates with a micromachined polyimide interferce layer for RFIC interconnects", *IEEE Trans.*, **MTT–49**, No. 5, 866–870 (2001).

6.12   M. N. Khan, A. Gopinath, J. P. G Bristow and J. P. Donnelly, "Technique for Velocity–Matched Traveling–Wave Electrooptic Modulator in AlGaAs / GaAs", *IEEE Trans.*, **MTT–41**, No. 2, 244–249 (1993).

6.13   R. G. Hunsperger, *Integrated Optics: Theory and Technology*, Springer–Verlag, Berlin and etc. (1991).

6.14   I. J. Bahl, "Application Notes. High-Q and Low-Loss Matching Network Elements for RF and Microwave Circuits", *IEEE Microwave Magazine*, **1**, No. 3, 64–73 (2000).

6.15   L. Knisevskaja, M. Tamosiuniene and V. Shugurov, "MSL on GaAs substrate", *Lith. Phys. J.*, **41**, No. 3, 247–250 (2001).

6.16   L. Knisevskaja, M. Tamosiuniene, V. Shugurov and V. Tamosiunas, "MSL on two–layer GaAS and SiO_2 substrate", *Lith. Phys. J.*, **41**, No. 3, 251–255 (2001).

6.17   J. Zheng, Y.–C. Hahm, V. K. Tripathi, A. Weisshaar, "CAD–oriented equivalent–circuit modeling of on–chip interconnects on lossy silicon substrate", *IEEE Trans.*, **MTT–48**, No. 9, 1443–1451 (2000).

6.18   L. Knisevskaja, M. Tamosiuniene, K.–F. Berggren and B. A. Galwas, Investigation of MSLs on two–layered semiconductor–dielectric substrate, *14th Intern. Conf. on Microwaves, Radar and Wireless Communications, MIKON2002, Poland, Gdańsk, May 20–22, 2002, Conf. Proceedings*, **2**, 480–483 (2002).

6.19   F. Schnieder and W. Heinrich, "Model of Thin–Film Microstrip Line for Circuit Design", *IEEE Trans.*, **MTT–49**, No. 1, 104–110 (2001).

6.20   L. Knisevskaja, M. Tamosiuniene, V. Tamosiunas and V. Sugurov, "Numerical investigations of MSLs with semiconductor and dielectric substrates", *High frequencies technology. Microwave. Electronics and electrical engineering, Technologija, Kaunas*, **44**, No. 2, 22–25 (2003).

6.21   E. Voges and A. Neyer, "Integrated–optic devices on LiNbO$_3$ for optical communication", *J. of Lightwave Technology*, **LT-5**, No. 9, 1229–1236 (1987).

6.22   H. Wulf and E. Voges, "Modeling and numerical simulation of electro–optic devices on LiNbO$_3$", *SPIE, Electro–Optic and Magneto–Optic Materials 2*, **1274**, 27–33 (1990).

6.23   I. Burneika, L. Knishevskaya and V. Shugurov, "The rigorous method for numerical analysis of electrodynamic characteristics of the open uniaxial anisotropic waveguide with an arbitrary cross–section", *Lith. Phys. J.*, **33**, No. 4, 179–185 (1993).

6.24   I. Burneika, J. McGeehan, L. Knishevskaya and G. Labutis, "Analysis of open microstrip lines on finite width anisotropic–dielectric substrates", *Lith. Phys. J.*, **32**, No 3, 353–358 (1992).

6.25   L. Knisevskaja, K.–F. Berggren, M. Tamosiuniene and V. Shugurov, "Computational Analysis of the MSLs on Lossy LiNbO$_3$ and LiTaO$_3$ substrates", *J. of Electrical Engineering*, **53**, No. 9/S, 57–60 (2002).

Chapter 7:

7.1   V. N. Korovin and G. G. Shishkin, "Nonreciprocal SHF devices using semiconductor plasma", *Foreign radioelectronics*, No. 3, 93–122 (1975).

7.2   L. V. Knishevskaya, A. K. Laurinavichius and V. K. Shugurov, "Investigation of the microstripline on the longitudinally magnetized layer semiconductor-dielectric substrate", *Lith. Phys. J.*, **21**, No. 5, 27–34 (1981).

7.3   L. V. Knishevskaya and V. K. Shugurov, "The coupled microstrip lines on the longitudinally magnetized semiconductor substrates", *Lith. Phys. J.*, **22**, No. 1, 71–78 (1982).

7.4   L. V. Knishevskaya and V. K. Shugurov, "The nonsimilar strip lines on the longitudinally magnetiz.ed semiconductor substrate", *Lith. Phys. J.*, **22**, No. 2, 35–39 (1982).

7.5   L. V. Knishevskaya, M. N. Kotov, A. K. Laurinavichius and V. K. Shugurov, "The characteristics' particularities of the microstrip line on the longitudinally magnetized semiconductor substrate", *In the book: Physics, Technology and Manufacture of the semiconductor devices, Materials of the republican conf., April 10–11, 1980, Vilnius*, **1**, 74–76 (1980).

7.6   L. V. Knishevskaya, M. N. Kotov, A. K. Laurunavichius and V. K. Shugurov, "Analysis of the microstrip line on the longitudinally magnetized semiconductor substrate", *Lith. Phys. J.*, **20**, No. 6, 47–55 (1980).

7.7   J. K. Furdina, "SHF–interferometer for the investigation of the helicon waves in semiconductor plasma", *Devices for the research*, No. 4, 75–81 (1966).

7.8   V. V. Nikolsky and T. I. Nikolskaya, "A stretch of the system of the coupled strip lines having different widths", *RaE*, **26**, No. 2, 277–285 (1981).

7.9   N. D. Maliutin, "Matrix parameters of the nonsimilar coupled strip lines with nonuniform dielectric", *RaE*, **21**, No. 12, 2473–2478 (1976).

7.10 R. A. Speciale, "Even and odd waves for nonsymmetrical–coupled lines in nonhomogeneous media", *IEEE Trans.*, **MTT–23**, No. 11, 897–908 (1975); R. K. L. Poon "Capacitance of a microstrip of unequal widths in a homogeneous medium", *Elec. Lett.*, **15**, No. 2, 44–45 (1979).

7.11 M. V. Amitin, A. S. Elifterev, T. M. Kazanskaya, Y. A. Lauchius, V. K. Shugurov and G. K. Yakovlev, "Investigation of the microstrip meander line on the ferrite substrate", *RaE*, **23**, No. 8, 1594–1603 (1978).

7.12 J. J. Green and F. Sandy, "Microwave characterization of partially magnetized ferrites", *IEEE Trans.*, **MTT–22**, No. 6, 641–645 (1974).

7.13 A. K. Stolyarov and V. S. Smirnov, "Magnetoacoustic phenomena in microstrip line with ferrite", *RaE*, **16**, No. 2, 434–437 (1971).

7.14 N. A. Semenov, *Technical electrodynamics*, Swiaz, Moscow (1973) (*in Russian*).

7.15 R. A. Pucel, D. J. Masse and C. P. Hartwig, "Losses in microstrip", *IEEE Trans.*, **MTT–16**, No. 6, 342–350 (1968).

7.16 M. E. Hines, "Reciprocal and nonreciprocal modes of propagation in ferrite strip line and microstrip devices", *IEEE Trans.*, **MTT–19**, No. 5, 442–451 (1971).

7.17 N. Krause, "Ein Verfahren zur Berecnung der Dispersion einer Mikrostreifenleitung auf gyrotropem Substrat", *AEÜ,* **B. 31**, H. 5, 205–211 (1977).

7.18 H. Ermert, "Guided modes and radiation characteristics of covered microstrip lines", *AEÜ*, **B. 30**, H. 2, 65 (1976).

7.19 J. Borburgh, "The behaviour of guided modes on the ferrite filled microstrip line with the magnetization perpendicular to the ground plane", *AEÜ*, **B. 31**, H. 2, 73–77 (1977).

7.20 Yu. N. Polukhin, N. M. Galdina and V. P. Brovyakov, "Analysis and experimental investigation of the fields in the strip transmission lines", *Proceedings of the 4–th Intern. Conf. on gyromagnetic electronics and electrodynamics. Warszawa*, 104 (1978).

7.21 A. S. Elifter'ev, J. A. Lauchius and V. K. Shugurov, "Microstrip line on the ferrite substrate at the transversal magnetization in the approximation of harmonic functions", *Lith. Phys. J.*, **20**, No. 3, 23–30 (1980).

7.22 V. V. Nikolsky and A. V. Druzhinin, "On influence of the asymmetrical strip line conductor thickness on the main wave propagation constant magnitude", *RaE*, **23**, No. 12, 2628–2632 (1978).

7.23 V. V. Nikolsky and A. V. Druzhinin, "Determination of the strip line eigenwaves taking into account the conductor thickness", *RaE*, **22**, No. 7, 1331–1340 (1977).

7.24 A. B. Dalby, "Interdigital microstrip circuit parameters using empirical formulas and simplified model", *IEEE Trans.*, **MTT–27**, No. 8, 744–752 (1979).

7.25　P. Reinhold and N. Stephen, "Calculation expression", *AEÜ*, **B. 29**, H. 3, 125–128 (1975).

7.26　E. Yamashita and K. Atsuki, "Analysis of thick–strip transmission lines", *IEEE Trans.*, **MTT–19**, No. 1, 120–122 (1971).

7.27　L. V. Knishevskaya and V. K. Shugurov, "Eigenwaves of the coupled microstrip line on the combined substrate having the longitudinally magnetized dissipative semiconductor layer", *Lith. Phys. J.*, **22**, No. 3, 27–36 (1982).

7.28　L. V. Knishevskaya and V. K. Shugurov, "Theoretical analysis of microstrip lines on layered substrate having longitudinally magnetized semiconductor", *Proceedings of the XII Europ. Microwave Conf. 82. Helsinki*, 753–758 (1982).

7.29　L. V. Knishevskaya and V. K. Shugurov, "Analysis of the electric field in the microstrip line on the ε–gyrotropic substrate", *In the book: the 5–th all union scientific–technical conf. Metrology in Radioelectronics, Moscow*, 298–301 (1981).

7.30　L. V. Knishevskaya and V. K. Shugurov, "Analysis of electromagnetic field of microstrip lines on ε– and μ–gyrotropic substrates", *Proceedings of the XIth Europ. microwave conf., Amsterdam, Netherlands, 7–11 September 1981*, 660–664 (1981).

7.31　L. V. Knishevskaya and V. K. Shugurov, "Microstrip lines on the dissipative substrates having double (ε, μ)–gyrotropy", *Proceedings of the VI Intern. conf. on gyromagnetic electronics and electrodynamics. Sofia*, 56–62 (1982).

7.32　A. R. Van de Capelle, P. J. Luypaert and K. U. Leuven, "A complete analysis of the discrete mode spectrum of open microstrip transmission lines", *IEEE MTT–S Int. Microw. Symp. Dig.*, 413–415 (1978).

Chapter 8:

8.1　L. V. Knishevskaya and D. J. Shegzhda, "Algorithm of computer analysis of the microstrip line on semiconductor substrates dispersion characteristics", *Mathematical and computer modelling in microelectronics. Proceedings of the III republican school–seminar. Palanga*, 95–100 (1986).

8.2　L. V. Knishevskaya and D. J. Shegzhda, "The rigorous solution of Maxwell's equations for the investigation of the open microstrip line eigenwaves spectrum by means of singular integral equations", *Lith. Phys. J.*, **26**, No. 2, 147–154 (1986).

8.3　A. N. Fedorov, N. N. Levina and G. S. Shchukina, "Eigenwaves of the open slot line", *RaE*, **26**, No. 7, 1414–1419 (1981).

8.4　A. N. Fedorov, N. N. Levina and N. A. Khametova, "Some results of the numerical investigation of the slot line and of the strip–slot lines", *RaE*, **28**, No. 7, 1284–1292 (1983).

8.5　L. V. Knishevskaya and M. M. Jarmalis, "Investigation of the open symmetric and asymmetric slot lines", *Electronic engineering, SHF Electronics*, **394,** No. 10, 18–22 (1986).

8.6 L. V. Knishevskaya and M. Jarmalis, "Theoretical and experimental analysis of open slot lines on semiconductor substrate with symmetrically and asymmetrically located conductors", *8ᵗʰ Colloquium on Microwave Communication. Budapest. Hungary*, 345–347 (1986).

8.7 D. A. Usanov and V. E. Orlov, "Coaxial connector of the rectangular waveguide and the transmission slot line", *Electronic engineering, SHF Electronics*, **323**, No. 11, 15–17 (1980).

8.8 D. A. Usanov, V. E. Orlov and A. A. Bezmenov, "The frame coupling elements in a waveguide", *Electronic engineering, SHF Electronics*, No. 3, 37–42 (1977).

8.9 J. Burneika, L. V. Knishevskaya and G. Labutis, "Investigation of the open micron sizes slot lines", *Proceedings of the conf. Physical electronics. Development of technical science in the republic, Kaunas*, 103 (1987).

8.10 M. Ikeuchi, M. Sawami and H. Niki, "Analysis of open–type dielectric waveguides by the finite–element iterative method", *IEEE Trans.*, **MTT–29**, No. 3, 234–239 (1981).

8.11 R. V. Gelsthorpe, N. Williams and N. M. Davey, "Dielectric waveguide; a low–cost technology for millimetre wave integrated circuits", *The Radio and Electronic Engineering*, **52**, No. 11–12, 522–528 (1982).

8.12 V. A. Karpenko, "The method of the rectangular dielectric waveguide computation", *RaE*, **29**, No. 8, 1494–1501 (1984).

8.13 V. A. Karpenko, Y. D. Stolyarov and V. F. Kholomeev, "Theoretical and experimental investigation of the rectangular dielectric waveguide", *RaE*, **25**, No. 1, 51–57 (1980).

8.14 E. A. J. Marcatili, "Dielectric rectangular waveguide and directional coupler for integrated optics", *The Bell System Technical Journal USA*, **48**, No. 7, 2071–2102 (1969).

8.15 K. Solbach and I. Wolff, "The electromagnetic field and the phase constants of dielectric image lines", *IEEE Trans.*, **MTT–26**, No. 4, 266–274 (1978).

8.16 L. V. Knishevskaya, F. H. Mukhametzyanov, A. N. Puzakov, V. K. Shugurov and G. K. Yakovlev, "The propagation of the electromagnetic waves in an open circular longitudinally magnetized ferrite waveguide", *Lith. Phys. J.*, **26**, No. 3, 296–304 (1986).

8.17 L. V. Knishevskaya, F. H. Mukhametzyanov, A. N. Puzakov, L. Shevlyakov and V. K. Shugurov, "The hybrid modes with odd azimuth number of the open circular longitudinally magnetized gyrotropic waveguide", *Electronic engineering, SHF Electronics*, No. 9, 6–12 (1985).

8.18 L. V. Knishevskaya and V. K. Shugurov, "The theoretical investigation of an open circular longitudinally magnetized layer gyrotropic–dielectric waveguide", *Lith. Phys. J.*, **28**, No. 3, 358–363 (1988).

Chapter9:

9.1 D. Polder, "On the theory of ferromagnetic resonance", *Phil. Mag.*, **40**, No. 300, 99–115 (1949).

9.2 P. Epshtejn, "Theory of propagation of the electromagnetic waves in a gyrotropic mcdia", *Uspechi fizicheskih nauk,* Moscow, Vol. 65, N 2, p. 283 – 311 (1958); A. G. Gurevich, *Super high frequency ferrites,* Fizmatgiz, Moscow (1960) (*in Russian*).

9.3 J. A. Kong, *Electromagnetic wave theory*, John Wiley & Sons, Inc., New York (1990).

9.4 G. T. Rado, "Theory of the microwave permeability tensor and Faraday effect in nonsaturated ferromagnetic material", *Phys. Rev.*, **89**, No. 2, 529 (1953).

9.5 *The reference book on design and construction of the SHF strip devices.* Edited by V. I. Volman, Radio and communication, Moscow (1982) (*in Russian*).

9.6 U. Crombach, "Analysis of single and coupled rectangular dielectric waveguides", *IEEE Trans.*, **MTT–29**, No. 9, 870–874 (1981).

9.7 L. V. Knishevskaya and V. K. Shugurov, "Computation of the dispersion characteristics of main and higher modes of open rectangular ferrite waveguide with metallized faced", *Proceeding of 7 Intern. conf. on microwave ferrites, former ČSSR,* 104–108 (1984).

Chapter 10:

10.1 S. W. Yun and T. Itoh, "Nonreciprocal wave propagation in a hollow image waveguide with a ferrite layer", *Proc. IEE*, **132**, pt. M, No. 2, 222–226 (1985).

10.2 S. W. Yun and T. Itoh, "A novel millimetre–wave isolator", *Proc. 13th European microwave conf., Nuremberg, West Germany*, 174–178 (1983).

10.3 M. Muraguchi and Y. Narbo, "Nonreciprocal device on open–boundary structures for millimetre–wave integrated circuits", *IECE Jap.Trans. B*, **64**, 604–611 (1981).

10.4 L. Knishevskaya and V. Shugurov, "Fundamental solution of waveguide Maxwell equations for transversely magnetized bigyrotropic medium", *Lith. Phys. J.*, **27**, No. 2, 172–183 (1987).

10.5 L. B. Felsen and N. Marcuvitz, *Radiation and Scattering of Waves*, Mir, Moscow (1978).

Chapter 11:

11.1 A. G. Roeder, N. E. Jensen and G. A. E. Crone, "Some European Satellite–Antenna Developments and Trends", *IEEE Trans. Antennas Propagation*, **38**, 9–21 (1996).

11.2 G. S. Smith, "Analysis of Miniature Electric Field Probes with Resistive Transmission Lines", *IEEE Trans. Microwave Theory and Techniques*, **MTT–29**, 1213–1224 (1981)

11.3 Y. Gao and I. Wolff, "A New miniature magnetic field probe for measuring three–dimensional fields in planar high–frequency circuits", *IEEE Trans.*, **MTT–44**, 911–925 (1996).

11.4 *New Fronts in Medical Device Technology.* Edited by A. Rosen and H. D Rosen, John Wiley and sons Inc., N.Y. and etc. (1995).

11.5 J. C. Lin, "Electromagnetics for the Heart", *IEEE Antenna's and Propagation Magazine*, **44**, No. 4, 103–105 (2002).

11.6 E. R. Adair and R. C. Petersen, "Biological Effects of Radio–Frequency/ Microwave Radiation", *IEEE Transactions*, **50**, No. 3, 953–996 (2002).

11.7 *Medical applications of microwave imaging.* Edited by L. E. Larsen and J. H Jacobi, IEEE Press Inc., N.Y. (1986).

11.8 *Biological effects of electromagnetic radiation.* Edited by J. M. Osepchuk, IEEE Press Inc., N.Y. (1984).

11.9 S. C. Mitchell, B. P. F. Lelieveldt, R. J. van der Geest, H. G. Bosch, J. H. C. Reiber and M. Sonka, "Multistage Hybrid Active Appearance Model Matching: Segmentation of Left and Right Ventricles in Cardiac MR Images", *IEEE Transactions on medical imaging*, **20**, No. 5, 415–423 (2001).

11.10 A. F. Frangi, W. J. Niessen and M. A. Viergever, "Three–Dimensional Modeling for Functional Analysis of Cardiac Images: A Review", *IEEE Trans. on medical imaging*, **20**, No. 1, 2–24 (2001).

11.11 D. Bleecker and G. Csordas, *Basic Partial Differential Equations*, VNR, N.Y. (1992).

11.12 P. Dennery and A. Krzywicki, *Mathematics for physicists*, Dover Publications, Mineola–N.Y. (1996).

11.13 L. Knisevskaja, K.–F. Berggren, V. Engelson and V. Shugurov, "Singular Integrals Method for Numerical Investigations of Microwave Scattering from Dielectric 3D–Structures", *Journal of Electrical Engineering*, **53**, No. 9/S, 11–14 (2002).

11.14 R. W. P. King and T. T. Wu, *The scattering and diffraction of waves,* Harvard University Press, Cambridge, Massachusetts (1959).

11.15 C. C. H. Tang, *Electromagnetic back–scattering from thin circular disks by a microwave pulse method. Scientific Report No 15*, Cruft Laboratory, Harvard University, Cambridge, Massachusetts (1958).

11.16 L. Knishevskaya, V. Tamoshiunas, O. Smertin, M. Tamoshiuniene and V. Shugurov , "The numerical analysis of diffraction characteristics of the microstrip reflecting 3–D structures", c*onf. Digest of 4th Intern. conf. on Millimeter and Submillimeter Waves and Applications, July 20–23 1998, San Diego, California, USA*, 152–153 (1998).

11.17 P. V. Sokolov, "Two dimensional periodic structure geometry influence on its frequency–selective properties", *Radiotechnica, Moscow*, **14**, No. 7, 31–34 (1996).

11.18 G. Z. Ajzenberg, S. P. Belousov, E. M. Zurbenko, G. A. Kliger and A. G. Kurashov, *VHZ Antennas*, Radio and Sviaz, Moscow (1985).

11.19 A. J. Greenspon, "Advances in Catheter Ablation for the Treatment of Cardiac Arrhythmias", *IEEE Trans.*, **MTT–48**, No. 12, 2670–2675 (2000).

11.20 C. Rappaport, "Treating Cardiac Disease with Catheter–Based Tissue Heating", *IEEE Microwave Magazine. Biomedical Applications*, **3**, No. 1, p. 57–64 (2002).

11.21 L. Knisevskaja, K.–F. Berggren, V. Engelson and B. A. Galwas, "Microwave Scattering by a Three–Dimensional Model Human Heart", *14th Intern. conf. on Microwaves, Radar and Wireless Communications, MIKON2002, Poland, Gdańsk, May 20–22, 2002, Conf. Proceedings, Gdansk, Telecommunications Research Institute*, **1**, 107–110 (2002).

11.22 F. A. Duck, *Physical Properties of Tissue*, Academic Press, London (1990).

11.23 M. Alison and R. J. Sheppard, "Dielectric properties of human blood at microwave frequencies", *Phys. Med. Biol.*, **38**, 971–978 (1993).

11.24 S. Gabriel, R. W. Lau and C. Cabriel, "The dielectric properties of biological tissues", *Phys. Med. Boil.*, **41**, 2251–2293 (1996).

APPENDIXES

A–1. THE ELECTROMAGNETIC SPECTRUM

Acronym	Classification	Frequency Range, f	Wavelength
THF	Tremendously High Freq.	300–3000 GHz	1 mm – 100 μm
EHF	Extremely High Freq. *(Millimetric waves/Microwaves)*	30 – 300 GHz	1 cm – 1 mm
SHF	Super High Freq. *(Centimetric waves/Microwaves)*	3 – 30 GHz	10 – 1 cm
UHF	Ultra High Freq. *(Decimetric wave)*	300 – 3000 MHz	1 m – 10 cm
VHF	Very High Freq. *(Metric waves)*	30 – 300 MHz	10 – 1 m
HF	High Freq. *(Decametric waves)*	3 – 30 MHz	100 – 10 m
MF	Medium Freq. *(Hectometric waves)*	300 – 3000 kHz	1 km – 100 m
LF	Low Freq. *(Kilometric waves)*	30 – 300 kHz	10 km – 1 km
VLF	Very Low Freq. *(Myriametric waves)*	3 – 30 kHz	100 – 10 km
ULF	Ultra Low Freq.	300 – 3000 Hz	$10^3 - 10^2$ km
SLF	Super Low Freq.	30 – 300 Hz	$10^4 - 10^3$ km
ELF	Extremely Low Frequency	3 – 30 Hz	$10^5 - 10^4$ km
TLF	Tremendously Low Freq.	< 3 Hz	$10^6 - 10^5$ km

A–2. SOME BASIC EQUATIONS FOR ELECTRODYNAMICS

Fundamental Electromagnetic Field Quantities

Field Quantity	Symbol	Unit
Electric field strength (or intensity)	$\vec{E}$	V/m (volt/meter)
Electric induction (Electric flux density, Electric displacement)	$\vec{D}$	C/m^2 (coulomb/meter^2)
Magnetic induction (Magnetic flux density)	$\vec{B}$	T (tesla)
Magnetic field strength (or intensity)	$\vec{H}$	A/m (Ampere/meter)
External constant magnetic field strength (or intensity)	$\vec{H}_0$	A/m (Ampere/meter)

Boundary Conditions between two linear dielectric materials:

$$E_{t1} = E_{t2} \qquad \frac{D_{t1}}{D_{t2}} = \frac{\varepsilon_1}{\varepsilon_2}$$

$$H_{t1} = H_{t2} \qquad \frac{B_{t1}}{B_{t2}} = \frac{\mu_1}{\mu_2}$$

$$D_{n1} = D_{n2} \qquad \varepsilon_1 E_{n1} = \varepsilon_2 E_{n2}$$

$$B_{n1} = B_{n2} \qquad \mu_1 H_{n1} = \mu_2 H_{n2}$$

Boundary Conditions between linear dielectric material (media 1) and a perfect metal (media 2):

1. On the side of perfect metal: $E_{t2} = H_{t2} = D_{n2} = B_{n2} = 0,$

2. On the side of dielectric: $E_{t1} = 0$, $\vec{a}_{n2} \cdot \vec{D}_1 = \rho_s$, $B_{n1} = 0$, $\vec{\alpha}_n \times \vec{H}_1 = J_s$,

where a_{n2} is the outward unit normal vector from medium 2 at the interface; ρ_s is electric charge density (coulombs/m^3); J_s is electric current density (amperes/m^2).

A–3. VECTOR DIFFERENTIAL OPERATORS

Rectangular coordinates (x,y,z):
$grad = \nabla$ (gradient); ∇ is a linear operator

$$\nabla\left(a\Phi+b\Psi\right)=a\nabla\Phi+b\nabla\Psi$$

$$\nabla\cdot\left(a\vec{D}+b\vec{G}\right)=a\nabla\cdot\vec{D}+b\nabla\cdot\vec{G}$$

$$\nabla\times\left(a\vec{D}+b\vec{G}\right)=a\nabla\times\vec{D}+b\nabla\times\vec{G}$$

$$\nabla\Phi=\hat{x}\frac{\partial\Phi}{\partial x}+\hat{y}\frac{\partial\Phi}{\partial y}+\hat{z}\frac{\partial\Phi}{\partial z}$$

$$div\vec{D}=\nabla\cdot\vec{D}\;;\quad \nabla\cdot\vec{D}=\frac{\partial D_x}{\partial x}+\frac{\partial D_y}{\partial y}+\frac{\partial D_z}{\partial z}$$

$$curl\,\vec{H}=\nabla\times\vec{H}$$

$$\nabla\times\vec{H}=\hat{x}\left(\frac{\partial H_z}{\partial y}-\frac{\partial H_y}{\partial z}\right)+\hat{y}\left(\frac{\partial H_x}{\partial z}-\frac{\partial H_z}{\partial x}\right)+\hat{z}\left(\frac{\partial H_y}{\partial x}-\frac{\partial H_x}{\partial y}\right)$$

$$\nabla\cdot\nabla=\nabla^2;\;\nabla^2=\Delta\;;\;\nabla\cdot\nabla\Phi=\Delta\Phi$$

$$\nabla^2\Phi=\frac{\partial^2\Phi}{\partial x^2}+\frac{\partial^2\Phi}{\partial y^2}+\frac{\partial^2\Phi}{\partial z^2}$$

$$\nabla\cdot\nabla\times\vec{D}=0\;;\quad \nabla\times\nabla\Phi=0$$

$$\nabla\times\left(\nabla\times\vec{D}\right)=\nabla\left(\nabla\cdot\vec{D}\right)-\nabla^2\vec{D}$$

$$\nabla\left(\Phi\Psi\right)=\left(\nabla\Phi\right)\Psi+\Phi\nabla\Psi$$

$$\nabla\cdot\left(\Phi\vec{D}\right)=\left(\nabla\Phi\right)\cdot\vec{D}+\Phi\nabla\cdot\vec{D}$$

$$\nabla\cdot\left(\vec{D}\times\vec{G}\right)=\left(\nabla\times\vec{D}\right)\cdot\vec{G}-\left(\nabla\times\vec{G}\right)\cdot\vec{D}$$

$$\nabla\left(\vec{D}\cdot\vec{G}\right)=\left(\vec{D}\cdot\nabla\right)\vec{G}+\vec{D}\times\left(\nabla\times\vec{G}\right)+\left(\vec{G}\cdot\nabla\right)\vec{D}+\vec{G}\times\left(\nabla\times\vec{D}\right)$$

$$\nabla\times\left(\Phi\vec{D}\right)=\left(\nabla\Phi\right)\times\vec{D}+\Phi\nabla\times\vec{D}$$

$$\nabla\times\left(\vec{D}\times\vec{G}\right)=\left(\nabla\cdot\vec{G}\right)\vec{D}-\left(\nabla\cdot\vec{D}\right)\vec{G}+\left(\vec{G}\cdot\nabla\right)\vec{D}-\left(\vec{D}\cdot\nabla\right)\vec{G}$$

Cylindrical coordinates (r, φ, z) :

$$\nabla \Phi = \hat{r} \frac{\partial \Phi}{\partial r} + \hat{\varphi} \frac{1}{r} \frac{\partial \Phi}{\partial \varphi} + \hat{z} \frac{\partial \Phi}{\partial z}$$

$$\nabla \cdot \vec{D} = \frac{1}{r} \frac{\partial (r D_r)}{\partial r} + \frac{1}{r} \frac{\partial D_\varphi}{\partial \varphi} + \frac{\partial D_z}{\partial z}$$

$$\nabla \times \vec{H} = \hat{r} \left(\frac{1}{r} \frac{\partial H_z}{\partial \varphi} - \frac{\partial H_\varphi}{\partial z} \right) + \hat{\varphi} \left(\frac{\partial H_r}{\partial z} - \frac{\partial H_z}{\partial r} \right) + \hat{z} \frac{1}{r} \left(\frac{\partial (r H_\varphi)}{\partial r} - \frac{\partial H_r}{\partial \varphi} \right)$$

$$\nabla^2 \Phi = \frac{1}{r} \frac{\partial}{\partial r} \left[r \frac{\partial \Phi}{\partial r} \right] + \frac{1}{r^2} \frac{\partial^2 \Phi}{\partial \varphi^2} + \frac{\partial^2 \Phi}{\partial z^2}$$

A–4. RELATIVE PERMITTIVITIES*

(The listed ones are average low–frequency values)

Material	Relative Permittivity (dielectric constant), ε_r
Vacuum	1.0
Air	1.0006
PTFE, FEP (Teflon)	22.1
Polypropylene	2.20–2.28
Polyethylene	2.3
Polystyrene	2.4–4.0
Paper	2.0–4.0
Rubber	2.3–4.0
Transformer oil	2.5–4.0
Hard rubber	2.5–4.8
Silicon Dioxide (SiO_2)	3.4–4.3
Quartz, Fused Silica (SiO_2)	3.8
Soil(dry)	3.0–4.0
Plexiglas	3.4
Bakelite	3.5–6.0
Wood (Maple)	4.4
Glass	4.0–10.0
Bakelite	5.0
Wood(Birch)	5.2
Porcelain, Diamond	5.7
Mica	5.0–6.0
Salt	5.9
Calcium fluoride(CaF_2) (at $f \sim 30\ GHz$)	6.486.51
Porcelain, Steatite	6.5
Glass–bonded mica	6.3–9.3
Aluminum Oxide, Alumina, Sapphire, (Al_2O_3)	7.5–10.5
Silicon (Si)	11.7–11.8
Gallium Arsenate (GaAs)	12–13.1
Germanium (Ge)	15.8
Wood (Oak), Methanol	33
Sea water	72
Microwave dielectrics D6, D10, D2O, D30, D36, D77, D88 (at $f = 5 - 10 GHz$)	6.5–90
Water (distilled)	80
Ice (-30°C)	99
Barium–strontium–titanite	7500

A–5 FUNDAMENTAL CONSTANTS of a vacuum (free space)

No	Constant	Symbol	Magnitude
1.	Permittivity	ε_0	$\sim \dfrac{1}{36\pi} \times 10^{-9}\,(F/m)$
2.	Permeability	μ_0	$4\pi \times 10^{-7}\,(H/m)$
3.	Wave Impedance	Z_0	$\sqrt{\dfrac{\mu_0}{\varepsilon_0}} \approx 120\pi \approx 376.73\,(\Omega)$
4.	Velocity of electromagnetic waves$_0$	c	$c = 1/\sqrt{\mu_0 \varepsilon_0} = 3 \times 10^8\,(m/s)$

A–6 INDEX OF TABLES

1. Table 6.1. The effective permittivity values of one–strip MSLs with dielectric substrates e

2. Table 6.2. T The wave impedance values of one–strip MSLs ($l/h = 2$, $\varepsilon_r^l = 12$)

3. Table 6.3. Values of the relative wavelength in one–strip MSLs ($\varepsilon_r^l = 2.55$, $\varepsilon_r^u = 1$)

4. Table 6.4. Values of the relative wavelength in one–strip MSLs ($\varepsilon_r^l = 2.55$, $\varepsilon_r^u = 1$)

5. Table 6.5. The wave impedance values of two–strip MSLs ($\varepsilon_r = 16$)

6. Table 6.6. Dependence of MSL sizes (Fig.6.2a) on transition attenuation constant α (dB) for fixed parameters $Z = 50$ Ω , $a/b = 1$ and $t/b = 0.0075$

7. Table 6.7. Dependence of the wave impedance Z Ω and of the transition attenuation constant α (dB) on ε_3 and b_1 for $w/b = 0.95$, $s/b = 1$, $a/b = 2$, $\varepsilon_1 = 10$, $\varepsilon_2 = 1$ and $t/b = 0.001$

8. Table 6.8. T The experimental data for the MSL shown in Fig.6.2a

9. Table 6.9. The experimental data for the MSL shown in Fig.6.2b

10. Table 7.1. The cutoff frequencies at different magnitudes of the external constant magnetic field strength H_0 for the meander line on the ferrite substrate

11. Table 8.1. Dependence of the longitudinal propagation constant h when we change the metal strip w_1 slightly

12. Table 8.2. Dependence of the longitudinal propagation constant h when we change the substrate length l_1 slightly

13. Table 8.3. Dependence of the longitudinal propagation constant h when we change the frequency slightly

14. Table 9.1. Parameters for three ferrite waveguides given in Fig.9.6.

15. Table 9.2. The testing of our algorithm on the external constant magnetic field H for a mirror ferrite waveguide

16. Table 9.3. The testing of our algorithm with respect to small changes of the mirror ferrite waveguide width l

A–7 INDEX OF FIGURES

Introduction

1.   **Fig. 1.** Cross–section of a strip waveguide line, where L_ε are contours bordering isotropic, anisotropic and gyrotropic media. L_m are contours bordering metal conductors and other designations.
2.   **Fig. 2.** Cross–section of a slot line, where designations are shown.
3.   **Fig. 3.** (a) Cross–section of two coupled $\ddot{\varepsilon}$ – gyrotropic rods; (b) designations.
4.   **Fig. 4.** (a) The configuration of a 3D structure, which is illuminated by outside microwaves; (b) designations.
5.   **Fig. 5.** The structure having a semiconductor – dielectric substrate with two metallic frames placed on it.
6.   **Fig. 6.** The model of the heart used in calculations: (a) the surfaces are approximated by triangles, (b) a view of ventricles and atria, (c) certain designations.
7.   **Fig. 7.** Designations used in the expressions for the field components.

Chapter 1

8.   **Fig. 1.1.** Waveguide transmission line with inhomogeneous filling.
9.   **Fig. 1.2.** Boundary conditions on the surface of ideal metal.
10. **Fig. 1.3.** Cross–section of a coaxial waveguide.
11. **Fig. 1.4.** Cross–section of a coaxial waveguide with layer filling.

Chapter 2

12. **Fig. 2.1.** Projections of a vector surface element.
13. **Fig. 2.2.** Designations for determining the electric field of a metal strip.
14. **Fig. 2.3.** Determinations for determining the eclectic field of two interacting metal strips.
15. **Fig. 2.4.** For determination of the capacitance of a metal strip.
16. **Fig. 2.5.** Cross–section of the symmetric stripline.
17. **Fig. 2.6.** The symmetric stripline capacitance dependence on the ratio u_1/d : curves 1, 2 and 3 were calculated by the formulae (2.21), (2.22) and (2.23) respectively.
18. **Fig. 2.7.** The designations on conformal mapping of one region into another.
19. **Fig. 2.8.** The dependences of the capacitance per unit length on the ratio u_1/d for a stripline with the shot–circuited end at the values ℓ/d : where curves 1, 2 and 3 are 0.1, 0.5 and 1.0 respectively.
20. **Fig. 2.9.** Topology of the symmetric stripline cross–section.
21. **Fig. 2.10.** On conformal mapping of one region into another.
22. **Fig. 2.11.** The dependence of the capacitance per unit length on the ratio p/d for an open end stripline at the following values of ℓ/d : the curves 1, 2 and 3 are 0.1, 0.5 and 1.0 respectively.
23. **Fig. 2.12.** The designations for determining the capacitance of a symmetric stripline with a flange end.

24.   **Fig. 2.13.** The dependence of the capacitance per the unit length on the ratio p/d for the stripline with a flange end for the following values of ℓ/d : the curves 1, 2 and 3 are 0.1, 0.5 and 1.0 respectively.
25.   **Fig. 2.14.** Cross–section of a stripline in a circular screen.
26.   **Fig. 2.15.** The dependence of capacitance per unit length on geometric parameters of the striplines shown in Fig. 2.14 (curves 1, 2) and Fig. 2.16 (curves 3, 4).
27.   **Fig. 2.16.** The capacitance of a symmetric stripline with a vertical strip.
28.   **Fig. 2. 17.** The plane of the complex variable $z = x + jy$.
29.   **Fig. 2.18.** The polygon in the plane of the complex variable $w = u + iv$.
30.   **Fig. 2.19.** The designation for determining the capacitance of a stripline in a rectangular waveguide.
31.   **Fig. 2.20.** Screened stripline with a vertical metal conductor.

Chapter 3

32.   **Fig. 3.1.** (a) The piecewise *arc L* with a corner point t_0 and (b) the smooth *arc L* .
33.   **Fig. 3.2.** (a) The piecewise smooth *arc L* , (b) the smooth *arc L* .
34.   **Fig. 3.3.** The plane cut running from the point a into ∞ .
35.   **Fig. 3.4.** A view of the plane cut running from the point b into infinity ∞ .
36.   **Fig. 3.5.** An explanation of Cauchy's type integral behavior at the ends of the integration contour.

Chapter 4

37.   **Fig. 4.1.** Transformation of the coordinates system x', y', z' to the system x, y, z .

Chapter 5

38.   **Fig. 5.1.** Cross–section a piece – wise uniform regular transmission line.
39.   **Fig. 5.2.** The designations when contour L is smooth L .
40.   **Fig. 5.3.** Cross–section of a piece–wise uniform multiconductor transmission line.
41.   **Fig. 5.4.** Helps explain a nonorthogonal basis of vectors in a two-dimensional space.
42.   **Fig. 5.5.** An electric circuit of line segments with distributed parameters loaded at points B and C by impedances Z_1 and Z_2 .
43.   **Fig. 5.6.** An electric circuit of line segments with distributed parameters loaded by impedances Z_1 and Z_2 .
44.   **Fig. 5.7.** Geometrically symmetric contours bounding metal and gyrotropic media.
45.   **Fig. 5.8.** When calculating matrix elements A_{ik} of geometric symmetry contours.

Chapter 6

46.   **Fig. 6.1.** Cross–section of a one–strip MSL: (a) with a magnetodielectric substrate; (b) with a layer dielectric substrate.
47.   **Fig. 6.2.** (a) Cross–section of a screened two–stripline MSL; (b) cross–section of an open two–stripline MSL.

48.   **Fig. 6.3.** The schematic of the director.

49.   **Fig. 6.4.** Cross–section of a MSL with the n–Si substrate and designations.

50.   **Fig. 6.5.** Dependence of the MSL complex impedance $\dot{Z}$ on the relative distance l/h between the metal strip and the substrate lateral edge. Our calculations are presented here by solid and dash lines and the experimental results with circles.

51.   **Fig. 6. 6.** Dependence of the complex wave impedance module of the MSL on the frequency when the specific resistivity $\rho = 0.3\ \Omega \cdot m$, $\rho = 1.0\ \Omega \cdot m$ and $\rho = 1.8\ \Omega \cdot m$ at the magnitude $l/h = 5$. Solid lines show our calculations and experimental data [6.9] is shown by circles.

52.   **Fig. 6.7.** Dependence of the attenuation constant h'' of the MSL on the frequency f for three specific resistivities: $\rho = 0.3\ \Omega \cdot m$, $\rho = 1.0\ \Omega \cdot m$ and $\rho = 1.8\ \Omega \cdot m$ at the magnitude $l/h = 5$. Our calculations are shown by solid lines and the experimental data [6.9] is shown by circles.

53.   **Fig. 6. 8.** Dependence of the real part Z' of the complex wave impedance of the MSL with the low ohmic $n–Si$ substrate on the specific resistivity ρ of the substrate material when the metal strip widths are $w/h = 0.4$ (the solid line) and $w/h = 0.82$ (the dash line).

54.   **Fig. 6.9.** Dependence of the imaginary part Z'' of the complex wave impedance of the MSL with the low ohmic $n–Si$ substrate on the specific resistivity ρ of the substrate material when the MSL sizes are: $t/h = 0.002$, $l = h$, $w/h = 0.4$ (the solid line) and $w/h = 0.82$ (the dash line).

55.   **Fig. 6.10.** Dependence of the real part h' of the MSL complex propagation constant on the specific resistivity ρ of the $n–Si$ substrate material when the MSL sizes are: $t/h = 0.002$, $l = h$, $w/h = 0.4$ (the solid line) and $w/h = 0.82$ (the dash line).

56.   **Fig. 6.11.** Dependence of the imaginary part h'' of the MSL complex propagation constant on the specific resistivity ρ of the $n–Si$ substrate material when the MSL sizes are: $t/h = 0.002$, $l = h$, $w/h = 0.4$ (the solid line) and $w/h = 0.82$ (the dash line).

57.   **Fig. 6.12.** Cross–section of a MSL with a two–layer substrate.

58.   **Fig. 6.13.** Dependence of the MSL complex wave impedance $\dot{Z}$ on the real part ε'_d of the relative permittivity of the upper dielectric layer at $\rho = 1.8\ \Omega \cdot m$.

59.   **Fig. 6.14.** Dependence of the MSL complex wave impedance $\dot{Z}$ on the imaginary part ε''_d of the relative permittivity of the upper dielectric layer at $\rho = 1.8\ \Omega \cdot m$.

60.   **Fig. 6.15.** Dependence of the MSL complex wave impedance $\dot{Z}$ on the upper dielectric layer thickness d.

61.   **Fig. 6.16.** The MSL with the $GaAs–SiO_2$ substrate and designations.

62.   **Fig. 6.17.** Dependence of the MSL complex wave impedance $\dot{Z}$ on the metal strip width w/h when its thickness $t/h = 0.004$ and $GaAs$ substrate width is constant $p/h = 2.6$.

63. **Fig. 6.18.** Dependence of the MSL wavelength λ and the effective permittivity ε_{MSL}^{ef} on the metal strip width w/h when its thickness is $t/h = 0.004$ and the *GaAs* substrate width is constant $p/h = 2.6$.

64. **Fig. 6.19.** Dependence of the MSL attenuation constant h'' and capacitance per unit length C on the metal strip width w/h when its thickness is $t/h = 0.004$ and *GaAs* substrate width is constant $p/h = 2.6$.

65. **Fig. 6.20.** Dependence of the MSL complex wave impedance $\dot{Z}$ on the relative distance l/h between the metal strip and the substrate lateral edge when sizes are: $w/h = 0.8$, $t/h = 0.004$ and *GaAs* substrate width p/h is not constant.

66. **Fig. 6.21.** Dependence of the MSL wavelength λ and effective permittivity ε_{MSL}^{ef} on the relative distance l/h between the metal strip and the substrate lateral edge when sizes are: $w/h = 0.8$, $t/h = 0.004$ and the GaAs substrate width p/h is not constant.

67. **Fig. 6.22.** Dependence of the MSL attenuation constant h'' and capacitance per unit length C on the relative distance l/h between the metal strip and the substrate lateral edge when sizes are: $w/h = 0.8$, $t/h = 0.004$ and the *GaAs* substrate width p/h is not constant.

68. **Fig. 6.23.** Dependence of the complex wave impedance $\dot{Z}$ of the MSL with the *GaAs–SiO$_2$* substrate on its upper dielectric layer thickness d/h when sizes are: $w/h = 1.2$, $t/h = 0.004$, $l/h = 0.7$.

69. **Fig. 6.24.** Dependence of the MSL wavelength and the effective permittivity ε_{MSL}^{ef} of the MSL with the *GaAs–SiO$_2$* substrate on its upper dielectric layer thickness d/h when sizes are: $w/h = 1.2$, $t/h = 0.004$, $l/h = 0.7$.

70. **Fig. 6.25.** Dependence of the MSL attenuation constant h'' and capacitance per unit length C on its upper dielectric layer thickness d/h when sizes are: $w/h = 1.2$, $t/h = 0.004$, $l/h = 0.7$.

71. **Fig. 6.26.** Dependence of the complex wave impedance $\dot{Z}$ of the MSL with the *GaAs–SiO$_2$* substrate on the metal strip width w/h when sizes are: $t/h = 0.004$ and the substrate width is constant $p/h = 2.6$.

72. **Fig. 6.27.** Dependence of the MSL wavelength λ and effective permittivity ε_{MSL}^{ef} of the MSL with the *GaAs–SiO$_2$* substrate on the metal strip width w/h when sizes are: $t/h = 0.004$ and the substrate width is constant $p/h = 2.6$.

73. **Fig. 6.28.** Dependence of the attenuation constant h'' and capacitance per unit length C of the MSL with the *GaAs–SiO$_2$* substrate on the metal strip width w/h when sizes are: $t/h = 0.004$ and the substrate width is constant $p/h = 2.6$.

74. **Fig. 6.29.** Dependence of the complex wave impedance $\dot{Z}$ of the MSL with the *GaAs–SiO₂* substrate on the relative distance l/h between the metal strip and the substrate lateral edge when sizes are: $d/h = 0.1$, $w/h = 1.2$, $t/h = 0.004$.

75. **Fig. 6.30.** Dependence of the MSL wavelength λ and the effective permittivity ε_{MSL}^{ef} of the MSL with the *GaAs–SiO₂* substrate on the relative distance l/h between the metal strip and the substrate lateral edge when sizes are: $d/h = 0.1$, $w/h = 1.2$, $t/h = 0.004$.

76. **Fig. 6.31.** Dependence of the attenuation constant h'' and the capacitance per unit length C of the MSL with the *GaAs–SiO₂* substrate on the relative distance l/h between the metal strip and the substrate lateral edge when sizes are: $d/h = 0.1$, $w/h = 1.2$, $t/h = 0.004$.

77. **Fig. 6.32.** The quasi–TEM wave electric field distribution in the MSL with the *GaAs–SiO₂* substrate when sizes are: $w/h = 1.2$, $l/h = 0.2$, $d/h = 0.1$, $t/h = 0.004$.

78. **Fig. 6.33.** A cross–section of a MSL with a two–layer *Al₂O₃–SiO₂* substrate and designations.

79. **Fig. 6.34.** Dependence of a complex wave impedance $\dot{Z}$ of a MSL with the *Al₂O₃–SiO₂* substrate on the upper substrate layer thickness d/h at the fixed MSL sizes: $w/h = 0.8$, $t/h = 0.004$, $l/h = 0.9$.

80. **Fig. 6.35.** Dependence of the MSL wavelength λ and the effective permittivity ε_{MSL}^{ef} of the MSL with the *Al₂O₃–SiO₂* substrate on the upper substrate layer thickness d/h at the fixed MSL sizes: $w/h = 0.8$, $t/h = 0.004$, $l/h = 0.9$.

81. **Fig. 6.36.** Dependence of the attenuation constant h'' and capacitance per unit length C of the MSL with the *Al₂O₃–SiO₂* substrate on the upper substrate layer thickness d/h at the fixed MSL sizes: $w/h = 0.8$, $t/h = 0.004$, $l/h = 0.9$.

82. **Fig. 6.37.** Dependence of the complex wave impedance $\dot{Z}$ of the MSL with the *Al₂O₃–SiO₂* substrate on the metal strip width w at the fixed MSL sizes: $d/h = 0.1$, $t/h = 0.004$, $p/h = 2.6$.

83. **Fig. 6.38.** Dependence of the MSL wavelength λ and effective permittivity ε_{MSL}^{ef} of the MSL with the *Al₂O₃–SiO₂* substrate on the metal strip width w/h at the fixed MSL sizes $d/h = 0.1$, $t/h = 0.004$, $p/h = 2.6$.

84. **Fig. 6.39.** Dependence of the MSL attenuation constant h'' and capacitance per unit length C of the MSL with the *Al₂O₃–SiO₂* substrate on the metal strip width w/h when sizes are: $d/h = 0.1$, $t/h = 0.004$, $p/h = 2.6$.

85. **Fig. 6.40.** Dependence of the complex wave impedance $\dot{Z}$ of the MSL with the *Al₂O₃–SiO₂* substrate on the relative distance l/h between the metal strip and the lateral substrate edges when sizes are: $d/h = 0.1$, $w/h = 1.2$, $t/h = 0.004$ and the substrate width p/h is not constant.

86. **Fig. 6.41.** Dependence of the MSL wavelength λ and effective permittivity ε_{MSL}^{ef} of the MSL with the Al_2O_3–SiO_2 substrate on the relative distance l/h between the metal strip and the substrate lateral edge when sizes are: $d/h = 0.1$, $w/h = 1.2$, $t/h = 0.004$ and the substrate width p/h is not constant.

87. **Fig. 6.42** Dependence of the attenuation constant h'' and capacitance per unit length C of the MSL with the Al_2O_3–SiO_2 substrate on the on the relative distance l/h between the metal strip and the substrate lateral edge when sizes are: $d/h = 0.1$, $w/h = 1.2$, $t/h = 0.004$ and the substrate width p/h is not constant.

88. **Fig. 6.43.** Cross–section view of a MSL with the $LiNbO_3$ substrate and designations.

89. **Fig.6.44.** Dependence of the real part Z' of the MSL complex wave impedance on the metal strip width w for two MSLs with different metal strip thicknesses $t/h = 0.024$ (the dotted line) and $t/h = 0.008$ (the solid line). Circles are the experimental values. The $LiNbO_3$ substrate width is constant $p/h = (2 \cdot l + w)/h = 2.6$.

90. **Fig. 6.45.** Dependence of the real part Z' of the MSL complex wave impedance on the displacement of the metal strip on the substrate when the $LiNbO_3$ substrate width is $p/h = const$, $w/h = 1.6$ and $t/h = 0.04$.

91. **Fig. 6.46.** Dependence of the real part Z' of the MSL complex wave impedance on the distance between the metal strip and the lateral edges of the substrate when $l_1 = l_2 = l$, $w/h = 1.6$, $t/h = 0.04$. The $LiNbO_3$ substrate width $p/h \neq const$.

92. **Fig. 6.47.** Dependence of the real part Z' of the MSL complex wave impedance on the metal strip thickness t when $p/h = (2 \cdot l + w)/h = 2.6$.

93. **Fig. 6.48.** Dependence of the real part Z' of the MSL complex wave impedance on the component ε_{zz} of the permittivity tensor when $\varepsilon_{xx} = \varepsilon_{yy} = 43$.

94. **Fig. 6.49.** The MSL with an anisotropic $LiNbO_3$ substrate and designations.

95. **Fig. 6.50.** Dependence of the MSL complex wave impedance $\dot{Z}$ on the metal strip width w/h when sizes are: $t/h = 0.04$ and the $LiNbO_3$ substrate width $p/h = 2.6$ is constant.

96. **Fig .6.51.** Dependences of the MSL wavelength λ and the attenuation constant h'' on the metal strip width w/h when sizes are: $t/h = 0.04$ and the $LiNbO_3$ substrate width is constant $p/h = 2.6$.

97. **Fig .6.52.** Dependence of the effective permittivity ε_{MSL}^{ef} and capacitance per unit length C on the metal strip width w/h when sizes are: $t/h = 0.04$ and the $LiNbO_3$ substrate width is constant $p/h = 2.6$.

98. **Fig. 6.53.** Dependence of the complex wave impedance $\dot{Z}$ of the MSL with the one – layer $LiNbO_3$ substrate on the relative distance l/h between the metal strip and the lateral substrate edges when sizes are: $w/h = 0.3$ and $t/h = 0.04$.

99. **Fig. 6.54.** Dependence of the MSL wavelength λ and the effective permittivity ε_{MSL}^{ef} of the MSL with the one – layer LiNbO$_3$ substrate on the relative distance l/h between the metal strip and the lateral substrate when sizes are: $w/h = 0.3$ and $t/h = 0.04$.

100. **Fig. 6.55.** Dependence of the MSL attenuation constant h'' and capacitance per unit length C of the MSL with the one – layer LiNbO$_3$ substrate on the relative distance l/h between the metal strip and the lateral substrate when sizes are: $w/h = 0.3$ and $t/h = 0.04$.

101. **Fig. 6.56.** Cross–section of the MSL with a two–layer *LiNbO$_3$–SiO$_2$* substrate and designations.

102. **Fig. 6.57.** Dependence of the complex wave impedance $\dot{Z}$ of the MSL with the two–layer LiNbO$_3$–SiO$_2$ substrate on the upper substrate layer thickness d/h at the fixed MSL sizes: $w/h = 1.0$, $t/h = 0.004$, $l/h = 1.0$.

103. **Fig. 6.58.** Dependence of the MSL wavelength λ and the effective permittivity ε_{MSL}^{ef} of the MSL with the two–layer LiNbO$_3$–SiO$_2$ substrate on the upper substrate layer thickness d/h at the fixed MSL size: $w/h = 1.0$, $t/h = 0.004$, $l/h = 1.0$.

104. **Fig. 6.59.** Dependence of the attenuation constant h'' and capacitance per unit length C of the MSL with the two–layer LiNbO$_3$–SiO$_2$ substrate on the upper substrate layer thickness d/h at the fixed MSL sizes $w/h = 1.0$, $t/h = 0.004$, $l/h = 1.0$.

105. **Fig. 6.60.** Dependence of the MSL complex wave impedance $\dot{Z}$ on the metal strip width w/h when sizes are: $d/h = 0.1$, $t/h = 0.004$ and the two–layer LiNbO$_3$–SiO$_2$ substrate width is constant $p/h = 2.6$.

106. **Fig. 6.61.** Dependence of the MSL wavelength λ and effective permittivity ε_{MSL}^{ef} on the metal strip width w/h when sizes are: $d/h = 0.1$, $t/h = 0.004$ and the two–layer LiNbO$_3$–SiO$_2$ substrate width is constant $p/h = 2.6$.

107. **Fig. 6.62.** Dependence of the attenuation constant h'' and capacitance per unit length C of the MSL on the metal strip width w/h when sizes are: $d/h = 0.1$, $t/h = 0.004$ and the two–layer LiNbO$_3$–SiO$_2$ substrate width is constant $p/h = 2.6$.

108. **Fig. 6.63.** Dependence of the MSL complex wave impedance $\dot{Z}$ on the relative distance l/h between the metal strip and the lateral substrate when sizes are: $w/h = 0.8$, $t/h = 0.004$ and the two–layer LiNbO$_3$–SiO$_2$ substrate width p/h is not constant.

109. **Fig. 6.64.** Dependence of the MSL wavelength λ and effective permittivity ε_{MSL}^{ef} on the relative distance l/h between the metal strip and the lateral substrate when sizes are: $w/h = 0.8$, $t/h = 0.004$ and the two–layer LiNbO$_3$–SiO$_2$ substrate width p/h is not constant.

110. **Fig. 6.65.** Dependence of the MSL attenuation constant h'' and capacitance per unit length C on the relative distance l/h between the metal strip and the lateral substrate when sizes are: $w/h = 0.8$, $t/h = 0.004$ and the two–layer $LiNbO_3$–SiO_2 substrate width p/h is not constant.

Chapter 7

111. **Fig. 7.1.** The open two metal strips MSL with gyrotropic substrate: (a) geometrically symmetric and (b) geometrically asymmetric.
112. **Fig. 7.2.** Dependence of absolute values of real and imaginary parts of the matrix $\dot{C} = C'_{ik} - iC''_{ik}$ elements for geometrically asymmetric MSL (Fig.7.1b).
113. **Fig. 7.3.** One metal strip MSL with a longitudinally magnetized semiconductor substrate.
114. **Fig. 7.4.** Dependence of the relative phase shift $\nabla\varphi$ (solid line) and attenuation h''_1 and h''_2 (dash lines) on the normalized width of the metal strip w/h .
115. **Fig. 7.5.** Dependence of the real part of the effective permittivity $\mathrm{Re}(\dot\varepsilon^{ef}_{MSL})$ (solid line) of the MSL (Fig.7.3) and the real part of effective permittivity helicon wave $\mathrm{Re}(\dot\varepsilon^{ef}_{hel})$ (dash line) on the value of the magnetic induction B .
116. **Fig. 7.6.** Dependence of the relative wave phase shift $\Delta\varphi$ on the magnetic induction change ΔB for three MSLs.
117. **Fig. 7.7.** Dependence of eigenwave propagation constants $\dot{h}_{1,2} = h'_{1,2} - ih''_{1,2}$ on the normalized distance s/h between the metal strips.
118. **Fig. 7.8.** Cross–section of the geometrically asymmetric MSL with the longitudinally magnetized semiconductor substrate (a) with metal strips of different widths and (b) with metal strips of different heights.
119. **Fig. 7.9.** Dependence of eigenwave losses h''_1 , and h''_2 in the asymmetric MSL (Fig. 7.1a) on ratio of metal strip widths w_2/w_1 .
120. **Fig. 7.10.** Dependence of eigenwave losses h''_1 and h''_2 in the asymmetric MSL (Fig. 7.1b) on the ratio of the metal strip thickness t_2/t_1 .
121. **Fig. 7.11.** The microstrip meander line on the ferrite substrate.
122. **Fig. 7.12.** Cross–section of a *regular* MSL having three interacting metal strips (where $w_1 = w_2 = w_3 = w$ and $s_1 = s_2 = s$).
123. **Fig. 7.13.** Dependence of the inductance matrix elements on the external constant magnetic field H_0 for the MSL (Fig.7.12) when the sizes are: $w/h = 0.7$, $t/h = 0.005$ and $s/h = 0.3$.
124. **Fig. 7.14.** Dependence of the propagation constant in the meander line (Fig.7.11) on the coordinate z : curve 1 was calculated for $H_0 = 0$, curve 2 was calculated $H_0 = +30 \cdot 10^{-4} T$ and curve 3 was calculated $H_0 = -30 \cdot 10^{-4} T$. The sizes of the meander line are: $w/h = 0.6$, $t/h = 0.005$, $s/h = 0.3$ and $l = 0.108\lambda_0$.
125. **Fig. 7.15.** The meander line phase shift dependence on the external constant magnetic field H_0 . Our calculations are solid lines and experimental results are

the dash lines with circles. The sizes of the meander line are: $w/h = 0.7$, $s/h = 0.3$, $t/h = 0.005$, $l = 0.1065\lambda_0$ and $h = 0.015\lambda_0$.

126. **Fig. 7.16.** Dependence of the phase shift in the meander line on the external constant magnetic field H_0. Our calculations are solid lines and the experimental results are the dash lines with circles. The sizes of meander line are: $w/h = 0.6$, $s/h = 0.3$, $t/h = 0.005$, $h = 0.0087\lambda_0$, $l = 0.1075\lambda_0$.

127. **Fig. 7.17.** Dependence of the phase shift φ_Σ on the wavelength ratio λ/λ_0 at the external constant magnetic field $H_0 = 60 \cdot 10^{-4} T$. Our calculations are a solid line and the experimental result is the dash line with circles. The sizes of meander line are: $w/h = 0.6$, $s/h = 0.3$, $t/h = 0.005$, $h = 0.0087\lambda_0$ and $l = 0.1075\lambda_0$.

128. **Fig. 7.18.** The results of calculations and experiments for the meander line with parameters $w/h = 0.7$, $t/h = 0.007$, $s/h = 0.5$ and $h = 10^{-3}$ m. Calculations are shown by the solid line and the experimental result is dash line with circles.

129. **Fig. 7.19.** The current density distribution on the lower strip side is designated as Roman No I and the upper strip sides are designated as Roman No II. The current density distribution on the metal plate correspond as: curve 1,2,3 when $H_0 = 0$; $H_0 = -120 \cdot 10^{-4} T$ and $H_0 = +120 \cdot 10^{-4} T$.

130. **Fig. 7.20.** Dependences of losses in metal conductors as a function of distances s at two corresponding wavelengths: $\lambda = \lambda_0$ (curve 1) and $\lambda = 1.2\lambda_0$ (curve 2). Our calculations are shown as solid lines and the experimental results are seen as dotted lines. The MSL sizes are: $w/h = 0.6$, $t/h = 0.005$, $l/h = 12$ and $h = 10^{-3} m$.

131. **Fig. 7.21.** Schematic structure to measure the phase characteristics of a meander line: 1 is a generator of the SHF signal; 2 is the coaxial T–junction; 3, 9, 11 are the coaxial valves; 4 is the expanding coaxial line; 5 is the attenuator; 6 is the indicator; 7 is the voltage source; 8 is the meander structure and 10 is the measuring line.

132. **Fig. 7.22.** Cross–section of (a) one metal strips MSL and (b) two metal strips MSL with the transversally magnetized ferrite substrate.

133. **Fig. 7.23.** Dependence of the propagation constant h' of the MSL (Fig. 7.22a) on the relative permittivity ε_r of the substrate material. Curves 1 and 2 were calculated according to the known formulae; Curves 3 and 4 we calculated by the SIE method. The normalized width of the metal strip: curves 1 and 3 correspond to $w/h = 0.5$ and curves 2 and 4 correspond to $w/h = 1$.

134. **Fig. 7.24.** Distribution of the electric field force lines in the MSL shown in Fig. 7.22(a) at the constitutive parameters $\mu_r = 1$ and $\varepsilon_r = 5$.

135. **Fig. 7.25.** Dependence of the magnetic field components on the transversal coordinate x for the MSL (Fig. 7.22a) with the parameters: $\varepsilon_r = 5$ and $w/h = 0.5$. Curve 1 shows H_x distribution; curve 2 shows H_z distribution and curve 3 shows H_y distribution. Our calculations are the solid lines and the experimental results of [7.20] are the dotted lines.

136. **Fig. 7.26.** Dependence of the phase shift on the external constant magnetic field for the MSL (Fig. 7.22a) with $w/h = 1.2$. Our calculations are the solid lines and the experimental results of [7.20] are the dotted line.

137. **Fig. 7.27.** Distribution of the electric field in a one metal strip MSL.

138. **Fig. 7.28.** Dependence of the propagation constant h' on the magnitude of the external constant magnetic field strength H_0.

139. **Fig. 7.29.** Cross–section of a one metal strip MSL with the longitudinally magnetized layer semiconductor–dielectric substrate.

140. **Fig. 7.30.** Dependence of (a) the complex wave impedance $\dot{Z}$; (b) the propagation constant $\dot{h}$ on the dielectric layer thickness d for the MSL shown in Fig. 7.29. Our calculations of the real parts of the values $\dot{Z}$ and $\dot{h}$ are the solid lines and the imaginary parts are the dotted lines. Curve 1 corresponds to the value of $n = 2.44 \cdot 10^{20}\,\mathrm{m}^{-3}$ and curve 2 corresponds to the value of $n = 2.0 \cdot 10^{20}\,\mathrm{m}^{-3}$.

141. **Fig. 7.31.** Dependence of (a) the complex wave impedance $\dot{Z}$ and (b) the propagation constant $\dot{h}$ on the relative permittivity ε_r of the upper layer for the MSL (Fig. 7.29). The solid lines are the real parts, the dotted lines are the imaginary parts of the values $\dot{Z}$ and $\dot{h}$. Curve 1 corresponds to the value of $n = 2.44 \cdot 10^{20}\,\mathrm{m}^{-3}$ and curve 2 corresponds to the value of $n = 2.0 \cdot 10^{20}\,\mathrm{m}^{-3}$.

142. **Fig. 7.32.** Dependence of (a) the complex wave impedance $\dot{Z}$ and (b) the propagation constant $\dot{h}$ on the frequency for the MSL shown in Fig. 7.29. Solid lines are the real part and dotted lines are the imaginary parts. Curve 1 corresponds to the value of $n = 2.44 \cdot 10^{20}\,\mathrm{m}^{-3}$ and curve 2 corresponds to the value of $n = 2.0 \cdot 10^{20}\,\mathrm{m}^{-3}$.

143. **Fig. 7.33.** Dependence of (a) the complex wave impedance $\dot{Z}$; (b) the propagation constant $\dot{h}$ and (c) the values of the relative phase shift $\Delta\varphi$ on the magnetic induction B for the MSL (Fig. 7.29). Our calculations are the solid line and the experimental results are circles.

144. **Fig. 7.34.** Dependence of (a) the effective permittivity $\dot{\varepsilon}^{ef}_{MSL}$ (b) the complex wave impedance $\dot{Z}$ and (c) the propagation constant $\dot{h}$ for the MSL (Fig. 7.29) on the normalized thickness t/h of the metal strip.

145. **Fig. 7.35.** Dependence of the losses h'' on the magnetic induction B in the one metal strip MSL (Fig. 7.29) (a) curve 1 corresponds $n = 2.0 \cdot 10^{20}\,\mathrm{m}^{-3}$, $t = 0.14 \cdot 10^{-3}\,m$; curve 2 corresponds $n = 2.0 \cdot 10^{20}\,\mathrm{m}^{-3}$, $t = 2 \cdot 10^{-5}\,m$ curve 3 corresponds $n = 2.0 \cdot 10^{20}\,\mathrm{m}^{-3}$, $t = 0.14 \cdot 10^{-3}\,m$; (b) curve 4 corresponds $n = 2.0 \cdot 10^{20}\,\mathrm{m}^{-3}$, $t = 0.14 \cdot 10^{-3}\,m$; curve 5 corresponds $n = 2.0 \cdot 10^{20}\,\mathrm{m}^{-3}$, $t = 2 \cdot 10^{-5}\,m$. (Calculations are the solid lines and the experiment results are the circles and points).

146. **Fig. 7.36.** Distribution of the MSL electric field strength $\vec{E}$ force lines at different moments of time for the one–strip line (Fig. 7.29) with parameters: $h = 0.5 \cdot 10^{-3}\,m$, $d = 0.25 \cdot 10^{-4}\,m$, $w = 10^{-2}\,m$, $t = 0.14 \cdot 10^{-3}\,m$, $\varepsilon_r = 9.8$ at $f = 34.02$ GHz.

147. **Fig. 7.37.** Cross–section of two metal strips MSL with the longitudinally magnetized semiconductor–magnetodielectric substrate.

148. **Fig. 7.38.** Dependence of (a) the propagation constants $\dot{h}_{1,2} = h'_{1,2} - ih''_{1,2}$; (b) the relative phase velocities $\beta_{1,2}$ and (c) the current coupling coefficients $\dot{p}_{1,2} = p'_{1,2} - ip''_{1,2}$ of the MSL eigenwaves on the relative permittivity ε_r of the upper substrate layer.

149. **Fig. 7.39.** Distribution of the electric field $\vec{E}$ vector polarization ellipses in the fixed cross–section of the MSL (the parameters correspond to Fig. 7.38) for the different values of the permittivity ε_r of the upper substrate layer.

150. **Fig. 7.40.** Dependence of (a) the propagation constants $\dot{h}_{1,2} = h'_{1,2} - ih''_{1,2}$; (b) the current coupling coefficients $\dot{p}_{1,2} = p'_{1,2} - ip''_{1,2}$ of the MSL eigenwaves on the relative permeability μ_r of the upper substrate layer.

151. **Fig. 7. 41.** Distribution of the electric field strength $\vec{E}$ polarization ellipses in a cross–section of the coupled MSL with the layer longitudinally magnetized semiconductor substrate for (a) the opposite excitation in phase of the metal strips and (b) the excitation of the metal strips in the same phase.

152. **Fig. 7.42.** Distribution of the electric field strength $\vec{E}$ force lines at fixed time moments $\Psi = \omega t$ for the opposite in phase excitation of the MSL metal strips.

153. **Fig. 7.43.** Distribution of the electric field strength $\vec{E}$ force lines at fixed time moments $\Psi = \omega t$ for the in phase excitation of the MSL metal strips.

154. **Fig. 7.44.** Distribution of the electric field strength $\vec{E}$ polarization ellipses in a cross–section of the coupled MSL with the layer semiconductor substrate. The two metal strips MSL potentials are: (a) $\varphi_1 = 1 \ and \ \varphi_2 = 0$; (b) $\varphi_1 = 0 \ and \ \varphi_2 = 1$.

155. **Fig. 7.45.** Cross–section of (a) one metal strip MSL and (b) two metal strips MSL with the longitudinally magnetized semiconductor–ferrite substrate.

156. **Fig. 7.46.** Dependence of (a) the complex wave impedance $\dot{Z}$ and (b) the propagation constant $\dot{h}$ on the signal frequency f.

157. **Fig. 7.47.** Dependence of (a) the complex wave impedance $\dot{Z}$ and (b) the propagation constant $\dot{h}$ on the normalized width w/h of the metal strip MSL.

158. **Fig. 7.48.** Dependence of (a) the complex wave impedance $\dot{Z}$ and (b) the propagation constant $\dot{h}$ on the concentration of charge carriers in the MSL substrate semiconductor material.

159. **Fig. 7.49.** Dependence of (a) the propagation constant $\dot{h}_{1,2} = h'_{1,2} - ih''_{1,2}$; (b) the current coupling coefficient $\dot{p}_{1,2} = p'_{1,2} - ip''_{1,2}$ of eigenwaves on the saturation magnetization of the MSL substrate ferrite layer.

160. **Fig. 7.50.** Dependence of (a) the complex wave impedance $\dot{Z}_{1,2} = Z'_{1,2} - iZ''_{1,2}$ and (b) the voltage coupling coefficient $\dot{q}_{1,2} = q'_{1,2} - iq''_{1,2}$ on the normalized ferrite layer thickness of the MSL.

161. **Fig. 7.51.** Dependence of the eigenwave propagation constant $\dot{h}$ on the electron mobility μ_{el} of the MSL substrate semiconductor layer.

162. **Fig. 7.52.** Dependence of the current coupling coefficient $\dot{p}_{1,2} = p'_{1,2} - ip''_{1,2}$ of the MSL (Fig. 7.45b) eigenwaves on the magnetic induction B value.

163. **Fig. 7.53.** Distribution of the electric field strength $\vec{E}$ force lines at different time moments $\Psi = \omega t$ in a cross-section of the MSL.

164. **Fig. 7.54.** Diagrams of polarization ellipses in a cross–section of the coupled MSL (a) the electric field strength $\vec{E}$; (b) the magnetic field strength $\vec{H}$ at the saturation magnetization of the ferrite $4\pi M_s = 396.8$ kA/m. The two metal strips potentials are: $\varphi_1 = 1$ *and* $\varphi_2 = 0$.

165. **Fig. 7.55.** Distributions of the electric field strength $\vec{E}$ (a) polarization ellipses and (b,c,d,e) the force lines at certain time moments $\Psi = \omega t$ of the MSL.

166. **Fig. 7.56.** Cross–section of the coupled MSL with the longitudinally magnetized substrate having double $(\ddot{\varepsilon}, \ddot{\mu})$ – gyrotropy.

167. **Fig. 7.57.** Dependence of (a) the propagation constant $\dot{h}_{1,2} = h'_{1,2} - ih''_{1,2}$ and (b) the current coupling coefficient $\dot{p}_{1,2} = p'_{1,2} - ip''_{1,2}$ of the MSL eigenwaves on the charge carrier concentration n in the substrate material.

168. **Fig. 7.58.** Dependence of (a) the propagation constant $\dot{h}_{1,2} = h'_{1,2} - ih''_{1,2}$ and (b) the current coupling coefficient $\dot{p}_{1,2} = p'_{1,2} - ip''_{1,2}$ of the MSL eigenwaves on the saturation magnetization M_s of the substrate material.

169. **Fig. 7.59.** Dependence of (a) the propagation constant $\dot{h}_{1,2} = h'_{1,2} - ih''_{1,2}$ and (b) the current coupling coefficient $\dot{p}_{1,2} = p'_{1,2} - ip''_{1,2}$ of the MSL eigenwaves on the magnetic induction B.

170. **Fig. 7.60.** Distributions of polarization ellipses in a MSL cross– section: (a) the electric field strength $\vec{E}$; (b) the magnetic field strength $\vec{H}$.

171. **Fig. 7.61.** Distributions the electric field strength $\vec{E}$ (a) polarization ellipses and (b,c,d) its force lines at certain time moments $\Psi = \omega t$ in the MSL with the layer $\ddot{\varepsilon} - (\ddot{\varepsilon}, \ddot{\mu})$ substrate.

Chapter 8

172. **Fig. 8.1.** Geometry of a waveguide cross–section.

173. **Fig. 8.2.** The integration contours.

174. **Fig. 8.3.** Geometry of a SL cross–section.

175. **Fig. 8.4.** Dispersion dependences of the longitudinal propagation constant h of the main mode (curve 1) and two higher modes (curves 2, 3) and the wavelength λ_{MSL} in the geometrically symmetric MSL with parameters: $d = 3.17 \cdot 10^{-3}$ m, $w/d = 0.96$, $\varepsilon_r = 11.7$. Here the solid lines are the calculated results and the circles are from [7.32].

176. **Fig. 8.5.** Dispersion dependences of the longitudinal propagation constant h of the main mode (curve 1) and two higher modes (curves 2, 3) for the MSL–A. The sizes are: $d = 10^{-3}$ m, $w = 2 \cdot 10^{-3}$ m, $l_1 = 0$, $l_2 = 2 \cdot 10^{-3}$ m, $t = 3 \cdot 10^{-6}$ m.

177. **Fig. 8.6.** Dispersion dependences of the longitudinal propagation constant h for the main mode (curve 1) and two higher modes (curves 2, 3) of the DML. The line sizes are: $d = 3.17 \cdot 10^{-3}$ m, $w = 3.043 \cdot 10^{-3}$ m, $l_1 = l_2 = 5 \cdot 10^{-3}$ m, $t = 3 \cdot 10^{-6}$ m.

178. **Fig. 8.7.** (a) the geometry of waveguides, (b) dispersion dependences of the main mode for waveguides.

179. **Fig. 8.8.** Dispersion dependences of the longitudinal propagation constant h of the main mode (curve 1) and the first higher mode (curve 2). The solid lines are our calculation and the circles are the results from [8.3], [8.4].

180. **Fig. 8.9.** Dispersion dependences of the main mode (curve 1) and higher modes (curves 2, 3) for the asymmetric SL. The line parameters are: $d = 10^{-3}$ m, $w_1 = 3 \cdot 10^{-3}$ m, $w_2 = 10^{-3}$ m, $l_1 = l_2 = 0$, $t = 3 \cdot 10^{-6}$ m. The solid lines are our calculations and the circles are the experiment [8.5], [8.6].

181. **Fig. 8.10.** Dispersion dependences of the main mode (curve 1) and three higher modes (curves 2, 3, 4) for the SL with narrow metal strips. The SL parameters are: $d = 10^{-3}$ m, $w_1 = w_2 = 2 \cdot 10^{-3}$ m, $l_1 = l_2 = 2 \cdot 10^{-3}$ m, $t = 3 \cdot 10^{-6}$ m. The solid lines are our calculations and the circles are the experimental results.

182. **Fig. 8.11.** Dispersion dependences of the SL with metal strips placed at the substrate edges for the main mode (curve 1) and two higher modes (curves 2, 3). The SL parameters are: $d = 10^{-3}$ m, $w_1 = w_2 = 2 \cdot 10^{-3}$ m, $l_1 = l_2 = 0$, $t = 3 \cdot 10^{-6}$ m. The solid lines are our calculations and the circles are experimental results.

183. **Fig. 8.12.** Dispersion dependences of the asymmetric SL with the protruding substrate edge for the main mode (curve 1) and the first higher (curve 2) mode. The SL parameters are: $d = 10^{-3}$ m, $w_1 = 10^{-3}$ m, $w_2 = 3 \cdot 10^{-3}$ m, $l_1 = 10^{-3}$ m, $l_2 = 0$, $t = 3 \cdot 10^{-6}$ m. The solid lines are our calculations and the circles are the experimental results.

184. **Fig. 8.13.** Dependence of the SL eigenmode number on the slot width s at $d = 100 \cdot 10^6$ m and $f = 37.5$ GHz.

185. **Fig. 8.14.** Dependence of the SL longitudinal propagation constants on the frequency at $d = 100 \cdot 10^{-6}$ m for the main mode and five higher modes.

186. **Fig. 8.15.** Dependence of the SL longitudinal propagation constants of the main mode and three higher modes (curves 1–3) on the frequency.

187. **Fig.8.16.** Dependence of the SL longitudinal propagation constants of the main mode and four higher modes (curves 1–4) on the frequency.

188. **Fig. 8.17.** Dispersion dependences for the main modes in the SL having different substrate widths at $d = 100 \cdot 10^6$ m.

189. **Fig. 8.18.** Dispersion dependence of the main modes for the SLs having different left metal strip widths.

190. **Fig. 8.19.** Dispersion dependences of the main modes of the SLs having different right metal strip widths.

191. **Fig. 8.20.** Cross–section geometry of the NSL.

192. **Fig. 8.21.** Dispersion dependences of the main and four higher modes (curves 1–4) for the NSL. The SL sizes are: $d = 10^{-3}$ m, $w_1 = 2 \cdot 10^{-3}$ m, $w_2 = 4 \cdot 10^{-3}$ m, $l_1 = l_3 = l_4 = 10^{-3}$ m, $l_2 = 2.1 \cdot 10^{-3}$ m and $s = -2 \cdot 10^{-3}$ m. The solid lines are our calculations and the circles are the experimental results [8.5], [8.6].

193. **Fig. 8.22.** Dispersion dependences of the main and four higher modes for the NSL. The SL sizes are: $d = 10^{-3}$ m, $w_1 = 3 \cdot 10^{-3}$ m, $w_2 = 2 \cdot 10^{-3}$ m, $l_1 = l_4 = 0.1 \cdot 10^{-3}$ m, $l_2 = 1.1 \cdot 10^{-3}$ m, $l_3 = 2.1 \cdot 10^{-3}$ m and $s = -10^{-3}$ m. The solid lines are our calculations and the circles are experimental results [8.5], [8.6].

194. **Fig.8.23.** Dispersion dependences of the main and four higher modes for the NSL with the nonoverlapping metal strip. The SL sizes are: $d = 10^{-3}$ m, $w_1 = 3 \cdot 10^{-3}$ m, $w_2 = 2 \cdot 10^{-3}$ m, $l_1 = l_4 = 0.1 \cdot 10^{-3}$ m, $l_2 = 3.1 \cdot 10^{-3}$ m, $l_3 = 4.1 \cdot 10^{-3}$ m and $s = 10^{-3}$ m. The solid lines are our calculations and the circles are the experimental results [8.5], [8.6].

195. **Fig. 8.24.** Dependence of the longitudinal propagation constants of the NSL eigenmodes on the frequency for different substrate thicknesses and overlapping of metal strips: (a) $d = 5 \cdot 10^{-6}$ m, $s = -20 \cdot 10^{-6}$ m; (b) $3d = 15 \cdot 10^{-6}$ m, $s = -20 \cdot 10^{-6}$ m; (c) $d = 5 \cdot 10^{-6}$ m, $s = -40 \cdot 10^{-6}$ m.

196. *Fig. 8.25.* Dependence of the longitudinal propagation constants of the NSL eigenmodes on the frequency for different substrate thickness and overlapping of metal strips: (a) $d = 5 \cdot 10^{-6}$ m, $s = -20 \cdot 10^{-6}$ m; (b) $d = 15 \cdot 10^{-6}$ m, $s = -20 \cdot 10^{-6}$ m; (c) $d = 5 \cdot 10^{-6}$ m and $s = -40 \cdot 10^{-6}$ m.

197. **Fig. 8.26.** Dispersion dependences of the longitudinal propagation constant of the main mode (curve 1) and three higher modes (curves 2, 3, 4) for the rectangular DW. The waveguide parameters are: $a = 15 \cdot 10^{-3}$ m and $d = 5 \cdot 10^{-3}$ m. The solid lines are our calculations and the circles are results of [8.10].

198. **Fig.8.27.** Dispersion dependences of the longitudinal propagation constant for the main (curve 1) and three higher modes (curves 2, 3, 4) of the strip DW. The waveguide parameters are: $d = t = l_1 = l_2 = 5 \cdot 10^{-3}$ m, $w = 10^{-2}$ m, $\varepsilon_{r_1} = 2.62$, $\varepsilon_{r_2} = 2.66$ and $\mu_r = 1$. The solid lines are our calculations and the circles are the results of [8.11].

199. **Fig. 8.28.** Dispersion dependences of the longitudinal propagation constant of the main mode (curve 1) and two higher modes (curve 2, 3) for the mirror DW. The waveguide parameters are: $d = 0.35 \cdot 10^{-3}$ m, $a = 0.7 \cdot 10^{-3}$ m and $t = 10^{-6}$ m. The solid ($\varepsilon_r = 11.8$) and the dash lines ($\varepsilon_r = 9.7$) are our calculations and the circles are the results from [8.12].

200. **Fig. 8.29.** Dispersion dependences of the longitudinal propagation constant of the main mode (curve 1) and four higher (curves 2–5) modes for the mirror demagnetized ferrite waveguide. The waveguide parameters are: $d = 10^{-3}$ m, $a = 2 \cdot 10^{-3}$ m, $t = 2 \cdot 10^{-5}$ m, $\varepsilon_r^f = 13.5$ and $\mu_r = 1$. The solid lines are dispersion curves for the open ferrite waveguide (mirror waveguide), the dash lines are for the cylindrical demagnetized ferrite waveguide with the radius d.

Chapter 9

201. **Fig. 9.1.** The geometry of a layer gyrotropic waveguide.
202. **Fig. 9.2.** Dispersion dependences of the main mode HE_{11} and two higher modes EH_{11} and HE_{12} for the layer ferrite–dielectric waveguide (Fig.9.1) at the layer thickness $d/r_f = 0.3$ and several ferrite magnetizations. $0 - M = 0$ (solid lines), $1 - M/M_s = 0.5$ and $2 - M/M_s = 0.7$ (dash lines).
203. **Fig. 9.3.** Dispersion dependences of the main mode HE_{11} and two higher modes EH_{11} and HE_{12} for the layer ferrite–dielectric waveguide (Fig.9.1) at the layer thickness $d/r_f = 0.5$ and several ferrite magnetizations. $0 - M = 0$ (solid lines), $1 - M/M_s = 0.5$ and $2 - M/M_s = 0.7$ (dash lines).
204. **Fig. 9.4.** Dependence of the differential phase shift on: (a) the ferrite magnetization at the dielectric layer thickness $d/r_f = 0.3$ and $\varepsilon_d = 9.6$; (b) on the dielectric layer thickness where symbol 1 corresponds to $\varepsilon_d = 9.6$, symbol 2 corresponds to $\varepsilon_d = 15$ and symbols I, II, III equates to $M/M_s = 0.3, 0.5, 0.7$ accordingly.
205. **Fig. 9.5.** Dependence of losses on the frequency for the waveguide with and without a layer. The solid lines correspond to the demagnetized ferrite M=0 and the dash lines correspond to the ferrite magnetization M/M$_s$= 0.7. Symbol 1 shows dependences for the ferrite waveguide with the dielectric layer $d/r_f = 0.3$ and $\varepsilon_d = 9.6$. Symbol 2 shows dependences for the ferrite waveguide without a dielectric layer.
206. **Fig. 9.6.** Experimental and theoretical dependences of the differential phase shift $\Delta\varphi$ modules on the material magnetization M/M_s for three ferrite waveguides. We designate with the circle the experimental results and with the solid line our calculations.
207. **Fig. 9.7.** The topology of the piece–wise uniform gyrotropic waveguide.
208. **Fig. 9.8.** The integration contour L_m surrounding different media.
209. **Fig. 9.9.** Dispersion dependences of the SL with the ferrite substrate at $4\pi M_s = 377$ kA/m. The sizes are: $w_1 = w_2 = 2 \cdot 10^{-3}$ m, $l_1 = l_2 = 2 \cdot 10^{-3}$ m, $d = 10^{-3}$ m, $s = 0.2 \cdot 10^{-3}$ m, $t = 2.5 \cdot 10^{-6}$ m.
210. **Fig. 9.10.** Dispersion dependence of the asymmetric SL with the ferrite substrate at $4\pi M_s = 377$ kA/m. The sizes are: $w_1 = 3 \cdot 10^{-3}$ m, $w_2 = 10^{-3}$ m, $l_1 = 0$, $l_2 = 2 \cdot 10^{-3}$ m, $d = 10^{-3}$ m, $s = 0.2 \cdot 10^{-3}$ m and $t = 2.5 \cdot 10^{-6}$ m.
211. **Fig. 9.11.** Dispersion dependences of the asymmetric SL with the ferrite substrate for the main mode (curve 1) and the first higher mode (curve 2). The sizes are: $w_1 = 3 \cdot 10^{-3}$ m, $w_2 = 10^{-3}$ m $l_1 = l_2 = 0$, $d = 10^{-3}$ m, $s = 0.2 \cdot 10^{-3}$ m and $t = 2.5 \cdot 10^{-6}$ m.
212. **Fig. 9.12.** Dispersion dependences (a) the outside wave number $\xi_\perp$ of the main mode (curve 1) and three higher modes (curves 2, 3, 4). (b) transversal

propagation constants $k_{\perp 1}$, $k_{\perp 2}$ of the main mode. The parameters are: the charge carrier concentration $n = 10^{18}$ m^{-3}, the magnetic field induction $B = 1$ T and the waveguide width $a = 10^{-3}$ m.

213. **Fig. 9.13.** Dispersion dependences of the longitudinal propagation constant h of the main mode (curve 1) and three higher modes (2, 3, 4). The parameters are: the charge carrier concentration $n = 10^{18}$ m^{-3}, the magnetic field induction $B = 1$ T and the waveguide width $a = 2 \cdot 10^{-3}$ m.

214. **Fig. 9.14.** Dependence of the longitudinal propagation constant h of the main mode on the free charge carrier concentration n. The parameters are: the frequency $fa = 0.03$ GHz·m, the magnetic field induction $B = 1$ T and the waveguide width $a = 2 \cdot 10^{-3}$ m.

215. **Fig. 9.15.** Dependence of the transversal wave numbers $k_{\perp 1}$, $k_{\perp 2}$ and the outside wave number $\xi_\perp$ on the free charge carrier concentration for the square plasma waveguide. The parameters are: $fa = 0.03$ GHz·m, the magnetic field induction $B = 1$ T and the waveguide width $a = 2 \cdot 10^{-3}$ m.

216. **Fig. 9.16.** The topology of (a) one - comb waveguide and (b) two – comb waveguide.

217. **Fig. 9.17.** Dispersion dependences of the main mode (curve 1) and two higher modes (2, 3) for the ferrite comb waveguide (Fig. 9.16a). The parameters are: $d = 10^{-3}$ m, $l_1 = w = t = 3 \cdot 10^{-3}$ m, $l_2 = 2 \cdot 10^{-3}$ m, and the ferrite magnetization is $4\pi M_s = 377$ kA/m.

218. **Fig. 9.18.** Dispersion dependences of the main mode (curve 1) and two higher modes (2, 3) for the two-comb ferrite waveguide (Fig.9.16b). The parameters are: $l_1 = l_2 = t_1 = w_1 = 3 \cdot 10^{-3}$ m, $d = s = t_2 = w_2 = 10^{-3}$ m and $4\pi M_s = 377$ kA/m.

219. **Fig. 9.19.** Dispersion dependences for a one–comb plasma waveguide (Fig.9.16a): (a) of the longitudinal propagation constant h of the main mode (curve 1) and three higher modes (2, 3, 4). (b) of the transversal propagation constants $k_{\perp 1,2}$ and of the outside wave number $\xi_\perp$. The parameters are: $l_1 = w = t = 3 \cdot 10^{-3}$ m, $l_2 = 2 \cdot 10^{-3}$ m, $d = 10^{-3}$ m, charge carrier concentration $n = 10^{18}$ m^{-3} and the magnetic field induction $B = 1$ T.

220. **Fig. 9.20.** Dependence for a one–comb plasma waveguide: (a) the longitudinal propagation constant h, (b) the transversal wave numbers $k_{\perp 1,2}$ and the outside wave number $\xi_\perp$ of the main mode on the charge carrier concentration. The parameters are: $l_1 = w = t = 3 \cdot 10^{-3}$ m, $l_2 = 2 \cdot 10^{-3}$ m, $d = 10^{-3}$ m, frequency $f \cdot d = 0.015$ GHz·m and the magnetic field induction $B = 1$ T.

221. **Fig. 9.21.** Dispersion dependences of the two–comb waveguide (Fig.9.16b): (a) the longitudinal propagation constant h of the main mode (curve 1) and three higher modes (2, 3, 4); (b) the transversal wave numbers $k_{\perp 1,2}$ and of the outside wave number $\xi_\perp$ of the main mode. The parameters are: $w_1 = l_1 = l_2 = t_1 = 3 \cdot 10^{-3}$ m,

$w_2 = s = t_2 = d = 10^{-3}$ m, the charge carrier concentration $n = 10^{18} \text{m}^{-3}$ and the magnetic field induction $B = 1$T.

222. **Fig. 9.22.** Dependence the two–comb plasma waveguide: (a) the longitudinal propagation constant h; (b) the transversal wave numbers $k_{\perp 1,2}$ and the outside wave number $\xi_\perp$ on the carrier concentration for the main mode. The parameters are: $w_1 = l_1 = t_1 = 3 \cdot 10^{-3}$ m, $w_2 = s = t_2 = d = 10^{-3}$ m, $l_2 = 2 \cdot 10^{-3}$ m the magnetic field induction $B = 1$T and the frequency $f \cdot d = 0.015$ GHz·m.

223. **Fig. 9.23.** The topology of the mirror ferrite waveguide.

224. **Fig. 9.24.** Dispersion dependences of the mirror ferrite waveguide. The parameters are: $d = 0.35 \cdot 10^{-3}$ m, $l = 0.7 \cdot 10^{-3}$ m, $t = 10^{-6}$ m, the saturation magnetization $4\pi M_s = 79.37$ kA/m, $\varepsilon_r = 9.8$ and the external constant magnetic field $H_0 = 39.68$ kA/m.

225. **Fig. 9.25.** Dependence of the longitudinal propagation constant h of the main mode (curve 1) and three higher modes (2, 3, 4) on the saturation magnetization. The parameters of the magnetized ferrite mirror waveguide are: $d = 0.35 \cdot 10^{-3}$ m, $l = 0.7 \cdot 10^{-3}$ m, $t = 10^{-6}$ m, $\varepsilon_r = 9.8$, constant magnetic field $H_0 = 39.68$ kA/m and the frequency $f = 70$ GHz.

226. **Fig. 9.26.** Dependence of the longitudinal propagation constant of the main mode (curve 1) and three higher modes (2, 3, 4) on the width of the magnetized ferrite mirror waveguide. The parameters are: $d = 0.35 \cdot 10^{-3}$ m, $t = 10^{-6}$ m, $\varepsilon_r = 9.8$, the saturation magnetization $4\pi M_s = 79.37$ kA/m, the external constant magnetic field $H_0 = 39.68$ kA/m and the frequency $f = 70$ GHz.

227. **Fig. 9.27.** (a) Dispersion dependences of the demagnetized ferrite waveguide: curve 1 for the main mode; curves 2 and 3 for the first and second higher modes; (b) dispersion dependences of the main mode propagating in the ferrite waveguide at several ferrite magnetization values. The parameters are: $d = 10^{-3}$ m, $l = 2 \cdot 10^{-3}$ m, $t = 20 \cdot 10^{-6}$ m and $\varepsilon_r = 13.5$.

228. **Fig. 9.28.** Dependence of the longitudinal propagation constant h of the main mode on the ferrite magnetization. The parameters of the mirror ferrite waveguide are: $d = 10^{-3}$ m, $l = 2 \cdot 10^{-3}$ m, $t = 20 \cdot 10^{-6}$ m, $\varepsilon_r = 13.5$ and $f \cdot d = 0.05$ GHz·m.

229. **Fig.9.29.** *Differential phase shift of the waveguide stretch. The solid line shows the calculated results and the circles show the measured results. The waveguide parameters are:* $d = 10^{-3}$ m, $l = 2 \cdot 10^{-3}$ m, $t = 2 \cdot 10^{-5}$ m, $\varepsilon_r = 13.5$ *and* $f \cdot d = 0.05$ GHz·m.

Chapter 10

230. **Fig. 10.1.** The waveguide topology.

231. **Fig. 10.2.** The sheet $\text{Im}\,\eta \prec 0$ for $b \prec 0$.

232. **Fig. 10.3.** The sheet $\text{Im}\,\eta \prec 0$ for $a \succ 0$, $0 \prec b \prec a^2$.

233. **Fig. 10.4.** The sheet $\mathrm{Im}\,\eta \prec 0$ for $a \prec 0$ and $0 \prec b \prec a^2$.

234. **Fig. 10.5.** The sheet $\mathrm{Im}\,\eta \prec 0$ for $a \succ 0$ and $b \succ a^2$.

235. **Fig.10.6.** The sheet $\mathrm{Im}\,\eta \prec 0$ for $a \prec 0$ and $b \succ a^2$.

236. **Fig. 10.7.** The function $\chi_{z_1}(\chi_x)$ cuts shown on the sheet $\eta(\chi_x)$, $\mathrm{Im}\,\chi_{z_1} \leq 0$, where C is the integration contour in the formula (10.19). This function $\eta(\chi_x)$ cuts start at the points $\chi_x = \pm 0.75 \pm i0.49$, the function $\chi_{z_1}(\chi_x)$ cuts start at the points $\chi_x = \pm 1$ and $\chi_x = \pm 0.648$.

237. **Fig. 10.8.** The function $\chi_{z_2}(\chi_x)$ cuts shown on the sheet $\eta(\chi_x)$ and $\mathrm{Im}\,\chi_{z_2} \prec 0$, where C is the integration contour in the formula (10.19). The function $\eta(\chi_x)$ cuts start at the points $\chi_x = \pm 0.75 \pm i0.49$, the function $\chi_{z_2}(\chi_x)$ cuts start at the points $\chi_x = \pm 1$ and $\chi_x = \pm 0.648$.

Chapter 11

238. **Fig.11.1.** (a) 3D structure (scattered body) used in our calculations, which was limited by the surfaces S_1, S_2, S_3; (b) the 3D structure surface was constructed of triangles.

239. **Fig 11.2.** (a) Circular metal disc; (b) back scattering cross–section dependence on kr for a circular metal disc. Solid line shows our calculations and the experimental data is shown by circles.

240. **Fig. 11.3.** The frequency selective structure with rectangular metal strip frames on its surface.

241. **Fig. 11.4.** The central frequency dependences (a, c, d) on substrate layer thicknesses d, d_1 and (b) on the relative permittivity ε_r of the upper dielectric layer.

242. **Fig. 11.5.** (a) polar coordinates; (b) the 3D structure of a sensor illuminated by microwaves $E_0(\vec{r})e^{-ik\vec{r}}$ and $H_0(\vec{r})e^{-ik\vec{r}}$.

243. **Fig. 11.6.** The transmission coefficient module $\left|\vec{E}^t\right|/\left|\vec{E}_0\right|$ dependence on polar angles Θ of the parallel– polarized incident EM wave.

244. **Fig. 11.7.** Description of the point position in a plane AOB.

245. **Fig. 11.8.** The model of the heart with the myocardium, the left and right atria with ventricles that are limited by the surfaces S_1, S_2, S_3 and designations.

246. **Fig. 11.9.** Dependences of the back scattering cross–section for the heart model.

247. **Fig. 11.10.** The normalized modulus of the scattered electric field as function of the distance X for a parallel–polarized incident microwave at f=10 GHz, φ=30°, θ=0°, ε_1=40–i15, ε_2=55–i7.

248. **Fig. 11.11.** The normalized modulus of the scattered electric field as a function of the distance X for a perpendicular–polarized incident microwave at f =10 GHz, φ =30°, θ=0°, ε_1=40–i15, ε_2=55–i7.

249. **Fig. 11.12.** The normalized modulus of the scattered electric field as a function of polar angle θ for a parallel–polarized incident microwave at f=10GHz, φ =30°, ε_1=40–i15, ε_2=55–i7.

250. **Fig. 11.13.** (a) A schematic longitudinal section of the heart model consisting of the myocardium, the atria and ventricles; (b) the same heart model with a catheter (the shaded rod) inside of its right atria.

251. **Fig. 11.14.** Distribution of the normalized electric field modulus on the x direction in three cross–sections of the heart model at the coordinates $Z = 2 \cdot 10^{-2}$ m (triangles), $Z = 3 \cdot 10^{-2}$ m (circles), $Z = 5 \cdot 10^{-2}$ m (squares) and coordinate $Y = 0$. The microwave catheter is made of dielectric material with the relative permittivity of 10.

252. **Fig. 11.15.** Distribution of the electric field modulus on the x direction in three cross–sections of the heart model at $Z = 2 \cdot 10^{-2}$ m (triangles), $Z = 3 \cdot 10^{-2}$ m (circles), $Z = 5 \cdot 10^{-2}$ m (squares) and $Y = 0$ when the microwave catheter is made of the metal.

A–8 INDEX

Absolute   135, 159, 191, 252
Absorption   279
Adjoin   14
Admittance   78, 135
Air   95, 112, 121, 124, 129, 136, 151, 188, 222, 270
Algebraic system   85, 236, 273
Algorithm   92, 117, 121, 151, 191, 193, 236, 237, 238, 250, 273, 279, 284
Alternating magnetic field   43
Aluminum Oxide   111
Amplitude   2, 5, 8, 43, 44, 54, 60, 61, 78, 79, 217, 226, 279
Analysis   9, 45, 85, 86, 133, 162, 224
Analytic   10, 16, 31, 34, 35
Analytical   9, 29, 191
Antiparallel   44
Anisotropic   1, 29, 41, 42, 50, 57, 69, 70, 86.121, 122, 124, 126, 132
Antisymmetric   55, 85
Antisymmetric tensor   41, 52
Angle   5, 6, 23, 30, 47, 48, 49, 50, 52, 62, 81, 84, 85, 136, 154, 189, 258-260, 278, 279, 282, 283
Angular   49, 221, 275, 280
Antenna   268, 271, 272, 273, 284
Approximation   4, 8, 9, 40, 43, 58, 60, 71, 85, 86, 112, 124, 133, 141, 142, 152-154, 175, 183, 184
Arbitrary   154, 185, 215, 216, 253, 267, 268
Argument   8, 21, 23, 24, 85, 217, 247, 257
Attenuation   77, 89, 90-96, 100-117, 122-132, 136, 148, 162, 164
Axis   1, 11-23, 43, 44, 48-52, 55, 62, 81, 217, 225, 257-264, 281, 282, 285, 286
Axes   19, 24, 45, 47, 48, 50, 52, 234, 260, 261, 262, 285
Azimuthal angle 278

Back waves   78
Back scattering cross–section   273, 274, 281
Band   153, 173, 202, 208, 221, 224, 239, 247
Bandwidth   202, 221, 274
Basic   4
Basis   73, 91, 117, 140, 252, 253, 283

Bessel   5, 7, 217, 226
Biomedical   268
Bigyrotropic medium   218, 267
Bloch equation   45
Blood substances   285
Bogoliubov-Krylov   39, 59, 60, 81
Boundary condition   3, 5, 8, 10, 38-42, 57-70, 105, 151, 153, 154, 185-188, 191, 196, 200, 216, 228, 230, 232, 270, 273
Boundless media   44
Branch   13, 34, 35, 37, 257-263, 267
Branch one-valued   257
Branch two-valued   257

Cable wave   1, 3, 4
Capacitance   6, 10, 16-28, 57, 65, 69, 71, 78, 101-130, 133, 150
Carrier   56, 133, 137, 158, 160, 163, 165, 166, 169, 173-175, 178-181, 241-243, 245-248
Cartesian   2, 42, 43, 45, 50, 225, 234, 254, 285
Catheter   271, 279, 283-287
Cauchy-Riemann   10, 11
Cauchy type integral   29, 30, 32, 33, 36-40, 189
Central   221, 222, 274-276
Characteristic impedance   89, 217, 269
Characteristic size   58
Charge   6, 12-16, 42, 54-60, 65-67, 70, 71, 133
Charge carrier concentration   137, 139, 158-166, 169, 173-176, 178-181, 241-243, 245-248
Circuit   18, 20, 78, 79, 80, 100, 284
Circular cylinder   24
Circular polarization
Coaxial   4, 6-9, 149, 150
Coaxial waveguide   4, 6-8
Coefficient   39, 40, 45, 47, 48, 59, 60, 65, 67, 68, 74, 75, 84, 85, 134, 135, 148, 166, 167-169, 175-177, 180-182, 191, 218, 219, 226-228, 236, 262, 267, 274, 277, 278
Complex amplitudes   2, 60, 143

Complex argument 217
Complex conjugate 52, 67, 68, 76
Complex coordinate 10, 60, 153
Complex function 10, 60
Complex number 85
Complex permittivity 279
Complex plane 15, 186, 255, 26
Complex potential 11, 60, 64
Complex variable 26, 29
Complex wave impedance 158, 159,
    160-164, 173-177
Concentration 55, 56, 95, 136, 137, 158,
    160-166, 169, 173-175, 178-181,
    241-243, 245-248, 282
Conductivity 42, 53, 54-57, 75
Conformal mapping 16, 17, 19, 21, 22,
    24, 26, 28
Conjugate 52, 67, 68, 73, 75, 76, 78
Connection between Matrices 69
Constant magnetic field 41, 43, 44, 50,
    53, 133, 137, 138, 140-146, 148-151,
    156-158, 163, 164, 169, 172, 182, 237,
    248-250, 252
Constitutive relations 41, 43, 44, 50, 53
Continuity 7, 37
Convergence 40, 265
Coordinate system 1, 4, 41, 48, 50, 52,
    55, 57, 260, 262, 263, 280, 285
Corner point 235
Coupled MSL 132, 133, 166, 169, 170,
    172, 175, 178, 179, 181
Coupling coefficient 134, 135, 166-169,
    175-177, 180-182
Christoffel–Schwarz formula 26
Criterion 184
Cross-section 178, 216, 253, 267, 273,
    274, 281, 284-287
Current 53-56, 60, 66, 68, 70-72, 74,
    77-80, 90, 92, 268
Current coupling coefficient 135, 166-169,
    176, 177, 180-182
Current density distribution 146-148
Cut 13-15, 29-37, 221, 247, 248, 255,
    257, 258, 260-265
Cutoff frequency 3, 146
Cylindrical 1, 4, 5, 214
Cylindrical coordinate 186
Cylindrical wave 1

Damping processes 44
Decibel 89, 91
Delta function 254
Density 6, 13, 29, 31-38, 53, 55, 59, 61,
    68, 71, 81, 84, 146, 147, 185, 216, 236
Derivative 2, 10, 11, 20, 45, 49, 50, 52,
    53, 80, 64, 151, 189, 228, 234, 235
Diagonal 40, 41, 44, 57, 81, 82, 85, 122,
    126, 142, 175, 191, 239, 244
Dielectric 81, 89, 100, 166, 214, 219-223,
    243, 253, 267, 268, 275-278, 284-286
Dielectric layer 8, 98, 99, 100, 105, 106,
    158, 159, 162, 166, 167, 193, 219-223,
    275, 276
Dielectric media 61, 67, 81
Dielectric substrate 86-88, 97, 100, 277
Dielectric waveguide 184, 212, 219-221,
    243, 253
Differential 2, 5, 6, 9, 39, 41, 58, 72,
    151, 185, 186, 216, 221-224, 226, 248,
    252, 254, 255
Differential equations 2, 5, 6, 9, 39, 72,
    216, 226, 254, 255
Diffraction 29, 268, 274, 279, 280, 284
Dipole 277
Dirac function 59
Dirac delta function 254
Disc 273, 274
Discontinuity 37, 63
Dispersion 185, 192-215, 216, 219-221,
    236, 239-251, 267
Dissipative 44, 45, 60, 139, 173
Distribution 8, 110, 111, 121, 146-149,
    151, 155-157, 166, 169-172, 177, 178,
    180, 182-184, 200, 207, 241, 282, 284,
    285-287
Divergence 41, 150
Double 83, 105, 172, 178, 179, 181

Edge effects 118
Effective permittivity 87, 101-116,
    122-131, 137, 163, 164
Effective mass 54, 56, 57, 133, 241, 245
Eigenmodes 193, 198, 201, 202, 210, 211
Eigenvalues 5, 72, 75, 76, 134
Eigenvector 72-75, 77
Electric field 3, 5-17, 59-70, 74, 93, 110,
    111, 121, 228, 229, 271, 273, 277,
    281-287

Electric induction 41, 42, 57
Electric wave 269
Electromagnetic energy 95, 99, 101, 109, 114, 120, 126, 127-129
Electromagnetic field 10, 39, 54, 58, 112, 152, 185, 254, 267
Electromagnetic wave 1, 78, 114, 120, 121, 124, 136, 142, 165, 173, 222, 267, 268, 280
Electrostatic 14, 61, 64, 65, 69, 70
Electrophysical 133, 158, 163, 173, 179, 186, 192, 219, 223, 241
Element 133-135, 141-143, 150, 152, 165, 175, 187, 191, 196, 129, 277
Energy 47-50, 74, 77, 95, 99, 101, 109, 112, 114, 120-128, 139, 144, 173, 202, 222, 247, 282, 284
Equilibrium 47, 50, 51, 53
Equivalent 45, 74, 112, 128, 146
Expansions 8
Experiment 66-68, 88, 91-94, 117, 118, 133, 137, 138, 142, 144-150, 155-157, 161-165, 193, 196-200, 206-209, 212-215
Extreme 128, 138, 159, 161

Factor 5, 29, 35, 38, 42, 60, 67, 72, 154, 189, 217, 266, 267
Ferrite 41-44, 47, 89, 132, 141-144, 146, 148-153, 156, 158, 172-181, 214, 219-224, 237-240, 243-253
Ferrite-dielectric waveguide 220, 221
Field distribution 8, 110, 155, 169
Finite leap 33
First kind 15, 26
Flange end 22, 23
Flow 283, 284
Force 154, 155, 166, 171, 178, 180, 183, 184
Force moment 42, 49, 50
Free member 273
Free space 269, 278
Frequency ranges 220
Friction force 44, 54, 56
Fourier 151, 255, 269
Fundamental solution 254, 267, 268

Gradient 269
Group velocity 77

Ground 66, 111
Gyrotropic media 40, 41, 83, 84, 132, 133

Hankel function 186, 187, 189, 219, 229, 247, 266, 267, 269, 273
Hermitian matrix 51, 53, 67, 68, 73, 75, 78
Harmonic function 10, 41, 48, 54, 56, 58, 150, 151, 153
Heart 279-287
Helicon wave 137, 161
Helmholtz's equation 9, 58, 186
Hilbert 44
Hoelder condition 32-34, 36, 38
Homogeneous 151, 154, 185, 191, 215, 226, 234, 285
Hybrid modes 214

Image 268
Imaginary 11-13, 20-25, 29, 47, 55, 76, 93, 96-106, 111, 113, 128, 130, 132, 135, 150, 158-160, 163, 167, 181, 185, 189, 217, 247, 257, 259-265, 281, 284
Incidence 277
Inductance 7, 14, 44, 62, 67, 69, 70, 72, 142, 143, 161
Induction 42, 43, 58, 66, 133, 136-138, 146, 161-169, 173, 177, 178, 181, 182, 241-243, 245-247
Infinite 13, 16, 45, 89, 105, 106, 116, 134, 150, 151, 196, 212, 265, 269, 273
Inhomogeneous 285
Integral Cauchy Type 29, 30, 32, 33, 36-40
Integral elliptic 16, 19, 28
Interference 138, 161, 285
Inverse matrix 71
Inverse tensor 57
Inverse transformation 19
Isotropic material 111, 122, 126, 175, 215
Isotropic substrates 41, 85, 86

Jacobi iteration method 40
Join 13, 14, 36, 38, 165, 261
Junction 150

Kernel 38, 191
Kirchhoff's low 78
Landau–Lifshits' equation 42

Lande factor 42
Laplace's equation 9, 10, 41, 58, 59
Laplacian 186
Light velocity 186
Linear equations 60, 81, 154, 226, 234, 273
Logarithmic 266
Longitudinal components 2, 4, 5, 8, 45,
  58, 60, 61, 68, 151, 152, 183, 185,
  186, 217, 225, 226, 234, 254, 265
Longitudinally magnetized 135, 138, 141,
  163-173, 181, 216, 219, 224, 239, 243,
  244, 248
Long line circuit 78
Loss angle 136
Losses 42, 44, 53, 60, 67, 73, 78, 97, 100,
  112, 126, 129, 133, 136, 139-141, 143,
  148-150, 159, 161-163, 165-168,
  173-181, 206, 220, 222-224, 261
Loss tangent 136
Lossless 44, 56, 71, 85, 257
Lossy 267, 286
Lower layer 86, 97, 101, 111, 115, 124,
  126, 178, 183, 275

Magnetic field 3, 6, 7, 10, 39-68, 112,
  133, 137, 138, 140-152, 155-158, 163,
  164, 169, 172, 178, 179, 182-185, 219,
  228, 237, 241-243, 245-250, 252, 253,
  272, 273
Magnetic field energy 112
Magnetic induction 42, 65, 133, 137, 138,
  146, 161-169, 173, 177, 178, 181, 182
Magnetic moment 42, 44, 47
Magnetic susceptibility 44-47, 51, 53
Magnetization vector 42, 44, 47, 49, 50
Magnetodielectric 61, 86, 148, 166, 168,
  187, 188
Magnetostatic 61, 64, 65, 69, 70
Material parameters 219, 237
Matching 150, 163
Matrix 39-53, 67-81, 84, 133-135, 142,
  143, 150
Maximum 20, 99, 109, 139, 144, 163,
  282, 285
Maxwell's equations 1, 39, 42, 60, 75, 151,
  152, 184, 185, 187, 215, 216, 217, 225,
  227, 253, 254, 267, 268, 270
Meander 141-148, 150

Measure 89, 91, 92, 138, 146, 148-150,
  161, 197, 200, 207
Micron size 201
Micron slot 193, 200
Microstrip reflecting structure 274
Microwave 93, 95, 97, 102, 252, 268,
  271, 274, 278, 279, 281-287
Minimum 47, 149
Mirror waveguide 201, 214, 249, 250
Model 173, 179, 274, 279-287
Module 85, 94, 224, 277, 278
Moment 42, 44, 47, 49, 50, 56
Monotonically 99
Multiconductor 60, 66, 71, 72
Myocardium 270, 280, 281, 284, 285

Neumann 5, 7, 219
Nonmonotonously 96
Nonplanar 193, 206, 209
Nonsingular kernel 38
Nonuniform 6, 29, 60
Normal component 13, 61-63, 65, 69

Odd 47, 88, 134, 139, 167
Ohm's low 53, 78, 79, 93
One-mode waveguide 243
One-valued branch 257
Open end 20, 22
Operator 75, 83, 151, 217, 269
Opposite bank of the function 264
Ordinary 163
Origin of the coordinate system 262, 263
Orthogonal 73, 74, 76

Parallel 1, 11, 20, 234, 278, 281-285
Parallel-polarized 281-283
Pattern 283
Perfect conductor 273
Permeability 41-45, 66, 69, 88, 141, 143,
  148, 150, 152, 168, 169, 172, 185, 217,
  237, 239, 243, 280
Permittivity tensor 41, 53-55, 117,
  120-122, 126, 133, 172, 175, 217, 237,
  239, 243
Perpendicular 274, 282, 283
Perpendicular-polarized 274, 282, 283
Phase 67, 96, 101, 138, 143, 169-171,
  221, 279, 253, 279

Phase shift   77, 136, 138, 141, 143-146, 149, 150, 156-171, 181, 189, 221-224, 248, 252, 253
Phase velocity   74, 167
Piece-wise   58, 66, 227
Planar   193, 196, 200, 206, 209
Plane wave   273, 277, 280-282
Plasma   133, 139, 163, 243-247
Plasma waveguide   243-247
Plemelj   29, 31, 39, 189, 191
Polar angle   277, 278, 282, 283
Polarization   144, 168-172, 178-184, 221, 222, 278, 279, 282
Polynomial   31, 48
Potential   3, 6, 10-16, 41, 60-72, 78, 81, 83, 86, 169
Power   31, 33, 34, 36, 47, 74, 77
Principal   189, 273
Product   47, 54, 55, 72, 73, 269, 271-273
Projection   4, 11, 43-45, 49, 50, 57, 62
Propagation constant   136, 138, 139, 142-144, 150-169, 173-177, 180-182, 85-197, 203-216, 217, 219, 224, 236, 239-242, 244, 245, 252, 254

Quasi-E -wave   226, 227, 240, 247, 255, 270
Quasi-H-wave   226, 227, 240, 247, 255, 270
Quasi-TEM   110-112

Radius   12, 13, 15, 24, 30, 187, 189, 214, 220, 222, 227, 273
Radiation   149, 268, 279
Reactive   163, 175
Receiving   204, 268
Reciprocity theorem   57, 65, 67
Rectangular   26, 27, 29, 45, 196, 206, 212, 214, 215, 275

Reflected electric field   273
Reflection   165, 274
Regular   1, 59, 141, 142, 185, 191, 200, 209, 216, 229, 267
Regular line   60
Relations   2, 43, 44, 47, 57, 70, 85, 133
Relative permeability   43, 88, 143, 169, 185, 280

Relative permittivity   86, 87, 92, 95, 97-101, 105, 110, 112, 128, 133, 152-167, 172, 185, 209, 220, 239, 241, 244, 275-281, 284, 286
Relaxation time   45, 54, 56
Representation   38, 57-59, 63, 64, 69, 185, 189, 216, 227, 254, 265
Resistance   146, 219, 237, 286, 287
Resistivity   93-98, 148, 196, 206
Resonance   44, 46, 56
Rest mass of electron   42
Riemann surface 258, 261, 262
Right   30, 37, 38, 61, 64, 68, 77-79, 111, 132, 133, 189, 190, 202-205, 254-257, 261-263, 267, 280, 284, 285
Rotation   49, 50-52

Sample   44, 47, 277
Saturated ferrite   152, 219
Saturation   42, 43, 50, 92, 137, 156, 173-182, 239, 243, 248-250
Scalar   2, 54, 56, 67, 72, 269, 279, 280

Scalar product   54, 72
Scattering   89, 207, 208, 268, 269, 273, 274, 280, 281
Second order   53, 186
Seidel iteration method   40
Semiconductor material   41, 136, 137158, 161-163, 175, 178, 179, 248, 252, 253
Semiconductor sensor   277, 278
Series   31
Sheets join   261
SIE method   29, 60, 86, 88, 124, 133, 146, 151, 154, 156, 162, 184, 185, 215, 216, 236, 252.-254, 266, 268, 273, 279, 287
Signal   74, 77, 92, 95, 100, 105, 112, 121, 126, 138-144, 150, 161, 163, 174, 185, 191, 193, 197197, 206, 223, 224, 274, 285
Silicon   92, 100, 111, 196, 206
Single-valued   30, 34, 35, 37, 266
Singling out of the singularity   266
Singular   216, 229, 266
Slot line   239
Sokhotsky-Plemelj   29, 31, 39, 235, 273
Source   58, 70, 216, 227, 268, 269, 272-274, 279, 284, 285

Space   47, 73, 95, 269, 274, 278
Specific resistivity   93-98, 148, 196, 206
Spectrum   202
Speed   44, 55, 56, 142, 149
Spherical   269, 273
Square waveguide   241
Static field   3, 6, 14, 64
Strong   42, 93, 95, 139, 165, 198, 208, 242, 278
Structure   29, 40, 41, 42, 141, 150, 154, 173, 236, 238, 267, 268, 270-278, 285
Superposition   198, 266, 270
Surface charge density   13
Surface integral   269, 279
Susceptibility   43-47, 51-53
Symmetric   17-25, 29, 41, 52, 55, 67, 83, 85, 133-135, 138, 155, 156, 167, 193, 194, 198, 215, 239, 277, 278
Symmetry   16, 47-49, 52, 83-85, 133, 148, 156, 241

Tangential component   3, 61-63, 66, 68, 69, 216, 218, 219, 228, 229
Tensor   133, 141, 142, 150, 152, 172, 175, 178, 183, 216, 217, 219, 225, 237, 239, 243, 252
TEM waves   110, 111
Theorem   57, 65, 67
Three-dimensional   268, 277, 279
Total losses   148, 149
Transformation   16, 19, 20, 24, 25, 41, 45, 47, 48, 59, 71, 83, 89, 255, 269
Transversal component   3-5, 8, 217, 225, 228, 234, 235, 266
Transversal propagation number   217, 219

Transversally magnetized ferrite   150, 151, 252, 253
Transmission   1, 58, 59, 65-67, 71-79, 85, 92, 100, 193, 211, 277, 278
Transition attenuation constant   89-91
Transverse electromagnetic   4
Triangle   270, 274, 279-281, 284, 286, 287
Two-comb waveguide   243, 244, 246, 248

Uniaxial   49
Uniform   5, 10, 47, 48, 54, 58-60, 66, 68, 72, 92, 216, 224, 227, 255
Uniform fillings   29, 58, 224
Uniform integral equations   59
Upper layer substrate   275

Vacuum   42, 58, 88, 95, 152, 185, 186, 217, 219, 269, 275, 281
Variable   16, 22, 26, 27, 29, 31, 37, 261, 266
Variation   90, 111, 130, 192
Vector-column   71, 72, 76
Vector product   55, 269, 271, 272
Velocity   74, 77, 167, 186, 275
Vertical strip   25
Voltage   71, 72, 74, 77-80, 150, 176
Voltage coupling coefficient   134, 175, 176

Wave equation   186, 217, 219, 227
Wave propagation   138, 139, 161, 176, 177
Wavelength   8, 58, 87, 88, 101-116, 123-131, 146, 149, 161, 193-202, 207, 210-213, 239, 250, 269, 273281, 285
Wave number   217-219, 226, 241, 243, 245-247, 269, 273, 281

 Appendixes

A–9 ACRONIMS:

CSWR	Current standing–wave ratio
2D	two–dimensional
3D	three–dimensional
D.C.	direct current
DML	MSL with dielectric substrate and without the metal plate
DW	dielectric waveguide
MSL / MSLs	Microstrip line / Microstrip lines
MSL–S	Symmetric microstrip line
MSL–A	asymmetric microstrip line
NSL / NSLs	nonplanar slot line / nonplanar slot lines
EH	Electro–Magnetic hybrid mode
EHF	Extremely high frequency
E–wave/mode	Electric wave/mode (known as transverse magnetic (TM) wave)
HE	Magneto–Electric hybrid mode
H–wave/mode	Magnetic wave/mode (known as transverse electric (TE) wave)
SHF	Super high frequency
SIE	Singular Integral Equation (one equation)
SIEs or SIE	Singular Integral Equations (two or more equations)
SL / SLs	Slot line / Slot lines
TEM	Transverse Electromagnetic
USSR	Union of Soviet Socialist Republics